SIGNS OF LIFE

SIGNS OF LIFE

How Complexity Pervades Biology

Ricard Solé

and

Brian Goodwin

A Member of the Perseus Books Group

Published by Basic Books,
A Member of the Perseus Books Group

Library of Congress Cataloging-in-Publication Data

Solé, Richard V., 1962–
Signs of life : how complexity pervades biology / Richard Solé and Brian Goodwin.
p. cm.
ISBN 0-465-01927-7 (alk. paper)
1. Life (Biology) I. Goodwin, Brian C. II. Title.
QH501.S63 2000
570–dc21 00-049825

Designed by *Nighthawk Design*

99 00 01 02 03 /10 9 8 7 6 5 4 3 2 1

To those days of my university student's life and to that old flat in Barcelona that I shared with the most strange and stimulating people. To those days that had the taste of the irrepeatable and to the friends that shared them with me: Joan Manel Solé, Adolfo Borraz, Juan Sanchez and the memory of Ricardo Labajos.
To those days of wine, roses . . . and chaos.

—Richard Solé

I acknowledge with gratitude many years of rich interaction with friends and colleagues at the Santa Fe Institute, particularly Stuart Kauffmen, Jack Cowan, George Cowan, Jim Crutchfield, Harold Morowitz and John Casti. And I recall with special pleasure my explorations into complexity theory in biology with two very bright, enthusiastic and lively young researchers who joined me at the Open University, who are now having fun pushing back the frontiers: Ricard Solé and Octavio Miramontes.

—Brian Goodwin

Contents

Preface

Grasping the nature of life is like catching a whirling eddy in a stream: the moment you have it in your hands it disappears and leaves you with the matter but not the form. For centuries biologists have been seizing life in different forms, holding it briefly, and then seeing it disappear as the collective belief, the energy that maintains the form, dissipates. We have had the organism as the expression of a vital principle, as a machine, as a complex chemical network, as a result of natural selection. None of these attempts to characterize life is wrong. Each gives a distinctive insight into the nature of organisms, but each is also limited. The latest vision sees the secret of life in the DNA coiled at the heart of every cell, organizing the dynamic activities of an organism like a conductor bringing coherent form from the orchestra. Now that we have looked deeply into this secret we see that it, too, is dissolving, or rather, exploding before our eyes. We are overwhelmed by what appears to be the sheer complexity of the information in the DNA and the problem of making sense of it. Information has meaning only within a context, and the living context still evades us.

Signs of Life is about this context, understood in terms of a new dynamics of living processes that has been taking shape in recent years. A remarkable burst of creativity in science is transforming traditional disciplines at an extraordinary rate, catalyzing movements whereby old boundaries are dissolving and newly integrated territories are being defined. The new vision comes from the world of complexity, chaos, and emergent order. This started in physics and mathematics but is now moving rapidly into the life sciences, where it is revealing new signatures of the creative process that underlie the evolution of organisms. A distinctive sign of life is the emergence of new order out of the complexities of its material foundations. The concept of emergence, once regarded by many biologists as a vague and mystical concept with dangerous vitalist connotations, is now the central focus of the sciences of complexity. Here the question is, How can systems made up of components whose properties we understand well give rise to phenomena that are quite unexpected? Life is the most dramatic manifestation of this process, the domain of emergence par excellence. But the new sciences unite biology with physics in a manner that allows us to see the creative fabric of natural process as a single dynamic unfolding (Figure 0.1).

Figure 0.1

The aim of this book is not to attempt a definitive treatment of the science of emergent phenomena, but to show how the new ideas are exerting their influence in a variety of biological areas. The consistent theme that runs throughout our treatment of such topics as genes and development, physiology, brain and behavior, organization in social insects, ecosystems, macroevolution, and the patterns of human culture is the understanding of biological processes in terms of complex dynamics from which emerge characteristic patterns of order. The objective is to show how scientists are thinking in this area and what tools are available for understanding the creative process. What we are seeing is the beginning of a science of emergent forms. This is a new biological frontier that will leave its mark on the life sciences and then transform into something else. But it is likely to have longer-term consequences on our view of science itself. It will become evident that the new understanding of complex processes takes us beyond the traditional scientific perspective of prediction and control of nature, to a relationship of participation in natural processes that are unpredictable, though still intelligible.

ONE

Nonlinearity, Chaos, and Emergence

In effect, there seems to be no end to the emergence of emergents. Therefore, the unpredictability of emergents will always stay one step ahead of the ground won by prediction. . . . As a result, it seems that emergence is now here to stay.

—J. Goldstein, *Emergence as a Construct*

Predictability and Control

One of the impulses behind science is the desire to gain reliable knowledge about the world so that we can control it. There is no denying the extraordinary success of this enterprise, which has given us electrical illumination, television, jet aircraft, antibiotics, computers, artificial pacemakers for the heart—the list goes on. The principle that underlies these technologies is an orderly and predictable relationship between cause and effect. More often than not the relationship is linear: the illumination from the light bulb changes in direct proportion to the amount the dimmer switch is turned; the power from the jet engines varies directly with the movement of the throttle. But the linear relationship between cause and effect in such processes holds only over a limited range; it always fails if that range is exceeded.

Nonlinearity does not mean that control is not possible. Plenty of nonlinear processes are orderly, predictable, and hence controllable. The pendulum of a grandfather clock follows a well-defined, stable cycle of nonlinear motion. Small extraneous impulses to the pendulum coming from mechanical disturbances of the clock, such as walking past it on a wooden floor, don't knock it from one periodic orbit to another, as would happen in a linear system in which impulses permanently change the motion (i.e., the whole of a linear system is the sum of its parts, which include extraneous perturbations). The pendulum of a clock always returns to the same stable cycle, a limit cycle, which is determined by the length and mass of the pendulum and the amount of energy added each cycle by the escapement mechanism. We control the period of the cycle simply by altering the length of the pendulum so that the clock can be adjusted to keep accurate time, as long as you remember to wind it up once a week.

This nonlinear mechanism has served for centuries as the paradigm of the mechanical world-view. The clock provides us with the two basic dynamic attractors that characterize so much of the controllable world: the stable point (point attractor) and the stable cycle (the limit cycle). Stop winding the clock, and it doesn't do something unexpected and strange, it runs down and stops at a stable state. Start a clock that has been wound up by giving the pendulum a small push from its rest position and the amplitude will grow, due to the energetic impulses from the escapement mechanism each cycle, until the pendulum reaches a steady cycle. Alternatively, if you give the pendulum a big push so that its initial amplitude is greater than the steady cycle, it will decrease until it reaches the same cycle that it attains starting with a small push. Either way the clock settles into its time-independent orbit, giving forth a steady, reliable tick-tock that is the heart of the intelligible, predictable, and controllable clockwork universe.

The motion of a clock becomes linear when the pendulum swings through very small angles. If we assume that there is no friction; we get simple harmonic motion. This is an ideal, not a real, state because any actual clock will experience friction. Yet this ideal has been enormously fruitful ever since Galileo, inspired by the swinging lamps in the cathedral at Pisa, was moved to envision perpetual periodic motion in a frictionless universe. He thus provided the foundations of an intellectual adventure that has led, among other strange and wonderful consequences, to the totally unexpected insights into the bizarre reality of quantum microstates where friction plays no role. The mathematization of nature, stimulated by the impulse to gain reliable knowledge of, and control over, natural processes leads in quite unanticipated directions. This book explores another consequence of following a route initiated by mathematics, which takes us beyond prediction and control but not beyond intelligibility.

Unpredictability with Intelligibility

The mathematics of this new adventure is one that describes the paradoxical world of complex processes. Here we encounter not simply nonlinearity but unpredictability, which comes in two forms. The first is the "sensitivity to initial conditions" associated with chaos, where tiny errors of measurement give rise to gross inaccuracies of prediction. This was first recognized by the great French mathematician and physicist Henri Poincaré (1854–1912) when he was studying the motion of three celestial bodies obeying Newton's laws of gravitational attraction. In his words, " . . . it may happen that small differences in the initial conditions produce very great ones in the final phenomena. A small error on the former will produce an enormous error in the latter. Prediction becomes impossible, and we have the fortuitous phenomenon."[1]

This unpredictability has a quite different source from that of a stochastic or random process, where the irregularity arises from the cumulative effects of a multitude of many extraneous influences. Chaos has its own intrinsic logic, and it can arise in a great diversity of nonlinear dynamic systems; it is a generic dynamic state. Hence we should expect that even the pendulum should be capable of manifesting chaos. And so it does, as illustrated in Figure 1.1. The upper curve shows the regular motion of a damped pendulum (i.e., with friction) that is driven by a periodic external force, a version of the escapement mechanism that drives a real clock. The variable is the angular velocity, the rate of change of the angle between the actual position of the pendulum at time *t* and its resting position (hanging vertically).

Figure 1.1b shows what happens when the periodicity of the driving force does not match the natural frequency of the pendulum, so that there is no stable state of the whole system: it becomes chaotic. Nevertheless, it is well-defined mathematically, and so the motion is deterministic: for any initial condition we may compute a unique trajectory. However, the slightest change of initial conditions results in a quite different trajectory, so that the behavior is unpredictable by any observer whose measurments are not perfect (i.e., any real observer). The importance of the observer in complex processes will come up again and again in this book.

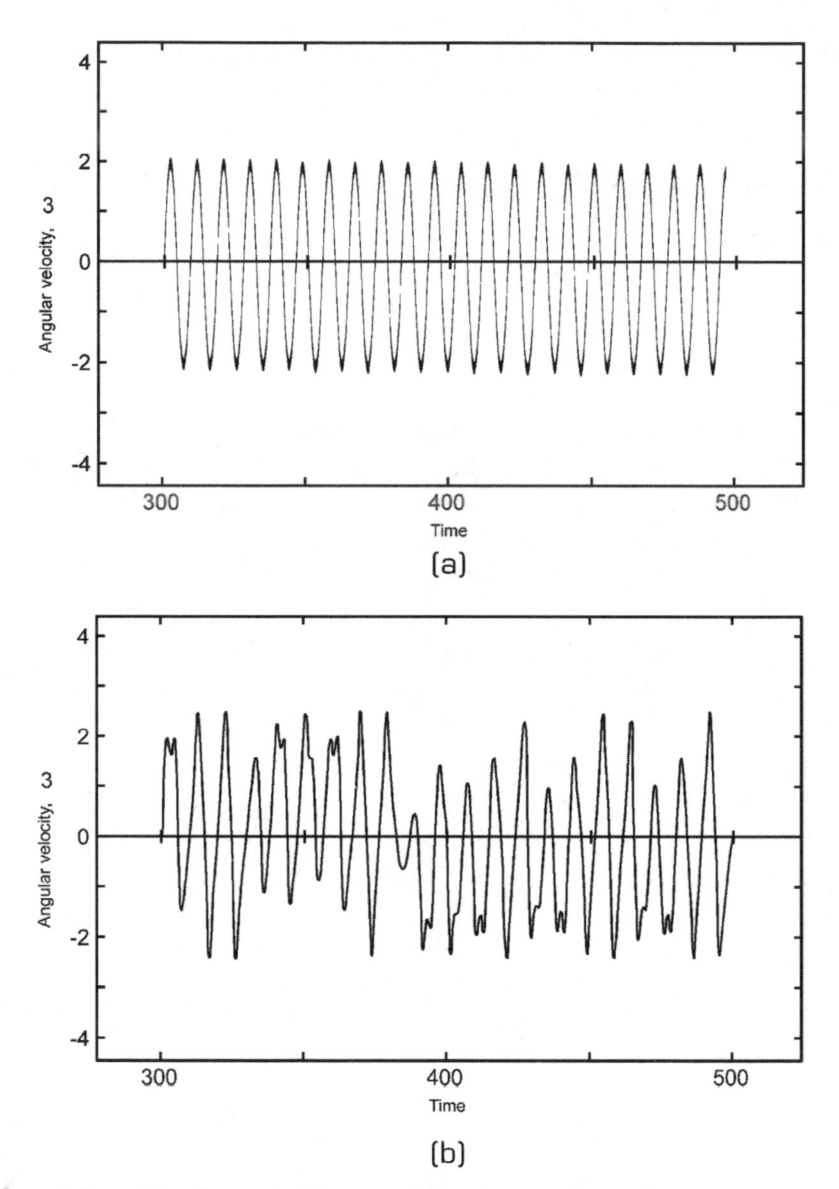

Figure 1.1 a. The damped, driven pendulum in periodic mode.
b. The same pendulum in chaotic mode.

What underlies the unexpected properties of chaos, making it simultaneously deterministic and unpredictable? Sensitivity to initial conditions means that the dynamic has a divergent aspect: some process is driving trajectories away from each other. Yet they remain bounded within a domain, as we can see from the constrained amplitudes of the angular velocity in Figure 1.1b. So there is a compensatory convergence that balances the divergence and results in a strange attractor: a region to which trajectories are attracted but within which they diverge.

The Intelligible Dynamics of Chaos

We now shift our attention from physics to biology, from a swinging pendulum to the growth of a population of organisms. In 1845, P.F. Verhulst derived a simple nonlinear differential equation to describe the growth of a population of organisms under limiting environmental conditions. The equation represents the consequence of two opposing influences: (i) the tendency of a population to grow exponentially (i.e., at an ever increasing rate and without bound); and (ii) the effect of a rate-limiting factor such as finite food supply or reduced fertility due to crowding, which keeps the population below a maximum size. These influences can be in balance so that the population grows to a particular size and then remains constant; and in some cases (see box 1) there can be an imbalance such that the population increases, then decreases, then increases again, producing a continuous oscillation; or the population size can change in a chaotic manner, increasing and decreasing in a perfectly deterministic but unpredictable pattern, like the pendulum in Figure 1.1b. The essential properties of the Verhulst equation can be studied by using a somewhat simpler expression called the logistic equation (see box 1) to describe these different patterns.

The Patterns of Population Growth

The equation derived by Verhulst took the form

$$\frac{dX}{dt} = \mu X\left(1 - \frac{X}{A}\right)$$

where X is population size, μ is the intrinsic growth rate, and A is the maximum size that the population can attain. For instance, we might be considering the growth of rabbits in a particular territory where the maximum number is 10,000, which is then the value of A. The tendency of the population to grow exponentially is given by the first term, μX, so that the larger the population, the greater the rate of growth. The second term, $(1 - X/A)$, means that as the population grows there is a self-limiting process (e.g., limited food, effect of population density on fertility, etc.), so that growth rate decreases. The parameter A can be removed by changing the variable to $x = X/A$, called the normalized population, with the result that the equation becomes

$$\frac{dx}{dt} = \mu x(1 - x), \quad \text{where now} \quad 0 < x < 1.$$

This is the well-known logistic equation. To get the actual population of rabbits X from x, just multiply by A. So if $x = 0.75$ and $A = 10{,}000$, then $X = 7{,}500$. A discrete version of the logistic equation is obtained assuming that the population is described by a discrete number x_n, n representing the nth generation. The assumption now is that there is synchronous reproduction of members of the population, or that the population census is taken once every generation cycle. This gives us the logistic map

$$x_{n+1} = \mu x_n(1 - x_n),$$

which relates the population size in the $(n+1)$th generation to that in the previous, nth, generation.

To study the behavior of successive generations described by the logistic equation (see Box 1), one has simply to specify a value of the parameter μ, representing the intrinsic growth rate of the population, and choose an initial value of the normalized population, x_0, where $0 < x_0 < 1$. Then x_1 can be calculated, which is in turn inserted into the equation to calculate x_2, and so on recursively. Suppose that we choose $\mu = 2.5$. This means that the population has the potential to grow to 2.5 times its initial size in one generation. But another influence, dependent on population size, reduces this rate, as described in Box 1. If we calculate the successive values of x_n we find that it converges to 0.6, irrespective of the initial value of x_0. In the example of a rabbit population with maximum size of 10,000, this means that the number of rabbits stabilizes at 6,000 ($0.6 \times 10{,}000$), independently of how many there were to begin with. This is called a point attractor (Figure 1.2a). With $\mu = 3.3$ the equation gives a continuous cycle between two different values (0.48 and 0.83), so that $x_{n+2} = x_n$ (Figure 1.2b): the rabbit population oscillates between 4,800 and 8,300 in successive generations. This is the discrete analogue of a limit cycle, similar to the pendulum revisiting the same values over and over. At $\mu = 3.5$ a more complex cycle occurs involving 4 distinct points, so that $x_{n+4} = x_n$ for large n (Figure 1.2c). With μ greater than about 3.57, successive values of x_n vary without any pattern or order; chaos has set in (Figure 1.2d). However, the values remain bounded within a domain that defines the chaotic attractor.

This whole sequence of different behaviors can be described by the familiar "march to chaos" through successive bifurcations from one type of solution to another as μ increases, as shown in Figure 1.3. At each bifurcation the number of interacting periodicities increases exponentially following the series 2, 2^2, 2^3, . . . , 2^n. Deterministic chaos sets in at the point where periodicity transforms into non-periodicity, where the number of interacting periodicities is such that no precise cycle occurs, and so predicting subsequent states becomes impossible. Here the apparently contradictory processes of simultaneous divergence and convergence have taken over to produce bounded but never repeating trajectories.

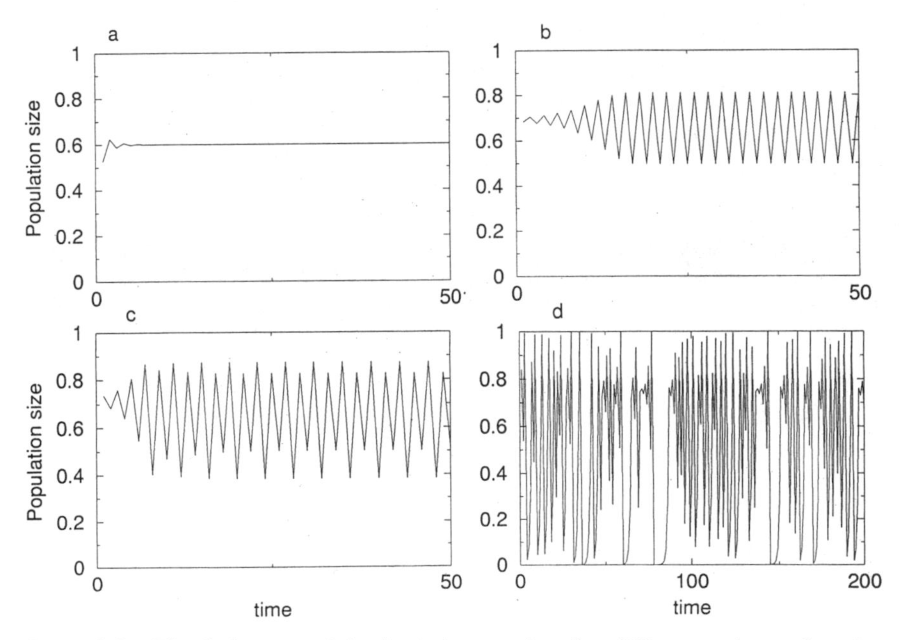

Figure 1.2 The behavior of the logistic equation for different values of μ, the intrinsic growth rate. a. $\mu = 2.5$ b. $\mu = 3.3$ c. $\mu = 3.5$ d. $\mu = 4.0$

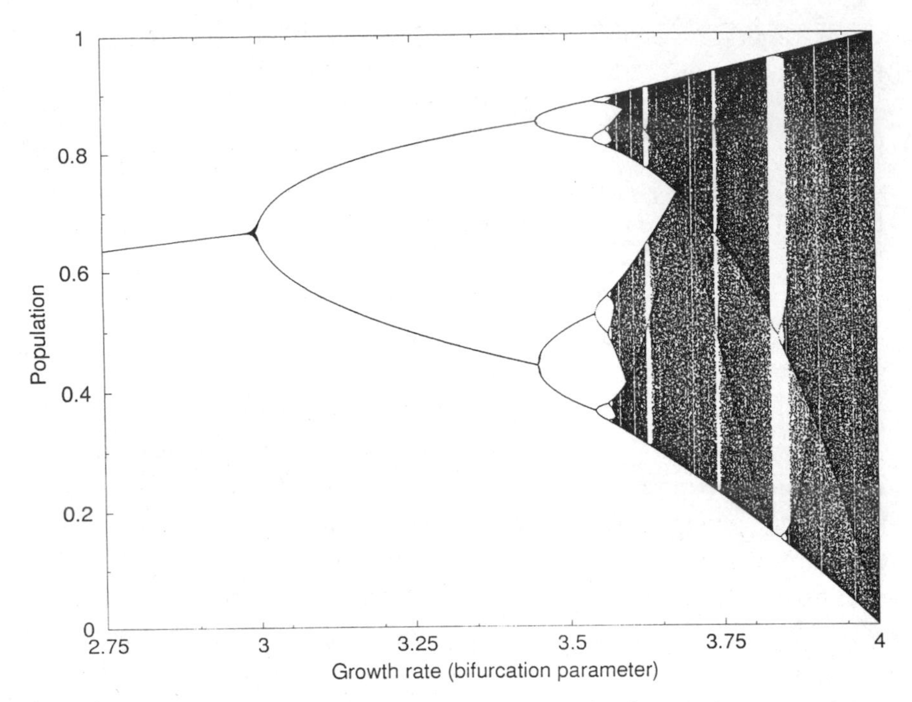

Figure 1.3 The march to chaos via bifurcations in the discrete logistic equation as μ is varied. The ordinate is the population size, and the curve shows the values taken by this variable as μ changes.

Stretching and Folding

These divergent and convergent processes are readily demonstrated with the logistic map. Taking $\mu = 4$, the equation is $x_{n+1} = 4x_n(1 - x_n)$. The maximum value of 1 for x_{n+1} occurs when $x_n = \frac{1}{2}$. The set of values of x_n in the interval $[0, \frac{1}{2}]$ map into x_{n+1} values in the interval $[0,1]$, while x_n values in the interval $[\frac{1}{2},1]$ also map to $[0,1]$ but in the reverse order, from 1 to 0. Thus the range of possible values of x_n, between 0 and 1, gets stretched to twice its length and folded back to give the picture shown in Figure 1.4. This stretching and folding is the key to divergence and convergence. The domain of change remains bounded, but initial values that start off close to one another diverge exponentially fast as stretching occurs, successive "lengths" of the iterated map of the interval $[0,1]$ expanding according to the series 1, 2, 4, 8, 16, ...

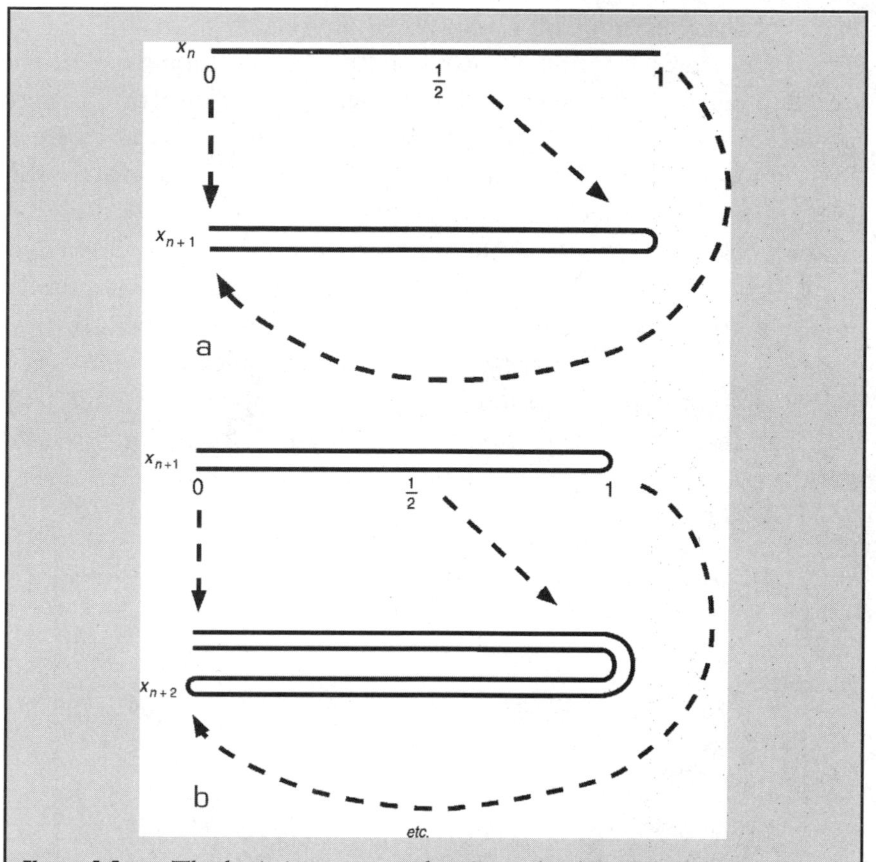

Figure 1.4 a. The logistic map stretches (0,1), the domain of x_n, to twice its length and folds it back so that $x_n = 1$ becomes $x_{n+1} = 0$. b. This is repeated again as x_{n+1} is mapped to x_{n+2}.

Making Sense of the Weather

The weather is a classic case of chaotic behavior. Its basic dynamics come from the behavior of the gases that make up the earth's atmosphere under the influence of the rotation of the earth and the sun's radiant energy. The equations of fluid dynamics can be used to describe the various patterns that can arise. In the early 1960s Edward Lorenz, working at MIT, was studying these possible patterns using a simplified version of fluid-dynamic equations and simulating their behavior on a computer. The variables in Lorenz's equations are related to velocity of movement of the air and to temperature. He discovered the strange and exotic flow pattern shown in Figure 1.5. The trajectories of motion are attracted to a form within which there is endless coherent flow on spiraling surfaces in three dimensions, part of which is shown in Figure 1.6. This remarkable mathematical form is the Lorenz attractor, an example of a strange attractor. A pair of nearby trajectories can separate so that one continues to spiral around the same center while its neighbor flips over to the other focus of rotation and spirals around it an indefinite number of times before flipping back. Small differences of initial condition thus become amplified, so that points initially on neighboring trajectories can quickly become very distant from one another. It took nearly half a century after Poincaré's initial observations before the mathematics of their formation was fully worked out.

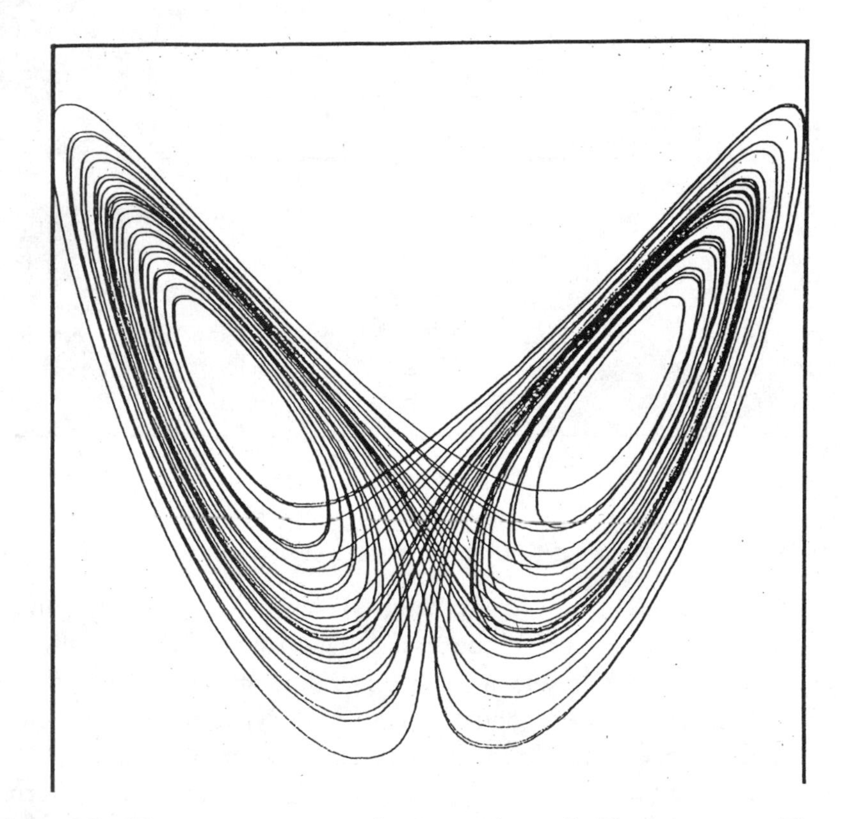

Figure 1.5 The Lorenz attractor showing its butterfly-like appearance. The attractor lives in 3 dimensions, so that trajectories do not cross one another.

Unpacking the Lorenz Attractor

The basic dynamics of these flows in the Lorenz attractor are organized around three singularities called saddle points. One of these is shown at the bottom center in Figure 1.6, while the other two are located at the centers of the outwardly spiraling trajectories. The saddle point at bottom center has trajectories flowing toward it in the vertical plane, as shown. Along the line that passes through this plane at right angles, trajectories travel in opposite directions away from this saddle point. They flow toward the other saddle points, around which they spiral a number of times before flipping over to the other spiral saddle. The result of the three saddle points yoked together with flows from one to the other is a stretching and folding of the surfaces containing the trajectories, as we saw in the case of the logistic equation, but now occurring in three dimensions. This results in continuous motion such that any pair of points

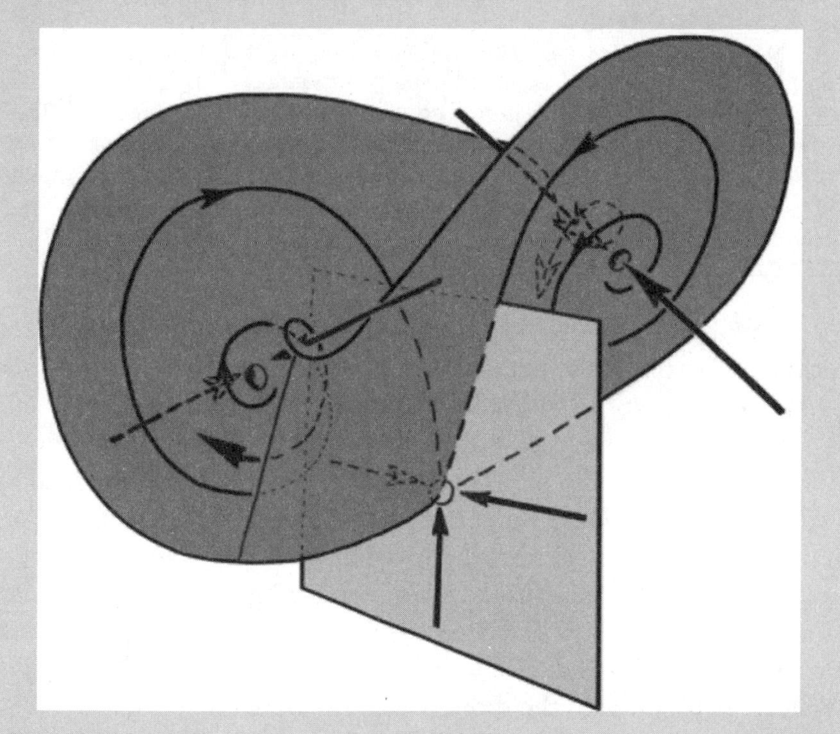

Figure 1.6 A three-dimensional view of the Lorenz attractor demonstrating the locations of the three saddle points that are yoked together to create the attractor. In the plane at the center is a radial saddle point, while at the center of each of the "butterfly wings" is a spiral saddle point.

that start as neighbors move apart exponentially fast along trajectories within the strange attractor.

It is the properties of saddle points—singularities that combine both inflows and outflows—that are the basic elements of strange attractors and account for simultaneous convergence and divergence of flows in these beautiful and strange structures. The converging trajectories are characterized by negative real parts of the characteristic exponents or Lyapunov coefficients, defining stable motion along a manifold, while the diverging trajectories have positive real parts of the Lyapunov coefficients, defining locally unstable motion.

To see this, consider the saddle point at the bottom center of the Lorenz attractor. The local flow with this saddle as the coordinates origin can be described as follows. Let the vertical plane be the *yz*-plane and the horizontal line the *x*-axis. Then the equations of motion of the flow are

$$\frac{dx}{dt} = x, \quad \frac{dy}{dt} = -y, \quad \frac{dz}{dt} = -z.$$

Trying the solutions $x = A\exp(\lambda_1 t), y = B\exp(\lambda_2 t), z = C\exp(\lambda_3 t)$, we find that the differential equations require that $\lambda_1 = 1, \lambda_2 = -1$, and $\lambda_3 = -1$ as the three Lyapunov coefficients for the local motion. The spiraling saddle points can be described locally by solutions defined by three Lyapunov exponents, two of which are complex conjugates with positive real part (the diverging spirals) and one real and negative (the inflowing trajectories).

The structure of the Lorenz attractor provides the clue to the curious properties of the weather. Because the atmosphere is a fluid and so obeys the equations of fluid dynamics, it is a deterministic system: its state at any moment is, in principle, calculable. Yet the nonlinear dynamics governing the weather include behavior that is described by a strange attractor. Here we encounter the unpredictability described by Poincaré where small errors in initial conditions give rise to very large errors in calculating expected outcomes. This is why the weather is outside our prediction and control. This type of behavior also limits our capacity to predict and control biological processes, yet it provides invaluable insights into the robust dynamics of the living state.

Reductionism, Self-Organization, and Emergent Properties

Earlier, we mentioned that unpredictability comes in two forms. The first arises from the sensitivity to initial conditions that is characteristic of strange attractors. The second one relates to emergent phenomena. Here the issue is whether one can predict the behavior of a system of many interacting components when the properties of the individual components themselves are understood. This clearly relates to the scientific strategy of reductionism: when faced with a complex system made up of many parts in interaction, study the properties of the parts in order to understand the whole. The question here is not the value of reductionist strategies in science; this is undeniable. But how do we discover the appropriate level at which to carry out the analysis into parts and how do we define the interactions that result in the higher-level behavior? Some simple examples illustrate the problem.

The properties of hydrogen and oxygen are both well described as elements and as molecules (H_2 and O_2) in their gaseous state. We understand their chemical behavior in terms of chemical bonding and the quantum mechanics of electron orbits. This knowledge, however, is not sufficient to predict the properties of their combination in the substance water (H_2O), even though the behavior of water can be recognized to be consistent with the properties of its constituents. To understand the properties of water we must observe its behavior in different contexts and discover equations that describe this behavior. These are called the Navier–Stokes equations, which were actually derived long before the quantum theory of chemical bonding was understood. The equations are based on liquid water's observed properties of cohesion, incompressibility, and fluidity. The same equations describe the behavior of compressible gases, and so these are the equations Lorenz used, in a simplified form, to study weather as consisting of the dynamic states of atmospheric gases.

The weather's dynamic patterns display order as well as chaos. A particularly striking instance of this is the pattern of air flow that produces the structures shown in Figure 1.7a in the sands of the Sahara. The sun heats the desert surface and hot air rises, while cold air descends from the upper layers of the atmosphere. The conflict is resolved by the emergence of orderly patterns of upward and downward air flows of the kind shown in Figure 1.7b. This is a very basic pattern of fluid flow under the action of an imposed temperature gradient, a pattern that holds for liquids as well as gases. The forms that define the flow are called Bénard cells (Fig. 1.7c). Combining a number of these to cover a surface gives us the picture shown in Figure 1.7d, which is the type of pattern of air flow that generates the dune structure seen in Figure 1.7a.

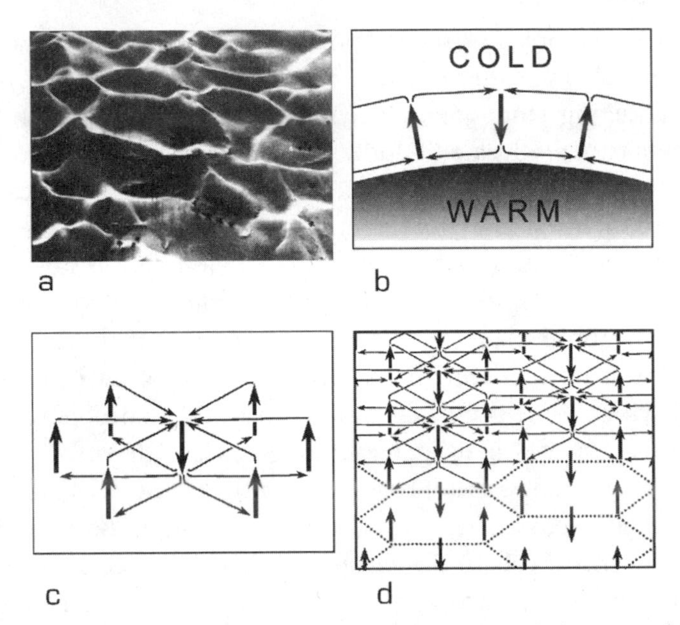

Figure 1.7 a. A roughly hexagonal arrangement of sand dunes in the Sahara reflecting upward and downward patterns of air flow caused by a thermal gradient, as in b. Within each hexagonal unit the air flows as shown in c, a number of these forming a coherent array of Bénard cells, shown in d.

These Bénard cells are more easily studied in the liquid state, where they were first seen in thin layers of liquid subject to a thermal gradient. It is very unlikely that anyone would have deduced such structures starting simply with the Navier–Stokes equations. This is because the behavior arises only under particular conditions: a thin layer of fluid and a thermal gradient resulting from a heat source at the bottom of the liquid layer. So we get the first, obvious, condition for an emergent phenomenon: there has to be an observer who is puzzled by the property being observed, because it is not obvious how it arises within a familiar context, in this case the usual behavior of liquids. Having observed the phenomenon we can then discover appropriate mathematical descriptions that capture the essence of the phenomenon.

Bénard cells are now understood to arise from the properties of the liquid state by a bifurcation in which, at a critical point, a self-organizing structure emerges. The immense number of degrees of freedom of individual molecules in liquids suddenly become organized into coherent flow patterns that dissipate heat more effectively than simple thermal conduction. A new emergent behavior, convection (transport of heat by mass movement) in self-organized spatial structures, becomes the dominant process at the critical point, and a new behavior emerges. This is initiated by an instability that arises as follows.

A locally hotter bit of fluid at the bottom of the thin layer, closer to the heat source, has lower density than the fluid above, so it will tend to rise. As it does so, it enters even colder regions, where the density is greater, so it keeps on rising. At first, the upward movement is counterbalanced by viscous dissipation and heat diffusion from the hotter fluid droplets to the environment. The hot droplet simply dissipates its heat to the surrounding liquid, and no organized flow pattern emerges. But when the temperature gradient exceeds a critical value, convection takes over from conduction as the dominant process of heat dissipation. The whole fluid layer then begins to self-organize into a coherent spatial configuration. First we see convection rolls of the type shown in Figure 1.8. Within each tube the fluid is rotating in a pattern as shown in Figure 1.9. Notice that the rolls turn together, like gears, so

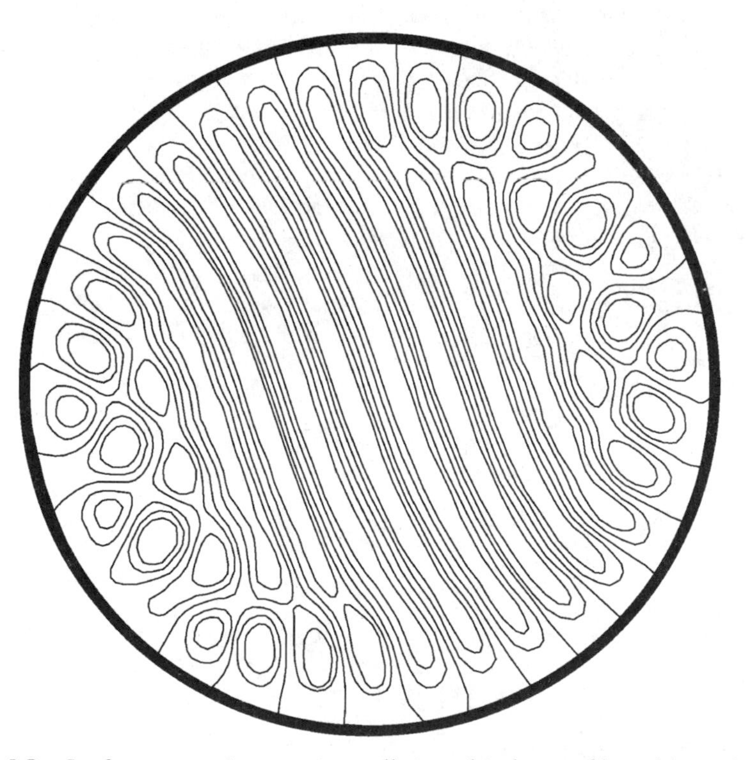

Figure 1.8 Surface view of convection rolls in a thin layer of liquid heated from below.

that there is smooth convective flow throughout. The geometry of the vessel determines the overall pattern, a circular dish producing both rolls and cells as in Figure 1.8. We can thus describe the emergent phenomenon in terms that are compatible with both the properties of the liquid state and the way heat is dissipated in liquids. This is where we can apply reductionist thinking, developing a description of a relatively complex phenomenon such as Bénard cells in terms of more elementary properties of liquids.

Order Parameters and Collective Variables

How does one find an appropriate mathematical description of the self-organizing Bénard cell pattern? There is no unique descriptor, and a convenient one that captures an essential aspect of the phenomenon has to be developed and agreed upon. Before the pattern emerges, the temperature decreases continuously across the fluid from bottom to top. At any distance in the fluid from the bottom the temperature will be constant. Once the roll pattern starts to emerge, temperature is no longer constant at a particular depth in the liquid layer: it will vary periodically, going from a maximum where the hot fluid is rising to a minimum where the cooler liquid is falling. The descending fluid heats up again as it gets closer to the bottom and the source of heat. So the emergence of a pattern can be described by the development of a periodic variation in temperature with a fixed wavelength, $\lambda_c = 2h$ (see Figure 1.9).

The equations describing this process, which now include both the Navier–Stokes equations and the equation of heat flow in liquids, can be

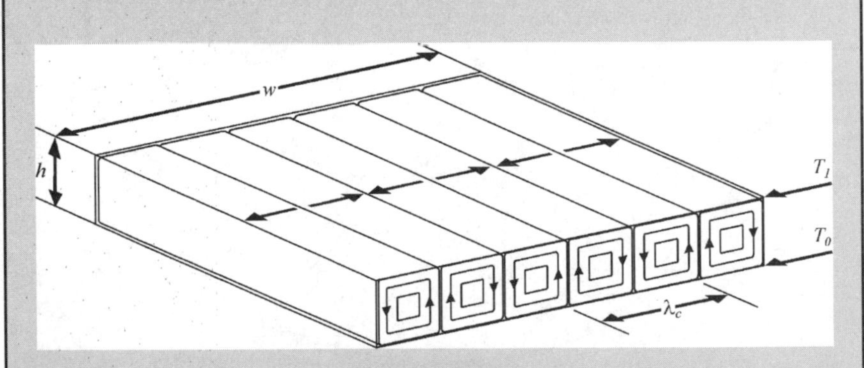

Figure 1.9 Schematic description of the flow pattern of liquid within the convective rolls shown in Figure 1.8, showing the coherent structure.

shown to go through a bifurcation at a critical temperature gradient β_C at which spatial patterns of convective flow begin to emerge. One of these, with a wavelength characteristic of the system, grows fastest and dominates the pattern (i.e., only one Fourier mode is active). This temperature periodicity can be described by

$$q(x, z) = A(\beta) \cos\left(\frac{2\pi x}{\lambda_C}\right),$$

where we are using the coordinates shown in Figure 1.9. The location in the fluid where the temperature variation is described is $z = h/2$, the center of the layer, for varying distances along the x-axis, and $A(\beta)$ is the amplitude of the periodic variation, which is zero when the temperature gradient is below the critical value β_C. As the gradient increases above this critical value, $A(\beta)$ begins to increase and is then limited to a maximum value by nonlinear effects. It is called an *order parameter*, which characterizes the emergence of an ordered pattern. The initially random microscopic fluctuations of the molecules self-organize at the critical temperature gradient β_C into a coherent macroscopic flow pattern described by the *collective variable* q.

As the temperature gradient is increased further, the rolls undergo a further bifurcation to produce hexagonal Bénard cells of the type shown in Figure 1.7c. Here there is periodicity in the spatial variation of temperature q in both the x and y directions. If the temperature gradient is increased further, other periodic modes grow in amplitude and contribute to the motion. The result is a series of bifurcations that describe greater and greater complexity of spatial pattern that is the precise spatial analogue of the "march to chaos" via period doublings in the logistic equation, ending in turbulence as the liquid analogue of chaos. Here liquids obey the deterministic chaos of a strange attractor of the type described in Figure 1.5.

Emergence, Consistency, and the Sciences of Complexity

Reductionism takes us from a complex phenomenon to more elementary properties of its components, giving us consistent explanations between descriptive levels. But rarely can we go from the properties of the constituent parts to a description of the whole. Understanding the properties of H_2 and O_2 does not allow us to predict the properties of H_2O; understanding the molecular properties of H_2O does not allow us to derive the Navier-Stokes equations; having the Navier–Stokes equations does not give us a prediction and description of Bénard cells.

In fact, unanticipated consequences of the Navier–Stokes equations are being discovered all the time. Self-organizing behavior emerges unpredictably in systems at different levels. We make it intelligible by recognizing how it is consistent with lower-level properties and by finding appropriate mathematical descriptors. But in doing this we don't *reduce* a whole to the properties of its parts and their interactions. Although there are cases where this has been achieved, in general there is a causal gap between one level of description and the next, which is covered by a mathematical relationship producing consistency between levels. Emergent properties provide the recognition that nature can be creative while denying the occurrence of miracles or inconsistencies.

Clarifying these issues has been a major task of the sciences of complexity, the study of nonlinear systems whose behavior, though regular and repeatable for any particular conditions, changes in unpredictable ways when circumstances are changed. It is relevant to look briefly at some of the different viewpoints in this discussion, for there is as yet no overall consensus about emergent properties.

There is a powerful school of scientific thought that regards all apparently emergent properties as epiphenomena, having no explanatory significance or causal efficacy. Emergent properties in this view do not influence lower-level behavior in ways that cannot be accounted for by the lower-level properties themselves. This position holds that reality is as described in classical mechanics, and reductionism is the only mode of satisfactory explanation: wholes can always be described in terms of the properties and interactions of their parts.

It is generally acknowledged that quantum mechanics is not subject to this kind of reductionism; it is fundamentally holistic, with emergent properties of wholes that cannot be described in terms of parts. Examples of this are legion, particularly in solid-state physics. In his 1994 book *A Career in Theoretical Physics*, Philip Anderson[2] listed supeconductivity, superfluidity, ferromagnetism, crystals, lasers, to name but a few. Perhaps the best-known property of quantum mechanics that takes it beyond the reductionism of classical mechanics is nonlocal connectedness, identified by Einstein, Podolsky, and Rosen[3] in 1935, sharpened by Bell[4] in 1966 in the form of his famous inequalities, and confirmed in experiments by Aspect and colleagues[5,6] in 1982. Nonlocal connectedness means that particles that are initially connected in particular ways within an atom or a nucleus preserve certain relationships even after separation, irrespective of distance or changes in the way the experimenter chooses to observe them.

Quantum states are entangled in ways that make a description of separate components and their local causal interactions impossible. The examples described by Anderson all relate to this fundamental property of quantum mechanics. He also identifies other irreducible properties, such as symmetry-breaking in quantum systems, which is the fundamental reason that there is something rather than nothing in our cosmos, that is, the reason why there are galaxies and stars and atoms rather than a uniform distribution of energy. Initial symmetry-breaking in cosmic evolution cannot be explained in terms of the interactions of parts, because at that stage of the cosmic drama there were no parts.

The occurrence of emergent, irreducible properties in quantum systems does not mean necessarily that they also occur in macroscopic systems. It is worth noting, however, that such emergent phenomena as superfluidity and superconductivity are macroscopic, so emergence is not restricted to microstates. Thus the nature of physical reality is not that described by classical physics, and there is no reason to assume a priori that all irreducible macroscopic emergent properties are necessarily derived from quantum properties. Many aspects of macroscopic systems have the appearance of irreducible emergent properties but may not be describable by quantum mechanics. Michael Silberstein[7] uses the term "radical emergence" to describe the position that there are properties of macroscopic systems, living or otherwise, that are intrinsically irreducible to interactions among component parts. He sees consciousness as a primary candidate for this in biology. Although this is currently a subject of intense interest, we don't discuss it in detail in this book, mentioning it only briefly in Chapter 5 in the context of brain dynamics.

We believe that reductionism is inadequate as the primary explanatory framework of science. Progress in understanding natural phenomena requires more than a study of parts in interaction. It often involves grasping relevant aspects of whole systems and finding appropriate mathematical descriptors that capture these properties. A central concept in the study of fluids, for example, is the hydrodynamic field whose properties are captured by appropriate mathematical relations (the Navier–Stokes equations) that define the behavior of these fields in time and space. The solutions of these equations under particular conditions, such as a thin layer of fluid heated from below, are then examined for their correspondence with the actual phenomena and used as tests of the mathematical model as an explanatory construct. This process requires constant movement from phenomenon to model and back again. High-level properties and properties of parts are both incorporated into the model. This aspect of scientific understanding, this continuous conversation between parts and wholes, between models and reality, is a major theme of this book.

The observer is crucial in all this. At any time a consensus may develop that a particular phenomenon, originally considered irreducible to a lower level of description, has been made reducible by the discovery of an appropriate lower-level description and the relevant pattern of interactions among components. This is part of the creative process of science. Only the process of scientific exploration and analysis can reveal which phenomena are "radically emergent" and which will turn out to be reducible. However, as stated by Jeffrey Goldstein in the journal *Emergence* and quoted at the beginning of this chapter, "the unpredictability of emergents will always stay one step ahead of the ground won by prediction. . . . As a result, it seems that emergence is now here to stay." All the examples presented in this book have required high-level concepts to describe and make them understandable, involving a comparative study and analysis of higher and lower levels in their resolution. In every case, pure reductionism was an inadequate strategy, irrespective of whether observers later reached a consensus that the phenomenon can be understood in terms of appropriate components and interactions.

Thus we can see parallels between chaos and emergent properties. Both involve unpredictability, though arising from different sources. With chaos, it is sensitivity to initial conditions that makes the dynamics unpredictable. With emergent properties, it is the general inability of observers to predict the behavior of nonlinear systems from an understanding of their parts and interactions. Once an emergent property is observed, we must first find the appropriate descriptors that capture its intelligible dynamic essence and then show how this is consistent with lower-level properties. As James Crutchfield[8] has put it, "It is rarely, if ever, the case that the appropriate notion of pattern is extracted from the phenomenon itself using minimally biased procedures. Briefly stated, in the realm of pattern formation 'patterns' are guessed and then verified." This is the creative work of scientists that parallels the intrinsic creativity of nature.

Emergent Phenomena in Biology

One of the clearest illustrations of both the value and the limitations of reductionism in biology arises from a remarkable process whereby a collection of single cells turns itself into a coherent multicellular organism. Here is self-organization in one of its most dramatic manifestations. The species in question is the rightly renowned cellular slime mold *Dictyostelium discoideum* (see Figure 1.10). For decades biologists have studied this and related species, whose life cycles have two quite distinct phases: (1) single cells that move about as amoebas, growing and mutiplying as they engulf and digest bacteria in rotting vegetation, but paying not the slightest attention to one another; and (2) multicellular

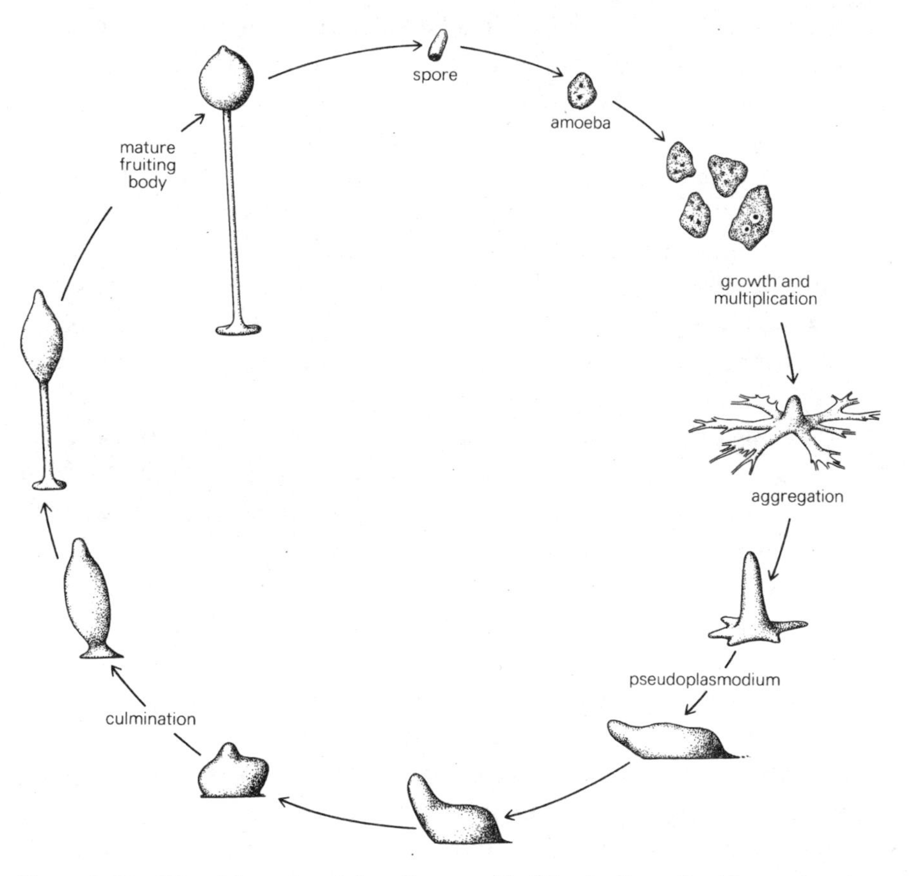

Figure 1.10 The life cycle of the slime mold, *Dictyostelium discoideum*, showing the stages of transformation of single vegetative cells into a coherently organized multicellular organism.

organisms consisting of a stalk and a cap of spores. The transition from one phase to the other is initiated by starvation. When there are no more bacteria to eat, the amoebas change their behavior in a very simple way that has remarkable consequences.

First, their metabolism changes under starvation conditions. They start to produce a molecule known as the glucose distress signal, cyclic adenosine monophosphate (cAMP), first recognized in starving bacteria. Produced within cells when glucose has run out, it prompts the cells to start metabolizing other food sources. But cellular slime mold amoebas do something different. Instead of keeping the molecule within the cell, they release it in little pulses. These have an effect on their neighbors: first, a pulse induces amoebas to move toward the origin of the signal; second, it stimulates them to release a pulse of the same chemical. The distress signal thus becomes a call for the amoebas to gather together and to make a multicellular structure that allows them to survive starvation. Between 10,000 and 100,000 cells collectively self-organize to generate a fruiting body that consists of a structural stalk (about one-third of the cells) on top of which sits a cap of spores (the remaining two-thirds of the cells). When moist conditions return, the spores germinate, releasing amoebas that then move about in search of bacteria. How do individual amoebas manage to organize themselves into a complex fruiting body by means of signaling, chemotaxis, and other types of interaction? Can this be explained simply by describing the cellular mechanisms involved?

The history of research on this problem is both intrinsically interesting and very informative on this issue. Detailed observations of hundreds of thousands of starving amoebas spread on a petri dish show that collectively they produce quite extraordinary spatial patterns that are difficult to explain simply on the basis of signaling and chemotaxis of individual cells. Some of these are shown in Figure 1.11. The patterns seen in Figure 1.11a–c are due to differences in the behavior of amoebas in the dish: the light bands occur in regions where amoebas are elongated and moving toward the origin of the pulsatile signal, while the dark bands occur in regions where the amoebas are not oriented and are moving more randomly. The signal always originates from the center of a spiral pattern and propagates outward from cell to cell, producing a traveling wave. There are clearly many such centers distributed over the dish. As the amoebas move toward the signal sources (a process called aggregation) they begin to break up into separate territories, each responding to signals that propagate from a particular center and ignoring the others (see Figure 1.11d).

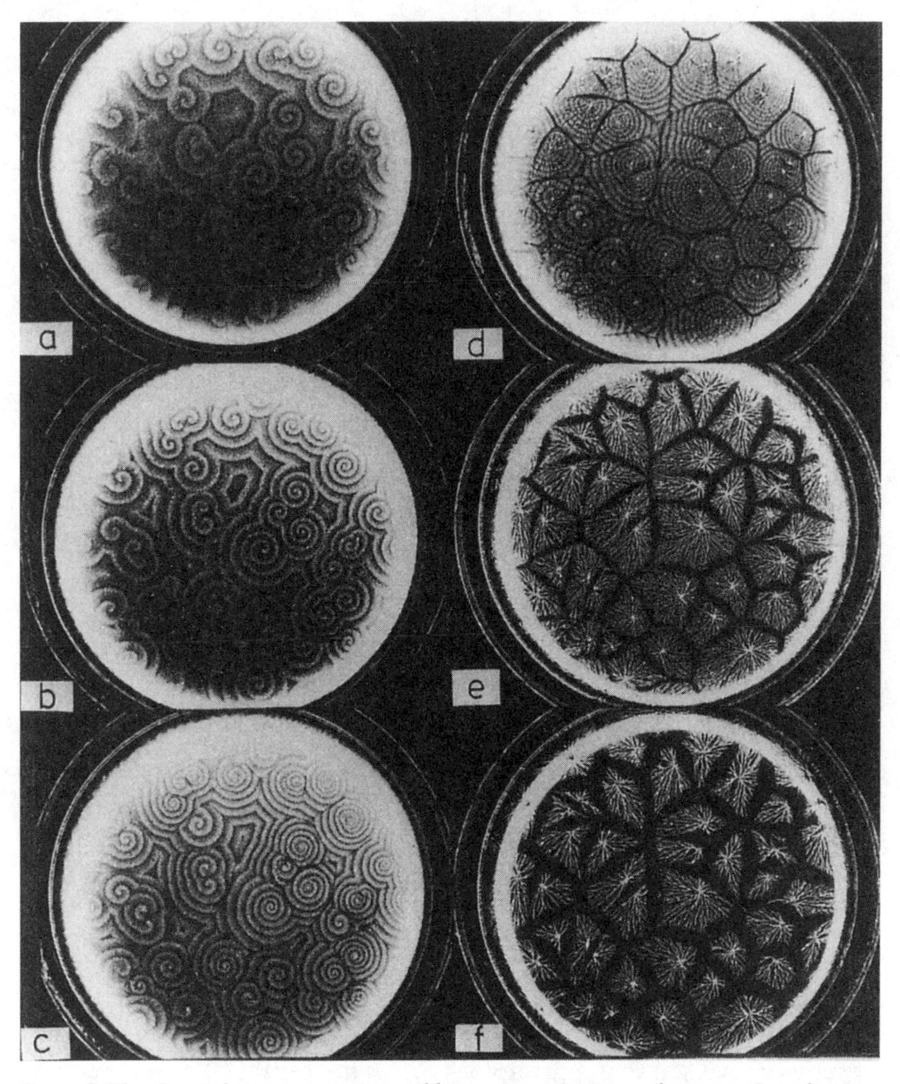

Figure 1.11 Spatial patterns generated by aggregating amoebae, a–c. Spiral structures arising from the movements of amoebae in response to propagating waves of cAMP, whose frequency increases progressively to give shorter wavelengths of the spirals. d. separate domains of aggregating amoebae form, each under the influence of a signaling center that is a reentrant cycle of stimulation. e, f. Separate aggregates break up into fanlike converging arrays due to an intrinsic instability of the interacting domains of cells.

The next step is the transformation of the separate territories, which initially consist of continuous lawns of amoebas, into fanlike arrays consisting of discrete streams of cells moving toward the aggregation center (Figure 1.11e,f). Each of the territories, consisting of thousands of cells and controlled by a single center, will form a single fruiting body. How are we to explain the spiral patterns seen in Figure 1.11a; and why do discrete streams form from continuous distributions of amoebas? Answering these questions requires us to move to a level of dynamical analysis that is above that of the single cell.

Excitable Media

During the 1960s and '70s there emerged a description of dynamic behavior in interacting units such as nerve or muscle cells, which have the property of responding to an electrical signal by generating a similar signal and propagating it to another connected cell. These became known as excitable media, and both the units and their collective behavior were seen to have certain properties. A necessary property of the units in order that signals should propagate in one direction only is the occurrence of a refractory condition after signal transmission, during which the unit is unresponsive to further signals. This was well known in nerve and muscle: the electrical action potential that propagates along the cell membrane leaves the membrane temporarily unable to respond again to another electrical impulse. It takes many milliseconds for the cell to recover its excitability. Another essential property is that the signal must decay or be destroyed rapidly, so that the medium does not get saturated and make all signals invisible.

When they created computer models of excitable media, researchers found that certain spatial patterns of signal transmission could occur. One of these is a spiral pattern such as we saw in aggrgating cellular slime mold amoebas. It was not initially evident how such spiral waves of activity arise in excitable media. Examination of the pattern of activity revealed that a group of cells at the center of a spiral acts as a self-exciting loop. Clearly, the circle of cells has to be large enough for each cell to have recovered from the previous excitation before the signal comes round again. Once this reentrant cycle is established, however, it can continue indefinitely; furthermore, it will tend to run at the maximum frequency possible for the system, determined by the refractory period.

These properties explain observations made on wave patterns in the slime mold. First, such patterns are possible only because, like nerve and muscle, amoebas have a refractory period: after they have responded to a cAMP signal, they are unable to respond again until after a recovery period that initially lasts several minutes. Also, they release an enzyme that rapidly destroys the signal, so that each pulse is transient. Yet the first waves to occur in a mass of starving amoebas are not spirals but concentric circles. The origin of the circular wave is a single cell at the center of the bull's-eye pattern. Such a pattern can occur only if the central cell is itself producing a periodic signal and so acts as a local pacemaker. This is what starving amoebas do.

Levels of Explanation

Not all cells simultaneously become periodic pacemakers. There is considerable diversity of metabolic state over the population, so the setting up of territories under the influence of different centers occurs gradually. But over time, the initial concentric wave pattern gradually transforms into spirals of the type shown in Figure 1.11a. The explanation for this came from the general study of excitable media, described above: the signal from a single pacemaker can travel around a circle of cells and close on itself, forming a self-exciting loop. Spiral wave forms arising from excitation waves traveling around a closed circle of cells tend to run at the maximum frequency, determined by the refractory period of the units. Once these self-exciting loops have arisen they will tend to take over from the concentric circle patterns, which run at lower frequencies because single, unstimulated cells will not "fire" as soon as they have recovered. Thus we see spirals gradually replacing concentric circles as the dominant signaling pattern. Collections of amoebas evidently acquire the properties of excitable media.

But a paradox emerged. Amoebas move toward the source of the signal. But as the signal travels past an individual amoeba, the signal looks the same from behind as it does from in front. How can amoebas distinguish a signal that is arriving from one that is leaving? If they can't, then they should first move forwards and then backwards, getting nowhere. This paradox was resolved by going back to the amoebas and asking a new question about their behavior as units. Do they become not only unable to respond to another signal after they have produced their pulse of cAMP, but also desensitized and unable to reorient or repolarize in a different direction after they have responded to an oncoming cAMP pulse? This appears to be the case, though the mechanism involved is not yet clear[9]. What we see here is a dialogue between a high-level analysis of interacting amoebas as excitable media and an analysis of the necessary properties of the cells themselves. One can take a reductionist position and argue that once the pattern of behavior of cells necessary for spiral wave formation and directed movement has been understood, the dynamics have been reduced to properties at the cellular level. This seems perfectly reasonable. It is important, however, to acknowledge the role of a higher-level analysis, using the global concept of an excitable medium, in clarifying the origin of the dynamics. In other cases this reduction has not yet occurred, and may never occur.

Consider the formation of separate streams of amoebas moving toward the aggregation centers. How are we to explain this? At the cellular level, it has been observed that this is the stage at which cells become adhesive, sticking together and so beginning to form a multicellular system that transcends the properties of a collection of cells. Is this the crucial cell-level step that initiates a new global level of order? Subsequently, cell adhesion and cell interactions trigger differential gene expression. Cells in some regions undergo transformation to states resulting in stalk formation; others become spores. Large-scale organization emerges.

The instability that transforms a uniform distribution of cells into branching streams arises not from cell-to-cell adhesion but from the combined effect of the repetitive cAMP pulse together with cell chemotaxis. Instead of moving along the uniform wave front defined by the propagating signal, the system shows a dynamical instability, with a tendency for cells to bunch together. This then enhances the instability, which grows as a spatial mode with a characteristic wavelength corresponding to the dimension of the streams.[9] Thus the streaming pattern is an emergent property of an excitable medium with the characteristics peculiar to cellular slime mold aggregation, explained by a dynamic analysis of the system as a continuum with specific properties of cell signaling and chemotaxis. Cell adhesion then consolidates this pattern and leads to cell differentiation. What we see here, then, is a sequence of symmetry-breaking events typical of morphogenetic processes.[10] The distinctive emergent forms of the living realm cannot be understood simply in terms of particular cell properties but require also the development of such theoretical concepts as excitable media.

From Control to Participation in Science

Earlier, we remarked that the significance of the observer would be a recurrent theme in our study of emergent phenomena. There is an unavoidable ambiguity in the very term used for our area of research: the sciences of complexity. We described this as the study of those systems in which there is no simple and predictable relationship between levels, between the properties of parts and of wholes. But what is the definition of complexity that is being used here? There is no objective definition because the very concept is relative to an observer. My complexity may be your simplicity. Chaos is simple from the point of view of the iterative computations used to generate it, but complex from the perspective of someone trying to predict the resultant sequence of states or describe the generic properties of the pattern generated. And it was not until the relevant order parameter was developed by observers that the simplicity of Bénard cell patterns could be systematically described and understood as an enormously constrained dynamic pattern relative to the degrees of freedom available to the molecular components of the fluid, from which perspective the system is very complex. Emergent phenomena are manifestly the result of descriptions by observers; that is, emergence is a construct, but it is taken to refer to a real process in nature.

Still, the same can be said of all scientific discoveries. So why are we stressing observers so forcefully in this context? It is to emphasize that with the sciences of complexity the process of understanding the world enters a new phase, with two primary characteristics. First, control of natural phenomena begins to slip out of the grasp of observers, both because sensitivity to initial conditions severely limits the possibilities for prediction and control and because emergent properties of complex systems are unpredictable from a knowledge of parts, so that deliberate manipulation of components brings unforeseen consequences. Second, these emergent properties can nevertheless be made intelligible in terms of appropriate descriptions of the processes involved, by using high-level concepts that capture their essential aspects. How observers choose to act in relation to natural phenomena that are thus revealed to be unpredictable but intelligible is clearly open and unprescribed.

The sciences of complexity show us that we are embedded in a world fundamentally different from that which has previously characterized modern science, with its emphasis on prediction and control of nature. We can clearly exercise whatever control remains possible in complex systems. But there are other options, such as participating rather than controlling, that is, recognizing that we can influence complex systems and proceeding cautiously with such influence because of the fundamental unpredictability of our actions and their consequences. We can no longer be naive observers who live outside the phenomena we manipulate. The properties of organisms, their development and health, the dynamic activities of brains and communities, the characteristic order of ecosystems, the patterns of evolutionary change, are processes in which we are directly involved. For better or worse, we participate in them, and of course, we would wish to participate wisely rather than irresponsibly.

TWO

Order, Complexity, Disorder

For the philosopher is right to say, that nothing thicker than a knife's blade separates happiness from melancholy

—Virginia Woolf, *Orlando*

Entropy, Chance, and Randomness

Stories of small events that trigger profound changes are without number. As an example, let us cite an event that completely changed the life of the molecular biologist François Jacob. In his amazing autobiography *The Statue Within*,[1] Jacob writes that he was looking for a supervisor whose work was oriented toward his personal interests. His ideal was to find a place in André Lwoff's lab, on the third floor of the chemistry building at the famous Pasteur Institute. "In his office," Jacob recalls, "I told him about my ignorance, my willingness, my desires. He fixed me for some time with his blue eyes. Tossed his head several times. His laboratory was already fully staffed. There was no place for me." Jacob was deeply disappointed, but over the winter he tried several times to see Lwoff. Each time he was refused. In June, his optimism fading away, Jacob decided to try for the last time. When he met Lwoff, "I find his eyes bluer than usual, the toss of his head more pronounced, the welcome warmer." Before Jacob started to say anything about his ignorance, willingness, or desires, Lwoff announced, "You know, we have just found the induction of the prophage." Jacob greeted this news with an "Oh" into which "I put all possible surprise, amazement and admiration while thinking 'What does it mean? What can a prophage be? What does it mean to induce a prophage?' " Lwoff looked at him and said, "Would you be interested on working on the phage?" and Jacob stammered, "That's just what I'd like to do." Lwoff appointed him to start in September, and Jacob ran into a bookstore to look for the magic words "phage" and "induction." The story ends with the Nobel Prize, which Jacob shared with Lwoff and the great Jacques Monod. Nobody knows what would have happened if Jacob had not decided to try one last time. Many years later, he still asked himself why he was accepted: "to this day, I still do not know the answer. What I do know, however, is that had I been he, I would surely not have accepted into my lab a fellow like me."

The idea that a random event can change history has been a great source of inspiration for both scientists and writers alike. We live in a universe with strong laws and much contingency. In our search for the laws of complexity we often find islands of randomness in an ocean of regularity, like the island of trickery, home of games and gambling, found by the travelers in Gargantua and Pantagruel (Figure 2.1). Even games like chess, where strategy is fundamental, involve many open possibilities in which players choose one essentially equivalent move over another. In biological evolution, complex forms arise from both randomness and dynamical phenomena. In economic markets, individual random decisions involving limited rationality can trigger large-scale events and perhaps financial crashes.

Disorder is thus part of nature. How can we understand the origins and implications of randomness? How can we measure it? The microscopic rule leading to random behavior in nature was explored by some of the most celebrated physicists, including Boltzmann and Einstein. Boltzmann introduced a macroscopic characterization of what he called "microscopic chaos," namely, entropy. The underlying motivation of Boltzmann's work was the search for the origins of time's arrow.[2] Contrary to our daily experience, with its many irreversible events, the physics of microscopic entities seems to suggest that there is no real time arrow in physics. The equations of motion have what is called *time-reversal symmetry*. But since macroscopic events must be the result of many microscopic steps, the arrow of time needed an explanation, one that Boltzmann wanted to obtain from microscopic events. The search for some characterization of time's asymmetry led to the discovery of Boltzmann entropy.

Figure 2.1 The travelers reach the island of Trickery, in Gargantua and Pantagruel (drawn by Dore).

Entropy (see Box 1) is a measure of disorder in a physical system. But it has been used in different contexts such as information theory and ecology. Ecologists often equate entropy with diversity. For an ecosystem, the highest diversity would correspond to a museum, where each species is represented by a single individual (all possible "states," here species, are represented with the same probability, as defined in

Entropy and probability

The entropy of a given system can be defined by means of a probabilistic description of its states. For a physical system, say a gas, where the energy of its molecules can take only a discrete number of states, we have thus a set of possible energies, and we can define the set of probabilities associated with the set of possible energy states. Let P_j be the probability of observing a molecule with energy E_j (with $j = 1, 2, \ldots, S$). The equation for entropy is

$$H = -\sum_{j=1}^{S} P_j \log P_j.$$

The entropy H is maximal when the system is most disordered: this occurs when all events have the same probability. Let us consider the following example. Imagine a set of 100 dice spread on a table. We can spend some time ordering the system by putting all dice with the 6 face up. Let $P(1), P(2), \ldots, P(6)$ be the probabilities of observing a one, two, . . . , six, respectively. At the beginning we have $P(6) = 1$ and $P(\neq 6) = 0$ and so $H = -\log(1) = 0$. This is a highly ordered state with minimum entropy. Now we start to shake the table. As we perturb the system, more and more dice rotate into a different number, and it is easy to see that eventually, all numbers will be roughly equally represented, i.e., we will have $P(i) \approx \frac{1}{6}$ for all faces. This gives

$$H = -\sum \frac{1}{6} \log\left(\frac{1}{6}\right) = 6 \times \frac{1}{6} \log\left(\frac{1}{6}\right) = \log(6) \approx 1.79,$$

which corresponds to the maximum entropy available for this system. Not surprisingly, the entropy increased as we shook the table. And we know from experience that further shaking never returns the dice to the initial ordered state. Some things occur spontaneously as time goes by; others never do. Strictly speaking, there is a very small probability that the dice will return to their initial ordered state sometime soon. But (for a large number of dice) this probability is virtually zero.

Box 1). At the other extreme, if only one species is present, the entropy is zero. Let us note, however, that none of these extreme cases is observed in real (and complex!) ecosystems.

Entropy and disorder grow together. But nature also abounds in ordered structures. From geology we know that most minerals are stable structures organized in different lattice configurations. Ice is "nothing but" water molecules regularly placed as a lattice and interacting through chemical and hydrogen bonds (though we cannot predict such a structure from the quantum-mechanical properties of the single water molecule). When we look at snow crystals on a window, we look at a highly ordered structure. But if we look closely, complexity always involves both order and disorder. Ecosystems show well-defined regularities, but populations often fluctuate wildly. The brain stores extraordinary amounts of information, but its activity, as revealed by an electroencephalogram, is far from regular. Our cities are large-scale structures with a long history, but their growth is often almost organic, and we know that they can eventually disappear.

Thus complexity is neither complete order nor complete disorder. To illustrate this rough idea, let us for a moment examine the patterns shown in Figure 2.2. Figure 2.2a shows a very ordered crystal-like structure. We can infer the whole pattern from just a very small part of it. Figure 2c shows a totally random pattern. It has been obtained by repeatedly tossing a coin. Each time we get a head, we plot a white square, and if a tail, we plot a black square. Although this pattern is entirely random, the same statistical features are reproduced over and over at all scales of observation.

But now let us examine Figure 2.2b. The structure is neither totally random nor completely ordered. We observe triangles of many different sizes, and in between we can see domains of disorder. We perceive some underlying structure at many different scales. This pattern is

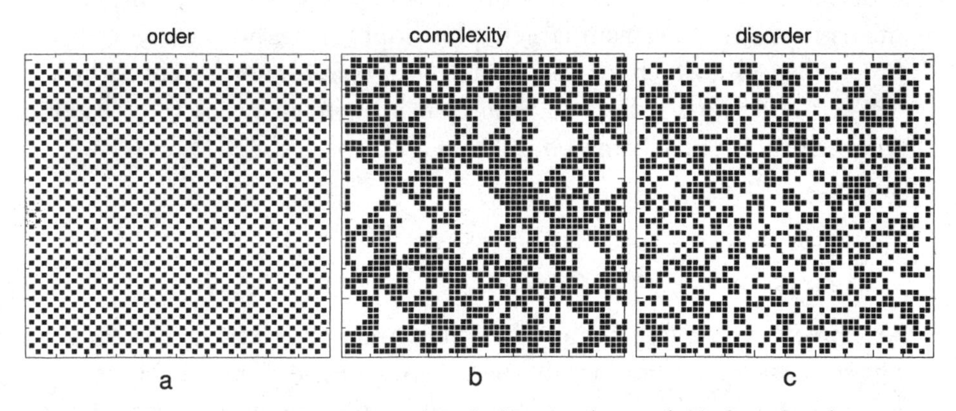

Figure 2.2 Three examples of (a) ordered, (b) complex, and (c) disordered patterns. The first is a low-entropy, regular lattice and a small part of it is equivalent to the whole. The last is a high-entropy pattern, but again a small part is (statistically) equivalent to the whole. The pattern in the middle has an intermediate entropy, but there are non-trivial structures at many different scales.

complex. It cannot be described in terms of a simple extrapolation of a few basic regularities. It displays nontrivial correlations that are not reducible to smaller, more fundamental units. This is, in fact, an example of a *fractal* object. The term "fractal" was first coined by Benoit Mandelbrot[3] to describe geometrical structures with features of all length scales; Mandelbrot made the surprising observation that nature abounds in such fractal, or self-similar, objects.

Figure 2.3 shows a clear example of this idea. Imagine that we pilot a spaceship to the moon. At some point in the travel we fall sleep, and we awake again when the spaceship is already approaching the moon's surface. Through our window, we see the image shown in figure 2.3, which is typical of the moon and other planetary surfaces lacking atmospheres. How far are you from the moon's surface? This question cannot be answered from simply looking at the image, because the moon's surface is a fractal object: we can see craters within craters at any scale of observation. There is no characteristic scale that we can use as a ruler. Like lunar surfaces, mountains, river networks, blood vessels, rainforest tree distributions, and cities are also fractal objects.[4]

Nature thus abounds in complex structures. But defining and measuring complexity is not a trivial problem. Entropy is not a good measure, since it increases with disorder (so a dead organism would be more complex than a living one). A properly defined complexity measure C should reach its maximum at some intermediate levels between the order of a perfect crystal and the disorder of a gas. This intuitive observation would remain largely philosophical without a theoretical framework with which to quantify it. In this chapter we present a whole class of phenomena involving the transition from order to disorder where the point of maximal complexity is sharply defined. This is where complexity lives.

Playing with Magnets

Among the most fascinating phenomena encountered in nature are phase transitions. There are two basic types: first-order (like the water–ice transition) and second-order (like that of magnetic materials; see below). The second group is particularly interesting for our analysis of complex systems. These transitions are called critical phase transitions, and the general class of phenomena arising from them are called critical phenomena. Of the many critical phenomena that have been analyzed

Figure 2.3 Moon's fractal landscape.

by physicists, one of the best known is the ferromagnetic transition.[5] A small piece of iron can tug on a paper clip at room temperature, but if we heat it to a high temperature, the magnetic force disappears.

Atoms of iron are themselves small magnets, which we can imagine as pointing "up" (+1) or "down" (−1). Each of these atoms occupies a node of a regular three-dimensional lattice and interacts only with its nearest neighbors. It is known that these atoms will spontaneously align in the same direction as their nearest neighbor (this is the configuration of lowest energy). At low temperatures, adjacent atoms couple together, creating large regions with the same orientation. This is illustrated in figure 2.4(a), where we see a unit (the white square) totally surrounded by neighbors of the opposite direction (black squares). At low temperature this unit will tend to switch toward the low-energy state by acquiring the same state as its neighbors. The magnetic fields created by all the regions add up to a well-defined global magnetization. The system is ordered.

At high temperature, atoms are constantly perturbed by thermal noise and can adopt either of the two orientations (figure 4b). Even in the most extreme case, when it is opposite to all its neighbors, the unit can switch or remain the same with equal probability. Local interactions cannot order the system even at small scales, and entropy wins out over (interaction) energy. The sum of the magnetic fields is zero, since half of them will be up (on average) and half down. The system is said to be disordered.

But what happens at intermediate temperatures? Let us consider the following classical experiment. We start with the iron at high temperature and slowly cool it down. We use some experimental apparatus to measure M, the magnetization. Clearly, M will be zero in the high-temperature domain. But something strange happens: at a critical temperature, M sharply increases to a nonzero value (see Figure 2.5). Suddenly, macroscopic order shows up. From a disordered phase where no correlations are present we move through a sharply defined boundary region where order and disorder coexist and the most remarkable features of many complex systems emerge from microscopic interactions.

To understand the nature of this phase transition, let us start with one of the most simple and celebrated models of statistical physics: the Ising model.[6] Instead of following a detailed description of iron atoms and their bonds, which necessarily involves a quantum mechanical formalism, a toy model, known as the Ising model, will be used as a theoretical approximation to this problem. In this model we replace the complicated iron atoms by binary variables, allowed to take two values only: $+1$ (up) and -1 (down). We will use a simplistic description of their interactions, almost a caricature of reality. At each time step we choose a given node, which will be up or down. With some *transition* probability the state of this virtual atom will switch to the opposite state. This transition probability will include the previous ingredients

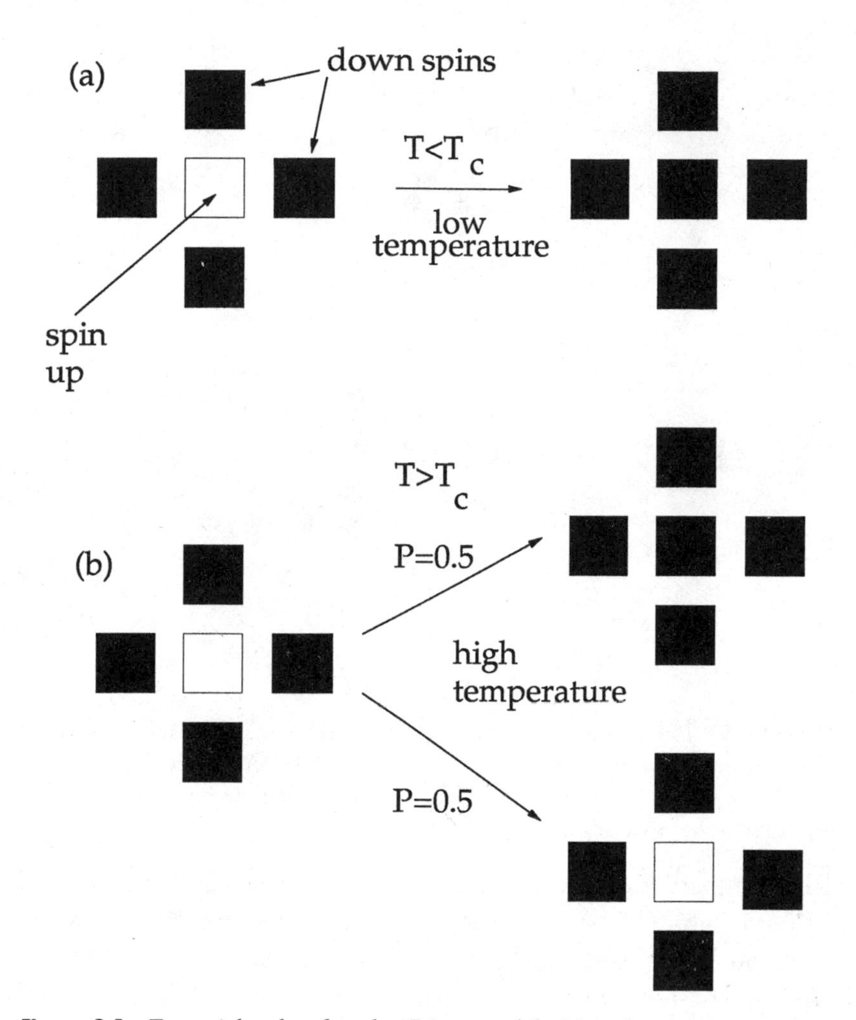

Figure 2.4 Essential rules for the Ising model: (a) at low temperatures (ordered phase) a given spin will get aligned with its neighbors; (b) at high temperature (disordered phase) the previous rule will fail and the spin, even if all its neighbors have the same orientation, will remain in the same state or switch to the other with equal probability.

(interaction and temperature) in a simplistic way. We shouldn't be surprised if this caricature fails to reproduce anything relevant.

But let us simulate this toy model in two dimensions using the temperature as a tuning parameter (Figure 2.6). At a high temperature T we see a random mixture of black and white squares (Figure 2.6). At low T we see an almost homogeneous lattice either black or white (Figure 2.6). But at the critical point something unexpected occurs: fractals spontaneously emerge. Figure 2.6 is a snapshot of the two-dimensional Ising model close to the critical temperature, clearly showing some

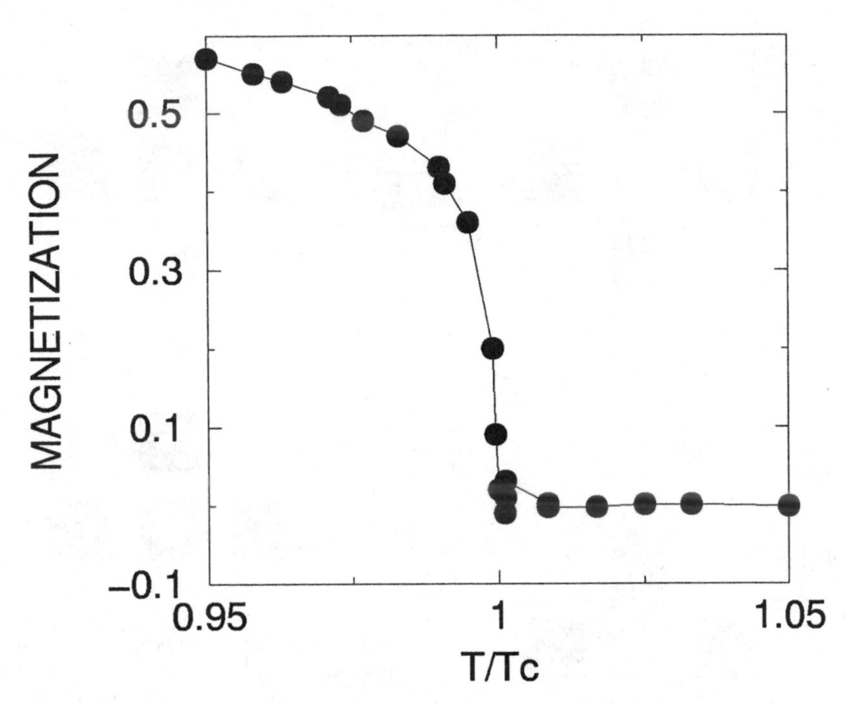

Figure 2.5 Phase transition obtained from experimental data for a two-dimensional magnetic system (see Back et al., 1995). At the critical temperature the system shows a sharp change linked to the phase transition from a non-magnetic to a magnetic state.

underlying order. We can see many different-sized clusters of both black and white squares. In fact, whatever scale we choose, we will always observe the same features: islands inside of islands inside of islands—a nested set of clusters over and over again. The Ising model close to the critical temperature spontaneously self-organizes into a fractal object. If you look at Figure 2.7 and start from an 800 × 800 lattice, and then look closely at a 400 × 400 subset, you will find the same basic pattern. If we measure the magnetization (here simply the sum of spin values) at different temperatures and plot its value, the plot (Figure 2.8) fully reproduces the experimental curve (Figure 2.5). A dramatic change occurs at $T = T_c$, the critical point of phase transition.

We can make several relevant observations. First, complexity seems to appear (in a nontrivial way) when T is tuned from high to low values, moving from order to disorder. Second, the sharp transition tells us that complexity arises at well-defined points. Third, the Ising model shows that even though our virtual atoms (and the real ones!) interact only with their closest neighbors, long-range correlations and large structures

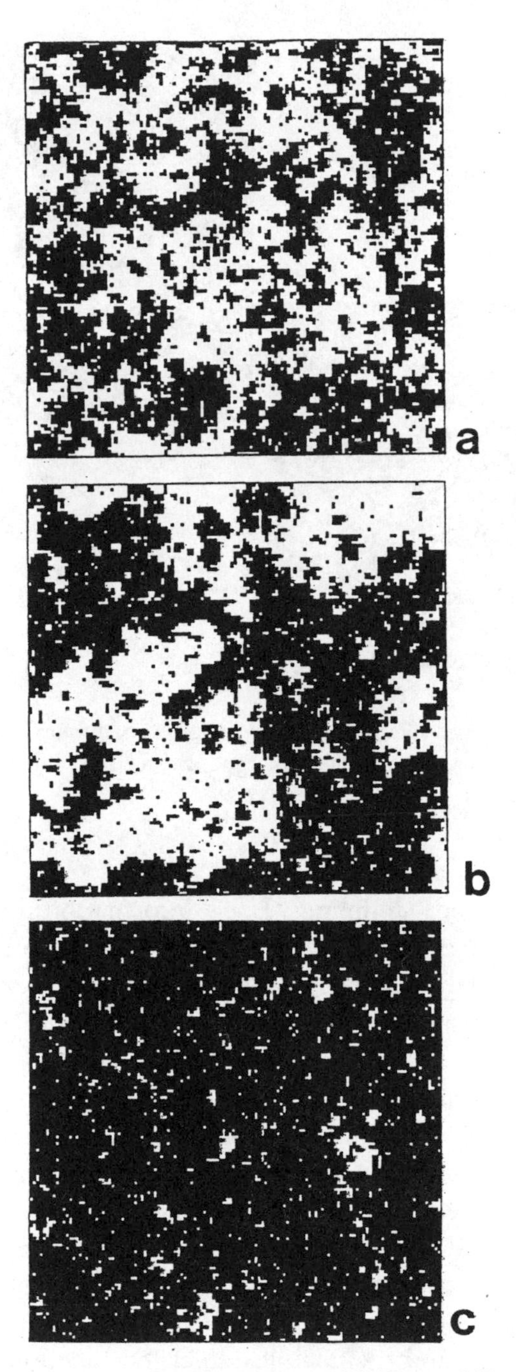

Figure 2.6 Phase transition from the Ising model. Three different temperatures are shown here associated to: (a) the disordered phase, (b) the critical point, and (c) the ordered phase.

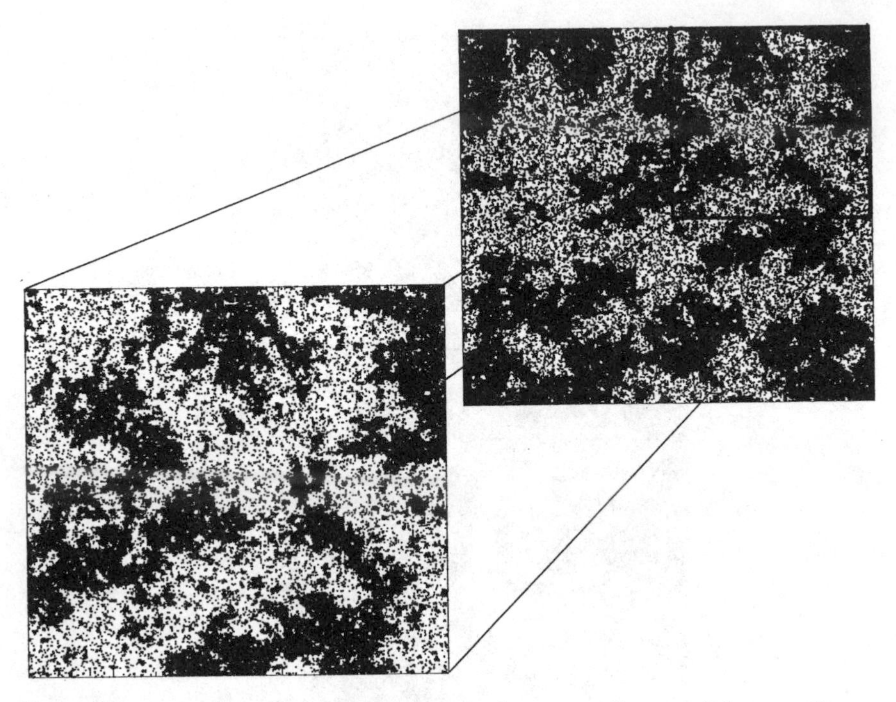

Figure 2.7 Fractals and phase transitions: when a small part of the two-dimensional lattice (at the critical point) is enlarged, we see the same properties reproduced at all scales.

emerge from the local couplings. Local information is somehow able to propagate through the whole lattice, creating structures that cannot be understood solely in terms of the properties of the units. Even in physics, reductionism often fails.

But there is still something more to be mentioned, something surprising, counterintuitive, and with tremendous consequences for our understanding of complex systems. The Ising model is not just a nice approximation to the critical behavior of a real piece of iron. It gives us a number of quantitative characteristics that we can measure and compare with reality. And now comes the surprise: the Ising model for the statistical behavior of the phase transition close to T_c gives *exactly the same values* as the experiment.[5,6,7] This implies that in order to understand the complex emergent phenomena arising at the critical point, we don't need a detailed description of the system. The reductionist approach assumes that a meticulous description of the parts is a necessary condition for a quantitative knowledge of the whole,

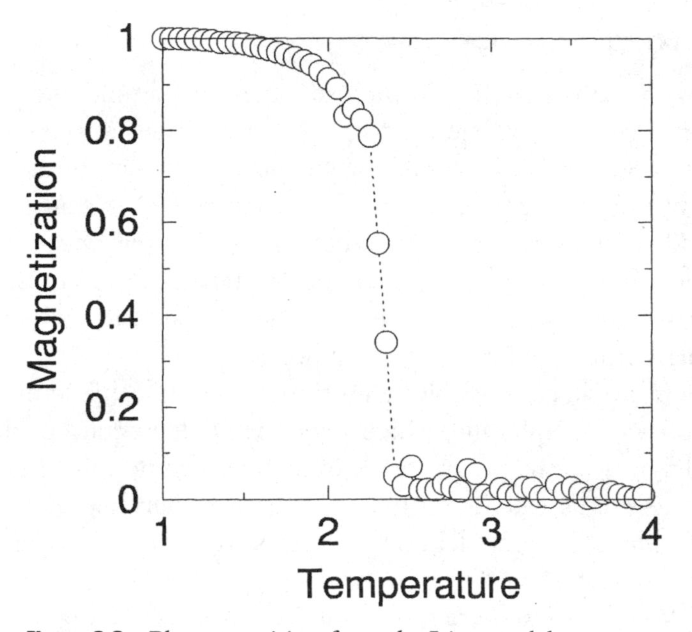

Figure 2.8 Phase transition from the Ising model.

yet here a very small number of features gives us an exact description of macroscopic behavior. This implies that disparate systems sharing some fundamental microscopic properties (essentially related to the symmetry and the dimension of the system) will behave in the same way close to their respective critical points. This idea is called *universality*. Simple rules can generate very complicated patterns of behavior, and the *interactions* among the different parts of a complex system—and not the detailed properties of their component parts—are the relevant part of the story. Universal behavior seems to be present in many natural systems, as well as in human societies, the economy, and the growth of cities. One of the most important challenges of complexity theory is to identify and characterize the different classes of rules responsible for their macroscopic behavior. These *universality classes* sometimes allow us to understand how complexity emerges in the universe.

Because of its simplicity and richness, the Ising model and a wide family of related models have been used as a theoretical framework in very different areas of science, often with extraordinary success. They have been applied to the distribution of galaxies in our universe, the astonishing adaptability of RNA viruses to their hosts, and the dark interiors of tropical rainforests.

Information and Complexity

Now, we need to measure complexity in some simple and intuitive way. Since complexity emerges from the interactions of the individual units, interactions must somehow be present in our measurement. When different parts of a system, for instance the molecules of a gas in equilibrium, do not exchange information, matter, or energy, then their global behavior can be understood in terms of the simple sum of separated entities. But once interdependencies appear, we require an appropriate measure involving these links.

There are many possible definitions of complexity, and here we present a very simple one, which needs only elementary probability. Consider an arbitrary system made of many different units. Let us take two of these units, call them i and j, and assume that they take only two values (as in the Ising model). If their states are indicated as S_i and S_j, at a given moment, then S_i will be either $+1$ or -1. The probabilities of finding S_i in these states are $P_i(+)$ and $P_i(-)$, respectively (and the same for S_j). Now, in order to introduce interactions, we need to take *both* simultaneously and consider their joint probabilities P_{ij}. For example, $P_{ij}(+, +)$ would be the probability of finding simultaneously both units in state $+1$. There are clearly four possibilities: $i+\ j+$; $i+\ j-$; $i-\ j+$; $i-\ j-$. This may be easier to grasp if you think of i and j as coins and $+$ and $-$ as heads and tails.

Let us now consider a very simple definition from elementary probability. If two events are independent, we have

$$P_{ij}(a, b) = P_i(a)P_j(b),$$

that is, the probability of finding the two units in two specified states is simply the product of the separate probabilities. The probability of two coins landing both heads, for instance, is $\frac{1}{2} \times \frac{1}{2} = \frac{1}{4}$ because the two coins flips are independent events.

But if the behavior of one unit modifies the state of the second one, then $P_{ij}(a, b) > P_i(a)P_j(b)$. This simply says that the probability of finding both i and j in some set of states is more than the product of the separate probabilities. It is as if the two coins were connected by a flexible spring that made them more likely to land both heads or both tails. So we can define a complexity measure (which we call *distance to independence*, D_I) by performing the sum of all the differences

$\delta = P_{ij}(a, b) - P_i(a)P_j(b)$. If the system is formed by parts that are independent, this measure will be zero.

In Figure 2.9 we have calculated this measure for the two-dimensional Ising model. The two extremes are easy to interpret. At high temperatures, the spins are completely independent: they behave like coin flips. At low temperatures, one of the states is predominant, and so we have, for example, $P_i(+) = P_j(+) \approx 1$ and $P_{ij}(+, +) \approx 1$. Meanwhile, $P_{ij}(-, +)$, $P_{ij}(+, -)$, and $P_{ij}(-, -)$ are all close to zero. All the terms in the definition of D_I cancel, and so again complexity is zero. But at intermediate temperatures none of these conditions are fulfilled, and in fact, this measure increases sharply close to T_c and peaks at this point, as expected from a well-defined complexity measure. This simple definition allows us to make clear what characterizes complexity: the emergence of interactions among different units and their conflict with randomness. Order and disorder find a compromise right at the critical point.

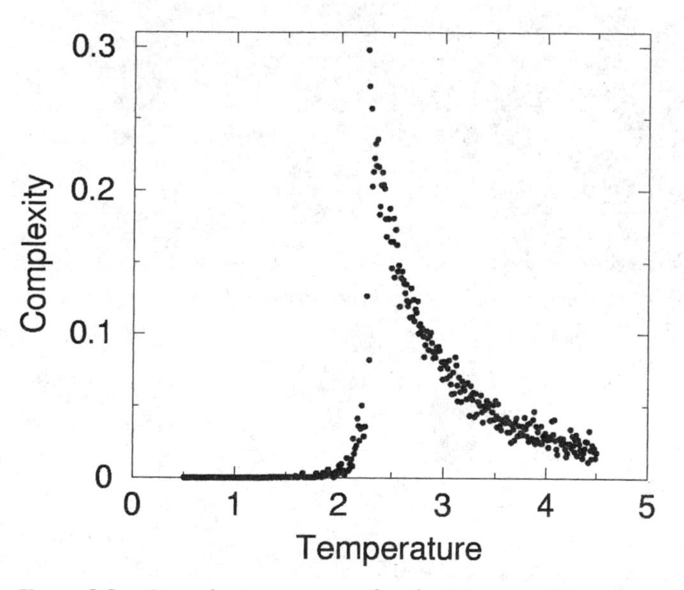

Figure 2.9 Complexity measure for the Ising model.

Renormalization

The critical point is a very special place. Above and below this point a system's behavior changes quickly and fractal structures disappear.

There is a powerful mathematical technique that allows us to find the critical point and also to characterize its statistical behavior.

Consider a square lattice with L nodes on a side. Each node can be occupied by a particle with probability p (and thus is empty with probability $1 - p$). Three examples of such a lattice are shown in Figure 2.10A–C for three different p values. The procedure followed is very simple. We choose a given p, say $p = 0.5$. At each node we flip a coin, and if it gives a tail, we place a particle at that node (and plot a black square). If not, we leave the site empty (white square). Let us call our occupied sites "trees." Imagine that a tree can be burned and that if one tree has a burning neighbor, it also starts to burn. Take the trees at the bottom of our two-dimensional lattice and burn them. The fire will propagate to the nearest trees. Clearly, if only a few trees are present, then the fire will go out after a few steps, while if the lattice is almost full of trees (large p), a large fire will propagate from the bottom

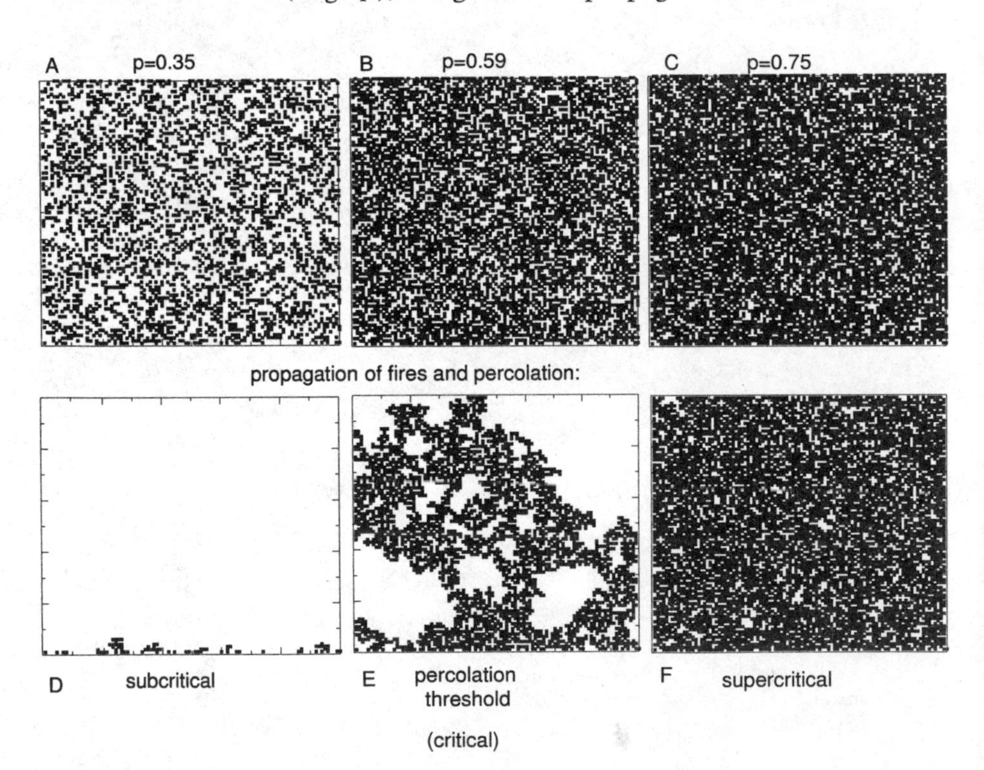

Figure 2.10 Percolation on a two-dimensional lattice. (A–C) three examples of the randomly generated patterns obtained by dropping "trees" at random locations with probability p, for three different probabilities; (D–F) results from the burning of the bottom row of trees.

to the top. Nothing interesting seems to occur for intermediate values of p. But surprisingly, randomness again is able to generate complex structures close to very special (critical) points.

If the whole set of points involved in a fire (the set of burned trees) is plotted as black squares, some underlying structures start to appear. In Figure 2.10 D–F we show some of these structures. For small p the fire does not propagate far, as expected. For large p the set of burned trees and the original set are not very different (Figure 2.10 C,F), since most trees have been burned, also as expected. But at some intermediate point (here $p = 0.59$) the web of burned trees becomes a large cluster with a very complex shape. In fact, this cluster is a fractal. If we zoom in on this object we cannot see clear differences: it is a self-similar set. In fact, this large object emerges at a critical point, known as the *percolation threshold*,[8] the lowest value of p at which the fire is able to propagate from bottom to top. Contrary to our expectations, this point is a sharp threshold: the percolating cluster is present for $p \geq p_c$ and never for $p < p_c$.

We can see this sharp transition by performing a couple of simple measures (Figure 2.11 a–b). First, we can calculate how many of the occupied sites are burned. The fraction of burning trees shows a sudden transition close to p_c, with a shape closely resembling the one that we obtained from the Ising model for magnetization (Figure 2.11a). And there is another way of characterizing this complex object: we can measure how long it takes for the fire to extinguish completely. One might expect something like an increasing function with p, since more trees means larger fires. But again this is not what we find: there is a sharp maximum at the critical point (Figure 2.11b). Below it, the fire cannot get far, and the time needed to burn the few trees involved in the fire is short. Beyond this point, the fire spreads without difficulty in a linear way: it needs about L steps, the distance in squares from top to bottom, to burn every tree. But why a maximum at p_c? We can understand this by looking at the percolation cluster. This object has a fractal shape. The fire must follow a very complex path, through many more than L squares, to burn every tree accessible to it. Somehow, the time required to burn all trees at the critical point is a measure of how difficult it is to describe the object. This is, in fact, a general feature of complexity: any quantity that measures a system's intrinsic details (such as path length) will show a sharp increase near the critical point.

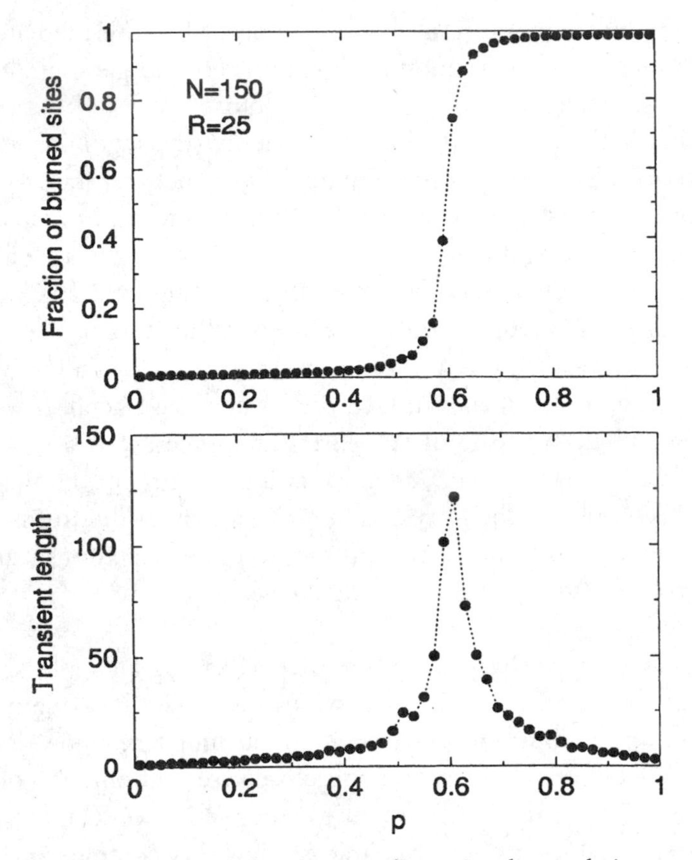

Figure 2.11 Phase transition for the two-dimensional percolation experiment shown in Figure 2.10. Upper plot: the fraction of burned trees displays a sharp increase close to a critical value, separating the phase where no propagation occurs to the percolation phase. Bottom picture: the time required to complete extinction of the fire.

There is a powerful and elegant tool that can be used to characterize this critical point and show how special it is. This mathematical tool is known as the *renormalization group*, and it was first introduced by the particle physicists Murray Gell-Mann and Francis Low. This sophisticated technique was deeply developed by the physicist Kenneth Wilson and has been a source of intense theoretical work in several areas of condensed matter physics and complexity theory. Although most of its areas of application involve rather difficult mathematical approaches, we can easily explain the main idea by using the Ising and the percolation models as clever examples.

Let us start with the Ising model. We have seen that this model shows a phase transition and that a fractal pattern emerges at a critical temperature, known as the *Curie temperature* T_c. In fact, at T_c the clusters of identical spins start to percolate through the system, and the fractal structures are in fact percolating clusters. Imagine that we take the critical point in two dimensions (see again Figure 2.7), using a very large lattice, and look at it from some distance. As we move away from the system we can no longer see the small details: the smaller scales are not available to our eye's resolution. Yet we observe the same pattern: percolating clusters with fractal behavior. On the other hand, if we take a low temperature, where most spins point in the same direction, as we move away from the system we see the lattice as a white (or black) homogeneous system. If we consider the Ising system at high temperatures, it looks grey, like a homogeneous object with no special features at any scale.

How, then, do we find these special points where nontrivial self-similar features emerge? Let us consider the following computer experiment. We start from a large $L \times L$ lattice. Instead of moving away from it, we will "renormalize" it. To do this, we can follow a simple rule (Box 2). Let us now describe the lattice as 3×3 blocks of spins. Then we replace each of these blocks by a single "renormalized" spin S', by applying a majority rule: if most spins in the block are up, then S' will be up. Otherwise, it will be down. Then we rescale the lattice spacing in such a way that the distance of the new spins becomes the same as before. This is not different from looking at the lattice at some distance: here, too, the small-scale details disappear. If some nontrivial, large-scale, self-similar feature is present, it will be preserved under the renormalization procedure. For $T < T_c$, even if some amount of structure is present (as small clusters) in the initial lattice, renormalization ultimately leads to a homogeneous system. The renormalization group is telling us that these structures are not preserved as we move away from the system. Likewise, for high temperatures T, even if we can see some small-scale structure, as the rule is applied over and over again the lattice becomes more and more disordered, finally looking like a set of random black and white squares. The renormalization group tells us that we are in the disordered phase, no matter how close we were to the critical point at the beginning. Only if we start at exactly $T = T_c$ will the successive renormalizations preserve the large-scale features of the fractal pattern. We can still see percolation, we can still see self-similarity: the critical boundary is a very special point, where complexity appears the same at all length scales.

Renormalization and Percolation

The renormalization group approach leads to simple ways of determining the percolation threshold by means of a simple mathematical transformation. As an example, let us consider the following problem: how to calculate the location of the critical point p_c using the idea of self-similarity (fractal behavior).

Imagine that we have again our two-dimensional lattice and that we look at the cluster formed by burned trees for different values of p (as shown in Figure 2.10D–F). A renormalized lattice is obtained by applying a simple procedure: let us take 2×2 blocks of sites. Several possible combinations of black and white sites are available (Figure 2.12 a). Let us imagine that these sites are metallic particles and that an electric current can move through two connected points from left to right (no current can take place through the diagonal). Then we consider a four-site square lattice mapping to a single (renormalized) site. All configurations of sites that span the cell from left to right are taken to be percolating and will map onto a single site on the new (renormalized) lattice (Figure 2.12a).

Using this rule we get a new, renormalized, lattice. The probability p' of finding an occupied site in the new lattice is easily obtained from the previous one. Let us assume that p is the probability of occupation in the original lattice. The probability of having all sites occupied in the 2×2 block is then p^4. If only three sites are occupied, we have probability $4p^3(1-p)$, since four possible ways of putting three particles in the 2×2 block are allowed; similarly, we have probability $2p^2(1-p)^2$ for the two-particle case. So the probability of having an occupied particle after renormalization is

$$p' = p^4 + 4p^3(1-p) + 2p^2(1-p)^2 = 2p^2 - p^4,$$

and we can repeat the renormalization by iterating the previous map. For the nth renormalization step we get

$$\begin{aligned} p(n+1) = f(p(n)) &= p(n)^4 + 4p(n)^3(1-p(n)) + 2p(n)^2(1-p(n))^2 \\ &= 2p(n)^2 - p(n)^4. \end{aligned}$$

This iterated map has three critical fixed points, obtained by using the polynomial $p = f(p)$, i.e., $p^4 - 2p^2 + p = 0$ (see Figure 2.12b). The following three physically meaningful points are obtained: $p^* = 0, p^* = 1$, and $p_c = 0.618$. The first two are the trivial ones, corresponding to totally empty and totally filled lattices, respectively. Clearly, these are two trivial cases where renormalization leads to the same pattern. But there is an additional point, p_c, where such self-similarity is also present. This point

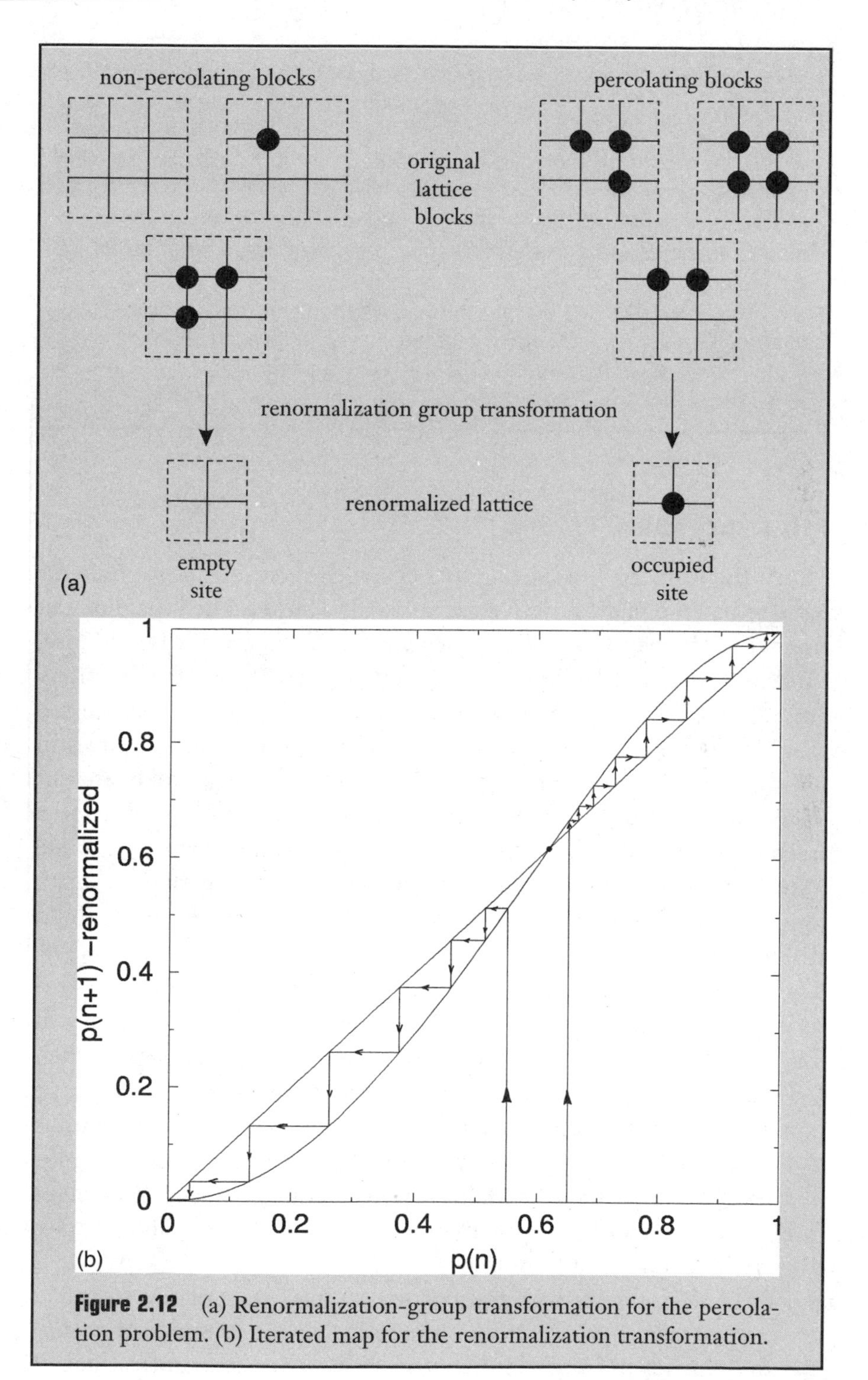

Figure 2.12 (a) Renormalization-group transformation for the percolation problem. (b) Iterated map for the renormalization transformation.

corresponds to the *critical point* of this system. It is very close to the exact value $p_c^{exact} = 0.593$ (as determined through extensive computer simulations on very large lattices), and we can see that it is unstable: starting from $p < p_c$, the iterates converge toward $p^* = 0$, and starting from $p > p_c$ they move to $p^* = 1$. In other words, unless we start at exactly the critical point, the starting lattice becomes more and more homogeneous as the transformation proceeds: no large-scale structure is preserved.

The renormalization group is a powerful tool: using a simple procedure, we can detect with rather good accuracy the presence and position of the critical point. And we see that it is a very special point: only if $p = p_c$ do we remain at criticality.

How to Get There

When the Emperor Joseph suggested that Mozart's *Abduction from the Seraglio* had too many notes, Mozart is said to have replied that none of them could be removed. All were necessary. Classical music is not noise; each piece has a delicate structure that requires concerted harmony among its different parts. But in fact, some scientists have suggested that classical music *is* a kind of noise: a special, ubiquitous, and still poorly understood type called $1/f$ noise.[4,9] This pattern, widely present in complex systems, is the time equivalent of a fractal pattern in space: a fractal pattern in time. And in fact, although we cannot remove a single note from a master's composition, it has been shown that Mozart's music is fractal, so in fact, we could remove many notes and it would still sound like Mozart!

An example of $1/f$ noise is shown in Figure 2.13, where the corresponding time fluctuations of the Ising model at criticality are shown for an 800×800 lattice. If we zoom in on the original signal, then a rescaled time series has the same basic appearance. You may be familiar with this type of signal from looking at market fluctuations in the newspapers. The intensity changes in earthquakes, fluctuations of species populations over time, Internet storms, large-scale changes in brain activity, and nest activity in some ant species are all examples of systems exhibiting $1/f$ noise.

The name comes from the particular properties of the so-called power spectrum, a powerful tool for exploring the presence of particular frequencies in time series. A power spectrum is simply a plot of the energy a system generates at various frequencies. If the system is

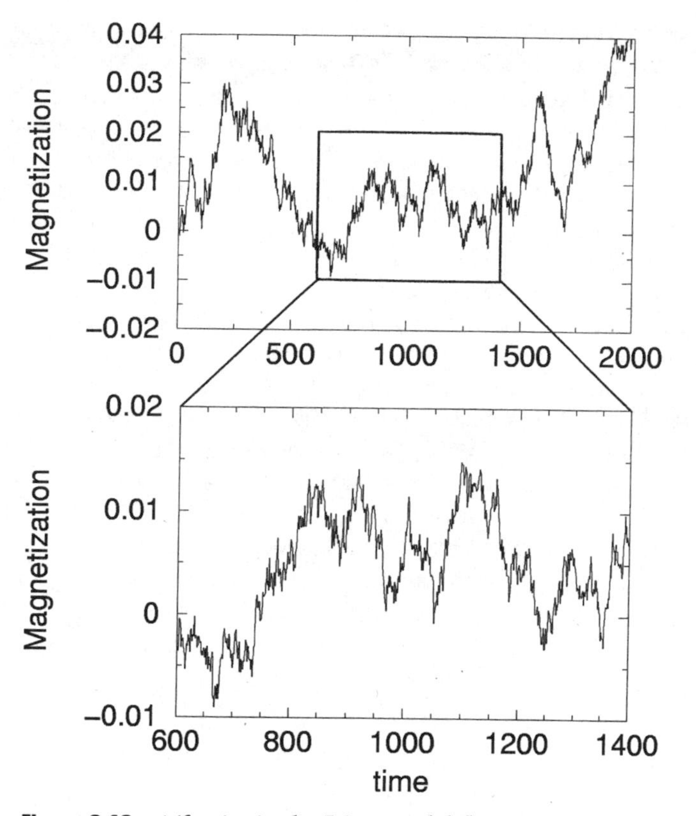

Figure 2.13 1/f noise in the Ising model fluctuations.

oscillating regularly, like a tuning fork, at a single frequency, then in the power spectrum plot we will see a single peak at that frequency. If two characteristic frequencies are at work, we will see two peaks. If one of them has a larger amplitude, then the power spectrum plot will show a higher peak at that frequency. What about random uncorrelated noise? Then no frequency is particularly favored, and we get a totally flat spectrum, known as white noise. But $1/f$ noise does something more interesting. Here all frequencies are represented, with a power law decay from the shortest to the largest. In other words, the power spectrum $P(f)$, which roughly measures the contribution of each frequency f to the overall series, is self-similar. There is no characteristic time scale: the dynamics are scale-free.

Power laws are an especially interesting feature of fractal patterns.[9] These power laws are widespread: they are obeyed by the distribution of city sizes, word frequencies, tree sizes in a rainforest, and the energy released in earthquakes (Box 3).

Self-Similarity and Power Laws

Power laws are very common in nature, from earthquakes or ecosystems to the distribution of matter in the universe. In figure 2.14 we show an example of the frequency of species $N(S)$ involving S individuals. This distribution, plotted in log-log scale, has been obtained from a sample of marine organisms, where the number of individuals of each species was measured. The result is that the frequency decays as a power: $N(S) = AS^{-\alpha}$, where A is a constant and α is the scaling exponent (for this specific example it is found that $\alpha \approx 1$). With this type of plot, the power law shows up as a straight line. This means that most species are represented by only one individual in the sample, but a few species can have a huge number of individuals. If the distribution were exponential, i.e. $N(S) = A\exp(-\gamma S)$, where γ is some constant, then we can easily show that no species with more than a given size would be observable.

To show that power laws are scale-invariant, we need only to see the effect of a scale transformation. Let us consider a linear transformation $S' = \beta S$, which we can see as a different scale of observation obtained by

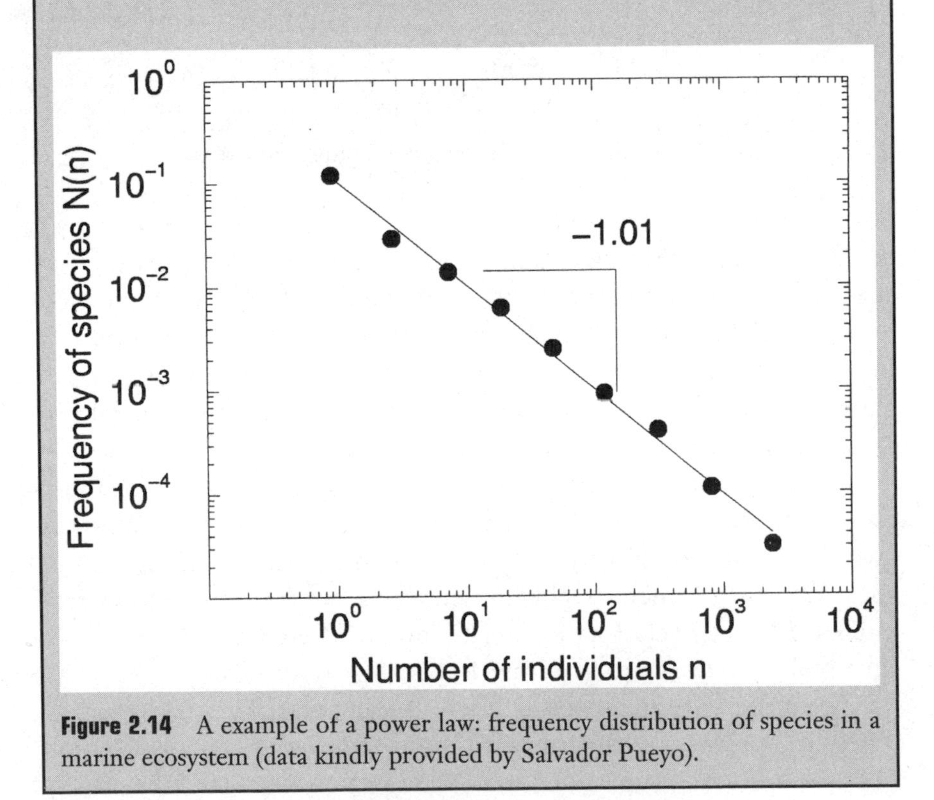

Figure 2.14 A example of a power law: frequency distribution of species in a marine ecosystem (data kindly provided by Salvador Pueyo).

rescaling S. In other words, if we look at a given size S, which we know follows a power law $N(S) = AS^{-\alpha}$, we want to see which distribution is followed by S'.

So we have

$$N(S') = N(\beta S).$$

Since S follows a power law, we have

$$N(S') = A(\beta S)^{-\alpha},$$

but we can write

$$N(S') = A\beta^{-\alpha} S^{-\alpha},$$

or in other words,

$$N(S') = A' N(S),$$

which is to say that the distribution is the same at all scales. It is interesting to mention that earthquakes also display self-similar fluctuations in the energy released over time and fractal distributions in space.

What is the dynamical origin of fractal structures? In 1987, a group of scientists led by the Danish physicist Per Bak, then at the Brookhaven National Lab, proposed a general scenario linking fractals with critical phenomena.[10,11] A key problem in a theory of fractals based on critical transitions is that in models like the Ising model, one must carefully tune a given parameter in order to determine the critical point. But in natural systems no one is tuning parameters: they organize themselves. This is particularly true in complex adaptive systems, like evolving populations or groups of traders in a stock market. In an adaptive system, rules change over time, these changes modify the system, and there is feedback from the system to the agents. Bak and his colleagues Tang and Wiesenfeld showed how such systems could reach critical states spontaneously. Their theory was called self-organized criticality.

How to get there? Their elegant solution involves the process of building a sandpile (Figure 2.15). The process is very simple: we

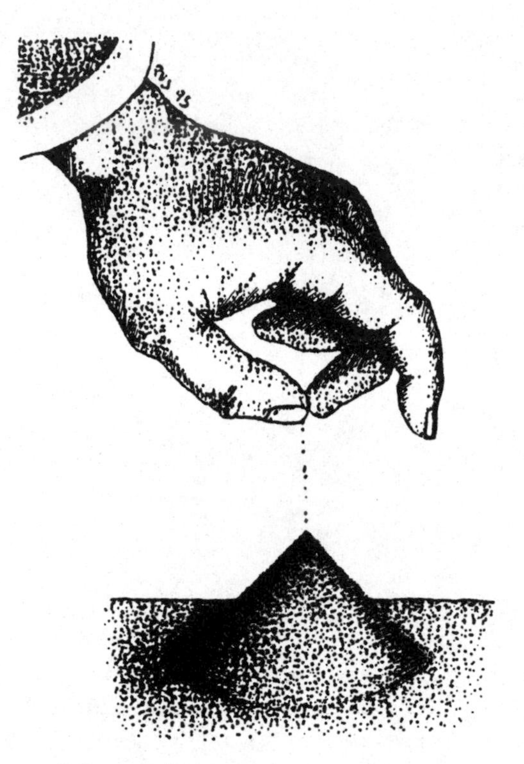

Figure 2.15 The sandpile experiment.

add sand slowly, one grain at a time. At the beginning of our experiment (preferably to be done on a sunny, windless day on a beach near Barcelona) the pile is relatively flat and the grains stay where they land. They are independent of each other, and their individual behavior is described in terms of gravity and friction. However, as we add more and more sand, the slope of the pile increases until it reaches a maximum, critical, angle. At this point, avalanches occur. Most avalanches will be small, involving only a few grains slipping down a short distance, but from time to time, very large avalanches will happen. In this regime, which is self-organized (we have not adjusted the pile to get the critical angle), the grains are no longer independent: a new behavior has emerged that cannot be extrapolated from the behavior of individual grains. Later, we will see that much the same thing happens to an ecosystem in which new species are slowly "injected," triggering avalanches of local extinction. No individual description of species allows us to understand this collective behavior.

The sandpile experiment is easily done in the lab by using grains of rice instead of real sand, and it shows that the size distribution of avalanches really does follow a power law. Bak's first experimental model, however, was not a real sandpile but a simple caricature of reality called *cellular automaton*. His strategy was the same one we used with the Ising model: use the simplest formal description able to capture the basic, universal features of the real system. The basic rules, given in Figure 2.16 (Box 4), are very simple and involve a slow driving process (one unit of virtual sand is added each step), a threshold (avalanches start when a given node becomes supercritical) and avalanche dynamics that are fast compared with the driving process. These ingredients are sufficient to define the state of self-organized criticality.

The sandpile cellular automaton shows a very rich range of patterns in both space and time. If we follow the time evolution of the sandpile, we get a highly fluctuating self-similar time series, with avalanches of all possible sizes (Figure 2.17a). By measuring how many avalanches involve a single grain, two grains, and so on, we find that they follow a power law distribution (Figure 2.17b). But we can also see fractals

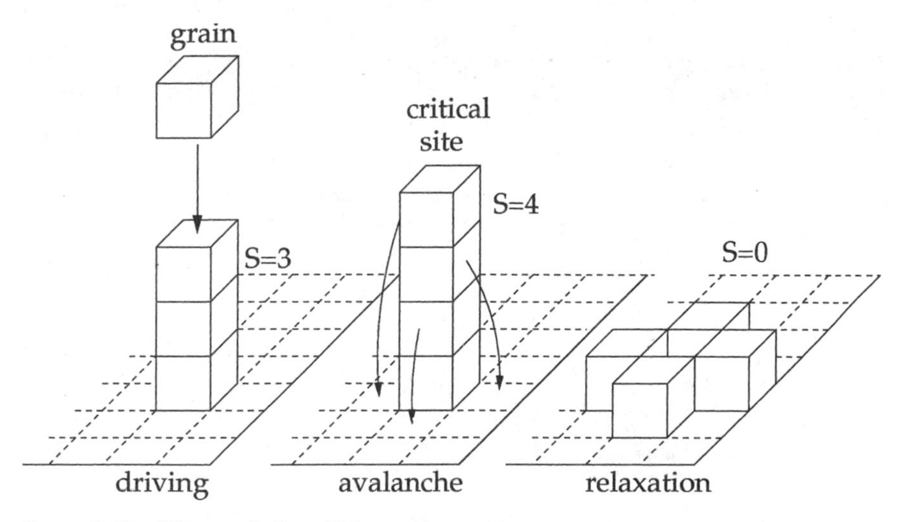

Figure 2.16 The sandpile cellular automaton.

in space. If we take the total set of grid locations involved in a single large avalanche, we see that this set forms a fractal object with a self-similar boundary. Most important, Bak's sandpile model shows a deep connection between spatial and temporal fractals: perhaps the origin of fractals is a dynamical proces. Maybe these two types of patterns are two faces of the same coin. It is interesting to see that, for example, earthquake epicenters are distributed in space in a fractal way, consistent with the observation that the energy released in earthquakes is self-similar in time.[12]

Nature abounds in power laws. And these scaling laws are also characteristic of critical states (but not exclusive to them; many natural phenomena displaying power laws involve very different mechanisms).

The Sandpile Cellular Automaton

We can simulate a sandpile by using a simple model. Consider a square $L \times L$ lattice. Each point is characterized by an integer value $S_i \in \{0, 1, 2, 3, 4\}$, which we call the height. The rules of the system are very simple: at each step, we drop a single grain of virtual sand at a randomly selected lattice point. Each time a grain is added, the local height increases by one unit (i.e., $S_i \rightarrow S_i' = S_i + 1$). If S_i' is less than $S_{max} = 4$, then nothing else happens, and we repeat this step after choosing a new random site. But if the local height is larger than $S_{max} (S_i > 4)$, then an avalanche starts to occur. Locally, the pile becomes unstable, and the four grains fall and become distributed over the four nearest lattice points (see Figure 2.16). However, as a consequence of the toppling of the grains to the nearest locations, one or some of them could also become critical. If this occurs, then a new toppling is produced, and a large avalanche can occur. Once the avalanche is finished (those grains reaching the limits of the lattice are lost) we start dropping sand again.

Measuring the avalanche size (in terms of the number of sites involved) we can plot the dynamics of this simulation, as shown in Figure 2.17a. We can appreciate a broad range of avalanche sizes, all of them produced by the same mechanism. The distribution of avalanche sizes is a power law (Figure 2.17b), i.e., the number of avalanches $N(S)$ of size S follows a distribution $N(S) = AS^{-\gamma}$, where A is a constant and $\gamma \approx 1.1$. Most events are small, but we only need to wait long enough to see events of any size.

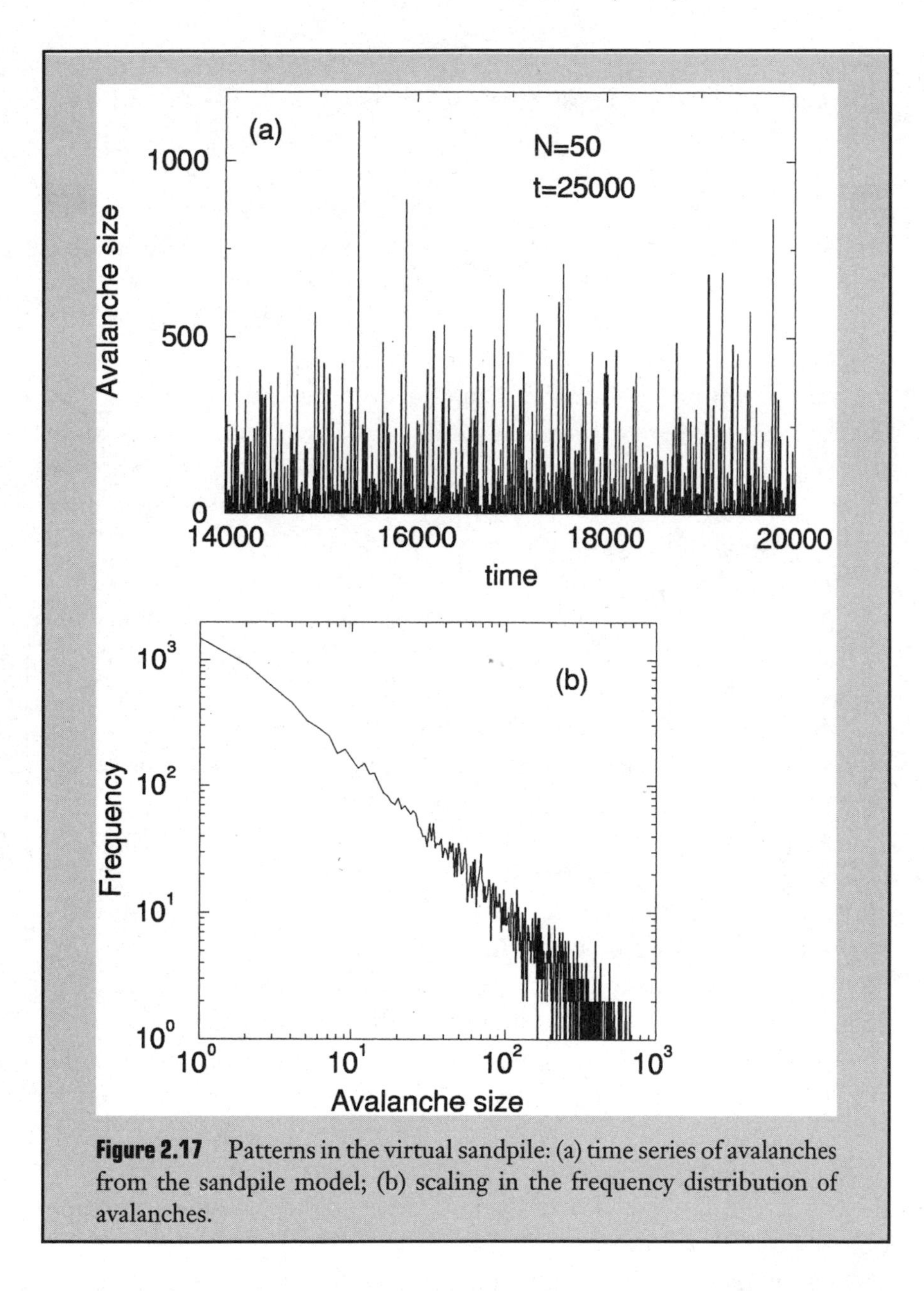

Figure 2.17 Patterns in the virtual sandpile: (a) time series of avalanches from the sandpile model; (b) scaling in the frequency distribution of avalanches.

In this book we will see several examples of systems exhibiting power laws and fractal behavior. Some are critical and some are not, but they all share some common features: they can be described through a very simple formal approach, they show a very small dependence (if any) on parameters, and the system response to small changes can be highly nonlinear.

Nonlinearity and collective behavior are characteristic features of complexity. They make complexity subtle, fascinating, and highly unpredictable. In fact, the Ising model reveals yet another interesting aspect of phase transitions, with significant consequences for evolutionary processes: the phenomenon of symmetry-breaking.

The Ising model can reach either of two ordered phases, with most units either up or down. All our rules are symmetric: there is no real distinction between "up" and "down," and any possible set of spins will behave in the same way if we invert all their values. Imagine that we start at $T < T_c$ from a random distribution of up and down units, such that their numbers are exactly the same. The dynamics of this system will, after some initial fluctuations, drive it toward one of the two possible attractors, corresponding to the two possible minimum energy states: either all up or all down. In spite of the symmetry between the states, the collective eventually breaks symmetry and chooses only one macroscopic state. There is nothing predictable in the final state of the system: it totally depends on the random fluctuations at the beginning.

If you should chance to visit the city of Florence and stop at the cathedral, you will see a beautiful example of how symmetry-breaking played a role in human society. The cathedral clock is rather peculiar: it was designed by Paolo Uccello in 1443, and it runs counterclockwise. At that time, no general convention for clock faces had yet emerged. But economic forces are highly nonlinear, and the use of a given resource by more and more people is a self-reinforcing phenomenon, introducing amplifications into the system that eventually break the symmetry in favor of one of the possible solutions. Perhaps if we were able to travel in time and introduce a large enough number of "counterclockwise" items into old Europe, before the current direction became dominant, our clocks would now run the other way.[13]

Symmetry-breaking plays a key role in how complex systems get organized. In early cosmic evolution, for instance, it was responsible for the virtual lack of antimatter in our present universe. We will need this phenomenon, together with chaos, nonlinearity, fractals, and power laws, to understand how a whole organism grows from a single cell, and it will also tell us some unexpected things about economics, history, and rationality. It may even be found to apply to the manner in which small events shape human destiny, as in Lwoff's decision to open the door of his laboratory to Jacob rather than keeping it closed, thus forever altering the history of biology.

THREE

Genetic Networks, Cell Differentiation, and Development

The development of all units—egg, embryo, regenerate, bud—all involve the problem of emergent form.

—N.J. Berrilll, *Growth, Development and Pattern*

Introduction

One of the continuing enigmas in biology is how genes contribute to the process of embryonic development whereby a coherent, functional organism of specific type is produced. How are the developmental pathways stabilized and spatially organized to yield a sea urchin or a lily or a giraffe? The problem here is that genes are themselves participants in the developmental process. They do not occupy a privileged position in making decisions about alternative pathways of differentiation. Yet they clearly constrain the possibilities open to cells: lilies do not make muscle or nerve cells, giraffes do not make the water-conducting elements of plants. How do genes act and interact within the context of cells so as to bring about these units of structure and function? How do cells act and interact within the context of the organism to generate coherent wholes, the different types of organism that populate the planet? It is not genes that generate this coherence, for they can only function within the living cell, where their activities are highly sensitive to context. The answer has to lie in principles of dynamic organization that are still far from clear, but that involve emergent properties that resolve the extreme complexity of gene and cellular activities into robust patterns of coherent order. These are the principles of organization of the living state.

We will not try to provide a comprehensive answer to this perennial problem of biological organization. In this postgenomic era we can see, but do not understand, the coordinated changes in the activities of thousands of genes involved in developmental processes.[1] Yet despite the extraordinarily complex patterns of gene transcription, cellular responses to differentiating stimuli are so well ordered that emergent properties are clearly involved. The strategy of this chapter, then, is to examine a few instances of developmental processes to see how genes participate in, but do not control, these phenomena.

Epigenetic Emergence

It is generally believed that cells with identical genomes in identical environments will lead identical lives. In a developing organism where the cells usually have the same genome, any changes of state that arise during differentiation are assumed to be caused by some spatial differences, however small. The egg may have a polarity that distinguishes, say, the animal from the vegetal pole, as in fish and amphibian eggs, giving rise to dorsal and ventral domains of the developing organism. Differences can also arise from an asymmetric environmental stimulus, as when light falls from above on the eggs of marine algae such as *Fucus* or *Acetabularia*. A leafy shoot (the thallus) grows toward the light, and the root (rhizoid) grows away from it. Even without an internal polarity or external heterogeneity, however, eggs have the capacity to spontaneously break their spherical symmetry and become polarized: they are excitable media.

Bacteria, because of their small size, have generally been assumed to conform to the principle that genome plus environment determines cell state. They, too, however, turn out to be capable of spontaneous diversification in uniform environments. In 1994, Elizabeth Ko and colleagues[2] studied activity levels of a growth enzyme, xylanase, and colony sizes in cultures of the common colon bacterium, *Escherichia coli*. They found that despite uniform culture conditions and identical genomes (no mutations occurred during the experiment) and uniformity of external culture conditions, individual cells developed different levels of enzyme activity and grew into colonies of different sizes.

Colonies grown from single cells continued to show large variations in enzyme activity throughout the thirty-day experiment. Cells varied in their growth rates as well, but they oscillated: more rapidly growing cells first decreased as a proportion of the whole culture as slower-growing cells increased, followed by the reverse. Thus there was no progressive accumulation of faster-growing cells, as one might expect from the dynamics of competition. The distribution of phenotypes remained at a constant average level in the culture, one type transforming into another and back again with a regular rhythm.

When Ko and her coworkers looked more closely at the colonies' patterns of enzyme activity, they found many different switching sequences between phenotypes with high and low levels. Cells descended from a colony with high enzyme activity could all retain the high level after a period of growth in liquid medium, or they could become heterogeneous, some progeny cells switching to low activity while others remained high. Cells from a low-activity colony could likewise either remain at low activities, or switch to high activity levels. And they could switch back and forth in any conceivable sequence.

These studies show that genotype and environment do not determine cell state in bacteria. Changes of state can occur spontaneously, without any defined internal or external cause. By definition, these changes are epigenetic phenomena: dynamic processes that arise from the complex interplay of all the factors involved in cellular activities, including the genes.[3] Epigenesis was Aristotle's term for the theory that developing embryos arise from the interaction of their emergent parts, in contrast to preformationism, the view that the adult organism preexists in the egg and simply grows. Genetic determinism in developmental biology is a type of preformationism. It regards the information in the genome as the sufficient cause of the developmental process that gives rise to a particular type of organism. This perspective leaves no room for emergent phenomena. However, the studies of Ko et al. tell us that even cells as simple as bacteria have dynamic properties that cannot be understood from a determinist point of view. A constant genome and uniform external environment do not themselves provide a causal basis for diversification of state in progeny. Still, there are ways of integrating Ko's observations with gene activity from an epigenetic perspective. One interpretion is that epigenesis displays properties similar to the sensitivity to initial conditions that characterizes deterministic chaos. The small differences of enzyme concentration that arise spontaneously between the progeny of a single cell can lead to significantly different enzyme levels in different individuals. The fluctuations are bounded but unpredictable.

The oscillation Ko found in growth rate suggests that some form of interaction between cells in liquid culture may result in a degree of temporal, and possibly spatial, order. This points to another level of epigenetic emergence. A striking example of the potential of bacterial colonies to produce complex spatial patterns is provided by the work of Eshel Ben-Jacob and his colleagues.[4] In these studies, Ben-Jacob grew *Bacillus subtilis* cultures on agar containing different amounts of water and nutrient. The bacterial colonies developed different spatial patterns exhibiting fractal structures that became progressively more branched and complex on poorer conditions of growth. Ben-Jacob et al. interpreted this to be a result of cooperative chemotactic behavior in the colony under adverse conditions, and showed how simple equations could capture the generic properties of the self-organizing patterns for different conditions. How do we explain such intriguing observations?

Dynamic Complexity in Cells

Kunihiko Kaneko and Tetsuya Yomo[5] recently constructed a model of cell growth and division based on networks of metabolic reactions that are collectively autocatalytic; that is, the reaction sequences produce all the components of the system so that it maintains itself and has the capacity to grow. Cells in this model are defined as closed networks within a membranous boundary. They take up nutrients from a shared medium through which they interact. Cell division occurs when a particular constituent, defined as a division factor, exceeds a threshold. Cells can also die from starvation. These networks have highly nonlinear dynamics, the most common being patterns of oscillation in the concentrations of the different chemical constituents. These oscillations have particular phase relations with the cycles of cell division.

At the beginning of growth in these computer cultures, the "progeny" of a cell passed through the same sequence of chemical oscillations as the "parent," and they all remained locked in synchrony. This collective stability persisted even after Kaneko and Yomo introduced small differences in the concentrations of chemicals in daughter cells to simulate the variations that inevitably accompany molecular processes. After several division cycles, however, the oscillations among the cells began to desynchronize. Clusters of synchronized cells arose, but chemical oscillations were phase-shifted between clusters. As a result, the clusters reduced competition for resources by what the authors call "time-sharing," a phenomenon that arose spontaneously and was not programmed. These cells had the same average concentrations of chemicals over the cell cycle, and the phase differences changed reversibly. No stable patterns of differentiation appeared.

With further divisions, cells emerged that had significantly different concentrations of some chemicals. These differences were stably inherited through cell division and became irreversibly fixed. Furthermore, when strong cellular interactions arose due to increased competition for resources, cells with particularly short division cycles emerged. These cells had large concentrations of the division factor and small concentrations of other constituents; they resembled certain types of cancer cells that arise under conditions of strong cell–cell interactions.[6] Such cells disturb the stable time-sharing of differentiating clusters, since they deplete the local supply of nutrients. Cell diversity is decreased by their presence. Kaneko and Yomo have named these processes of spontaneous differentiation of replicating biological units "isologous diversification." The term emphasizes a generic capacity of initially identical units to differentiate through interactions. More recent studies[7,8] have demonstrated the robustness of this process and the capacity of adhesive cells to form spatially organized aggregates.

Despite its biological simplicity, this model reveals a number of properties that are very suggestive of real cell behavior in culture. Clearly, there is a significant correspondence between the behavior of model cells and the bacterial cell dynamics observed by Ko and her colleagues. In fact, it was Kaneko's earlier theoretical studies of the dynamics of coupled chaotic elements[9] and clustering of states[10] that prompted Ko's experiements. What we find especially interesting is the way in which temporal and spatial order emerges from weakly interacting cells having chaos-like properties (sensitivity to initial conditions). As they diverge from their initial synchrony, cell states stabilize in new patterns. This is clearly reminiscent of early embryos, where there is initial synchrony of cell division followed by heterogeneous clusters of synchronized groups of cells that undergo progressive differentiation to form the main components of the emerging complex organism. The basic dynamics of development are perhaps a direct consequence of a spontaneously divergent pattern of cell change, tempered by local interactions that stabilize groups of cells into codeveloping clusters. These clusters define the emerging subfields that constitute the partially autonomous, spatially coherent modules of the emerging complex organism. Chaos and order live together and generate unexpectedly robust patterns of emergent organization.

The role of genes within this dynamic context is considerably less than full control and determination of the developing organism. Genes do not have to generate differences between groups of cells by amplifying small, programmed initial differences and then controlling specific pathways of cell differentiation. Instead, they have the simpler task of stabilizing generic patterns of emergent complexity in these multicellular systems. Producing reliably repeating patterns of development from fertilized eggs remains a significant task, and precisely how genes accomplish it is still far from clear. However, the recognition that spontaneous divergence of small differences within clusters of weakly interacting cells can generate ordered complexity is an important insight.

Robust Metabolism

The metabolic pathways of cells are a tangled web that for many years remained a black box of impenetrable complexity. With great ingenuity biochemists penetrated parts of this darkness by clarifying particular reaction sequences, such as the fundamental energy-generating processes by which a cell turns glucose into H_2O, CO_2, and the biological carrier of chemical energy, ATP. But the means whereby metabolic reaction sequences were regulated, so as to respond to cellular demands for energy and for the key building blocks of proteins and nucleic acids, remained a mystery. Genetic analysis was the key that unlocked this domain. But the particular mind-set that accompanied the discoveries resulted in a conceptual error that we are still struggling to overcome.

Bacteria have a metabolism every bit as complex as that of higher organisms, but it is easier to study. The discovery of mutations in bacteria that made them unable to produce particular key metabolites proved to be an immensely useful tool. With it the complex skein of metabolic pathways that all organisms use to generate molecules from simple precursors was unraveled. The metabolic tangle gradually resolved itself into a set of reaction pathways that transform one relatively stable molecular species into another. These stable metabolites—amino acids, nucleotides, sugars, lipids, ATP, and so on—are the building blocks of macromolecules—such as proteins, nucleic acids, polysaccharides, and structures such as membranes—and the energy sources used for biosynthesis. They are regulated at relatively constant levels by control processes that reveal the extreme sophistication of cellular activities. It was found that the stable metabolite at the end of a reaction sequence has an inhibitory action on the enzyme that catalyzes the first step of that reaction sequence. This creates a negative feedback loop: the end product is regulated at a concentration set by the sensitivity of the inhibitory response. The whole of the metabolic network can then be partitioned into partially autonomous reaction sequences, each controlled by feedback inhibition from the end product. This picture became more complicated with the subsequent discovery of positive feedback, and of crosstalk between pathways whereby they influence one another. The logic, however, seemed clear: the flux of metabolites along metabolic pathways is regulated by rate-limiting steps that act as throttles on flow, controlled by the concentration of the reaction sequence's end product. This view came to prominence during the 1960s and has prevailed ever since. It fits very well with the logic of assembly-line production in factories, and it is the way we would design control in a complex flow system. But is this, in fact, how cells regulate their metabolic activities?

Over the past two decades it has become clear that despite the importance of feedback processes, this picture lacks a fundamental property, one that changes the whole conceptual scheme for metabolic control. A basic biological phenomenon throws the theory of control by rate-limiting steps into question. This phenomenon is genetic dominance: for the vast majority of genes in diploid organisms (having two sets of genes), one normal copy of a gene is as good as two. (Bacteria, being haploid, do not have this property.) One would expect that reducing the concentration of a rate-limiting enzyme by one-half would significantly alter flux through the pathway. Since there are many rate-limiting steps in the metabolic network of any cell, feedback control theory leads us to expect that many heterozygotes, with only one functional gene, should have a significant difference of phenotype compared with a homozygote with two normal genes. For similar reasons, an increase in gene copy number from, say, two to three should again strongly influence flux along a metabolic pathway and so alter the phenotype by unbalancing flows. This does not happen.

Metabolism is remarkably unaffected by changes in gene copy number, so long as the genes are not completely absent or inactive. The scientists who provided an explanation for this published their results in the 1960s and '70s, but it was not until the 1980s that their work began to be recognized for its fundamental insights into the nature of metabolic control in living organisms.

What follows is a brief description of the kinetic analysis carried out by Henrick Kacser and James Burns.[11] This is the core of their insight into metabolic control. They studied the integrated kinetics of sequences of enzyme-catalyzed reactions that convert one stable metabolite into another, and the ways in which genes influence the flux through the pathway via the enzymes they produce. The kinetic analysis is described in Box 1. It shows that change of gene activity has very little effect on overall flux through the pathway because the controlling influences of genes are distributed relatively uniformly over all the steps in the metabolic sequence.

Distributed Control in the Metabolic Network

Kacser and Burns examined flux through a metabolic pathway described by the following scheme:

$$S_1 \overset{E_1}{\rightleftharpoons} S_2 \overset{E_2}{\rightleftharpoons} S_3 \ldots\ldots\ldots\ldots\ldots \overset{E_n}{\rightleftharpoons} S_{n+1}$$

The expression for the rate of conversion of S_i into S_j is given, according the Haldane modification of the Michaelis–Menten equation, by

$$v_i = \frac{V_i/K_{mi}(S_i - S_j/K_i)}{1 + S_i/K_{mi} + S_j/K_{mj}} \quad (1)$$

Here v_i is the rate of conversion, V_i is the maximum rate at which the conversion can occur, determined by the concentration of enzyme E_i and the rate constants; K_{mi} is the Michaelis constant for S_i, K_{mj} is that for S_j, and K_i is the equilibrium constant for the reaction of S_i into S_j.

Most enzymes are known to operate under conditions where substrates are nonsaturating, so that $S_i << K_{mi}$ and $S_j << K_{mj}$. Hence the denominator in (1) can be replaced by 1. At steady state, all the velocities v_i become equal to the overall flux rate for the reaction sequence, which we write as F. A sequence of equations can now be written, involving a rearrangement of terms as follows:

$$\frac{K_{m1}F}{V_1} = S_1 - \frac{S_2}{K_1},$$

$$\frac{K_{m2}F}{V_2K_1} = \frac{S_2}{K_1} - \frac{S_3}{K_1K_2},$$

$$\frac{K_{m3}F}{V_3K_1K_2} = \frac{S_3}{K_1K_2} - \frac{S_4}{K_1K_2K_3},$$

$$\cdots\cdots\cdots\cdots$$

$$\frac{K_{mn}F}{V_nK_1K_2}\cdots K_{n-1} = \frac{S_n}{K_1K_2\cdots K_{n-1}} - \frac{S_{n+1}}{K_1K_2\cdots K_n}.$$

Adding these together, we see that all terms on the right-hand side cancel in pairs except the first and the last. The result is

$$F\left(\frac{K_{m1}}{V_1} + \frac{K_{m2}}{V_2K_1} + \cdots + \frac{K_{mn}}{V_nK_1K_2\cdots K_{n-1}}\right) = S_1 - \frac{S_{n+1}}{K_1K_2\cdots K_n}.$$

The metabolites S_1 and S_{n+1} are the pools of stable metabolites at the beginning and end of the reaction sequence that are held relatively constant, so we can replace the right-hand side by a constant C. Furthermore, each V_i is the product of a rate constant k_i and the concentration of enzyme E_i. So each term within brackets on the left-hand side can then be written as $1/e_i$, which is a composite term for the genetically determined parameters that contribute to the overall flux. The final result is a strikingly simplified relationship describing how control is exercized by the components of a metabolic sequence:

$$F = \frac{C_n}{(1/e_1 + 1/e_2 + \cdots 1/e_n)}. \quad (2)$$

This demonstrates that control of flux through a metabolic sequence is not exercized by a key initial step in the chain but is distributed over all enzymes in the pathway. We can now examine the implications of this equation for differences in the concentration of any enzyme such as the difference between enzyme concentrations in a homozygote dominant and a heterozygote, the former having two active copies of a gene, while the latter has only one.

Focusing our attention on flux as a function of the concentration of any specific enzyme in the reaction sequence, say e_r, let us write this as a variable E, while all the other enzymes are taken to be constant, the sum of these terms being C_e. Then equation (2) takes the form

$$F = \frac{C_n}{C_e + 1/E}.$$

Rearranging, this becomes

$$F = \frac{C_n E}{1 + C_e E}.$$

With F plotted as a function of E, the result is shown in Figure 3.1. The curve has an initial slope of C_n and rises to the asymptotic value of C_n/C_e. The relationship between F and E is hyperbolic. For curves of this type, it is evident that if the homozygous normal level of enzyme is near the maximum value, as indicated, then a reduction of E by half will have a relatively small effect on the flux rate. Experimental evidence shows that enzymes normally function in the region of the curve where the slope is small.

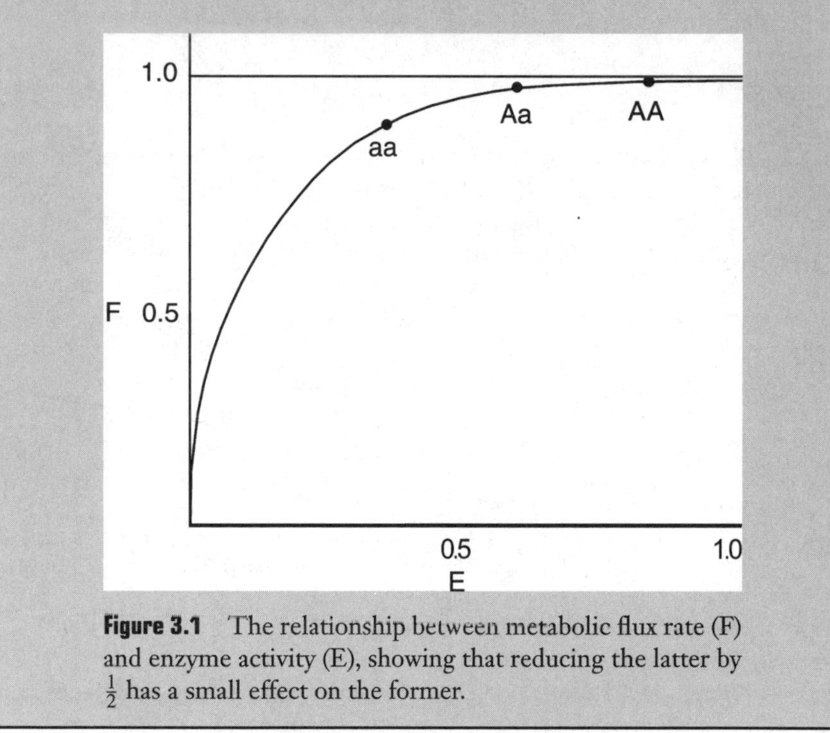

Figure 3.1 The relationship between metabolic flux rate (F) and enzyme activity (E), showing that reducing the latter by $\frac{1}{2}$ has a small effect on the former.

The result is that flux rates, and hence metabolic phenotypes, are insensitive to quite large changes of enzyme concentration. At the same time, the phenomenon of dominance is explained as a generic property of metabolic control processes. Kacser and Burns's comment on this result was, "The effect on the phenotype of altering the genetic specification of a single enzyme, with which the problem of dominance is concerned, is . . . unpredictable from a knowledge of events at that step alone and must involve the response of the system to alterations of single enzymes when they are embedded in the matrix of all other enzymes." The basis for this view was a systems perspective on biological processes that emphasized the need to go beyond the parts to the whole, a perspective Kacser had endorsed as early as 1957:

> The belief that a living organism is "nothing more" than a collection of substances, albeit a very complex collection of very complex substances, is as widespread as it is difficult to substantiate. . . . The problem is therefore the investigation of *systems*, i.e., components related or organized in a specific way. The properties of a system are, in fact, "more" than (or different from) the sum of the properties of its components, a fact often overlooked in zealous attempts to demonstrate "additivity" of certain phenomena. It is with these "systemic properties" that we shall be mainly concerned.[12]

This statement, from an appendix Kacser wrote to a book by C.H. Waddington, *The Strategy of the Genes* (1957), makes very clear Kacser's adherence to the principle that unexpected properties emerge from complex wholes, a position also espoused by Waddington in his epigenetic view of organismic development. The startling result that emerged from this systemic treatment of metabolic control has taken many years to become accepted, but it is now acknowledged as a milestone in understanding the emergent properties of biological metabolism. The general principle of distributed control over the components of a pathway does not preclude additional specific controls, as in the many known instances of feedback regulation. These, however, add fine tuning to an underlying robust sytem that is insensitive to extensive genetic perturbation. This example of an emergent "robust metabolism" is just one instance of the general phenomenon of robust dynamic order in organisms.[13]

Genetic Networks

So far, we have assumed that gene activities regulating concentrations of enzymes, e_i, are constant. The time scale of metabolism was regarded as fast enough, relative to genetic change, that the latter could be ignored. What if we now consider genetic change as a quasi-autonomous dynamic process and investigate the possibility of robust emergent behavior in patterns of gene interactions? Do any unexpected dynamic properties arise? Given that a higher organism such as a frog or a mouse or a human has on the order of 50,000–100,000 genes, one might expect that anything is possible, and there are no generic properties that characterize the dynamic patterns of gene activities in developing organisms. In fact, however, there are unexpected constraints on these patterns. Over the past few decades Stuart Kauffman[14] has developed a sophisticated, multifaceted approach to the study of genetic dynamics in both development and evolution. This research focuses on the self-organizing, emergent properties of systems involving the interactions of thousands of genes.

Kauffman was first inspired by the observations of François Jacob and Jacques Monod[15] on the regulatory dynamics of prokaryotic genes. These could be approximated as binary switches, genes being either on or off. Control of gene state is determined by the activities of other genes (inducers, repressors) that could also be approximated as binary variables, on or off. Restricting the number of control signals acting on any single gene to 2 (connectivity $K = 2$), Kauffman explored the dynamics of genetic networks in which genes can send signals to any gene (including themselves), so that all possible interaction patterns are allowed. The response of each gene to its two inputs was determined by random assignment from the set of sixteen Boolean functions of two variables.

There are four different combinations of input pairs, listed in each table of Figure 3.2 at the left, and the responding gene responds to the input pairs by going either off (0) or on (1), resulting in sixteen possible patterns of response. Genes are selected at random from this set and interact with random couplings. What could possibly emerge that is of interest from such networks of randomly coupled, randomly assigned genetic networks?

To get a sense of the dynamics of these genetic networks, let us start with a very simple example, such as the three interacting genes whose states are designated A, B, and C in Figure 3.3a. The first table defines how gene A responds to the combinations of values of the input signals from A and B, while the other two tables do likewise for the other two genes. The interactions are selected randomly. For the particular case described here, gene A feeds back to itself and has an input from B; B receives inputs from A and C; and C receives inputs from A and B. Each gene is assigned a Boolean function chosen at random from the set of sixteen shown in Figure 3.2. The ones chosen here are numbers

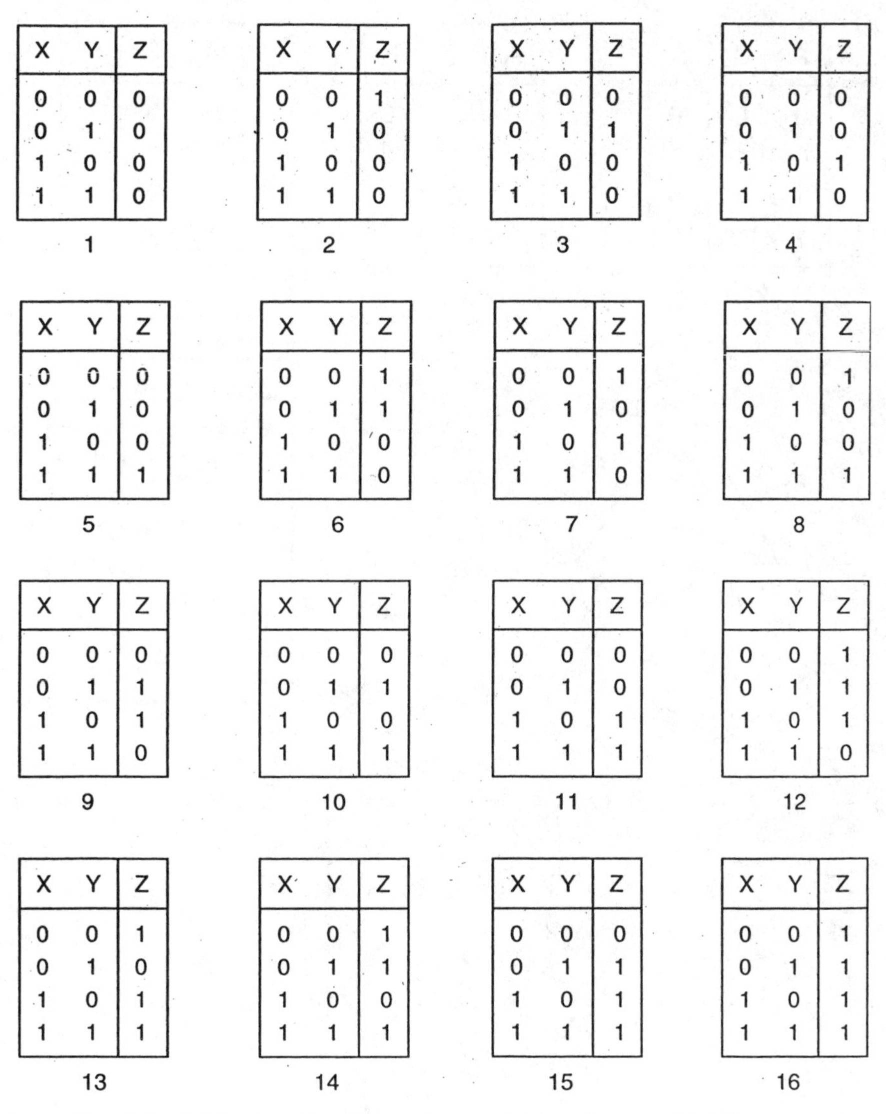

X	Y	Z
0	0	0
0	1	0
1	0	0
1	1	0

1

X	Y	Z
0	0	1
0	1	0
1	0	0
1	1	0

2

X	Y	Z
0	0	0
0	1	1
1	0	0
1	1	0

3

X	Y	Z
0	0	0
0	1	0
1	0	1
1	1	0

4

X	Y	Z
0	0	0
0	1	0
1	0	0
1	1	1

5

X	Y	Z
0	0	1
0	1	1
1	0	0
1	1	0

6

X	Y	Z
0	0	1
0	1	0
1	0	1
1	1	0

7

X	Y	Z
0	0	1
0	1	0
1	0	0
1	1	1

8

X	Y	Z
0	0	0
0	1	1
1	0	1
1	1	0

9

X	Y	Z
0	0	0
0	1	1
1	0	0
1	1	1

10

X	Y	Z
0	0	0
0	1	0
1	0	1
1	1	1

11

X	Y	Z
0	0	1
0	1	1
1	0	1
1	1	0

12

X	Y	Z
0	0	1
0	1	0
1	0	1
1	1	1

13

X	Y	Z
0	0	1
0	1	1
1	0	0
1	1	1

14

X	Y	Z
0	0	0
0	1	1
1	0	1
1	1	1

15

X	Y	Z
0	0	1
0	1	1
1	0	1
1	1	1

16

Figure 3.2 The 16 Boolean functions of two variables: there are 16 different output sequences (Z) for the 4 combinations of input values of X and Y, where all variables are binary (taking values 0 and 1 only).

3, 5, and 14 for genes A, B, and C, respectively. The interactive genetic network that this defines is shown in Figure 3.3b.

With the network now fully determined, we can study its dynamics. This requires working out the state transitions that the network undergoes for each of its possible states. These states are the triplets (A, B,

A	B	A
0	0	0
0	1	1
1	0	0
1	1	0

A	C	B
0	0	0
0	1	0
1	0	0
1	1	1

A	B	C
0	0	1
0	1	1
1	0	0
1	1	1

a

A

B C

b

T	T+1
ABC	ABC
000	001
001	001
010	101
100	000
011	101
101	010
110	001
111	011

c

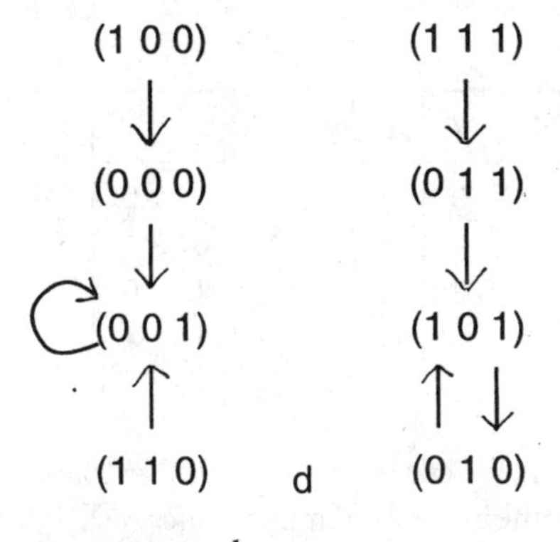

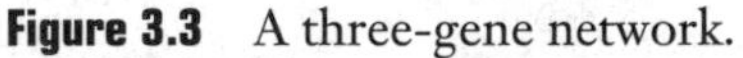

Figure 3.3 A three-gene network.
a. The Boolean functions defining the three genes.
b. The pattern of interactions in the network defined by the three genes.
c. The state transition table describing the sequential changes of state of the three genes from time T to $T+1$.
d. The kimatograph of the network, showing how the states change one into the other. One set of states terminates in the point (001), while the other ends in the cycle involving states (101) and (010).

C) in which A, B, and C take the numbers 0 or 1, corresponding to the off or on states of the three genes. Starting with (000), the Boolean function for gene A tells us that if A = 0 and B = 0, then A takes the value 0 at the next time step. Using the functions for B and C, we find that B goes to 0 while C goes to 1 when their inputs are both 0. Hence network state (000), all genes off, goes to (001): a and B off while C switches on. Working through all eight possible states of the network ($2 \times 2 \times 2 = 8$, each gene having a choice of two states), the result is the state transition table shown in Figure 3.3c. This can now be converted into a dynamic picture called a kinatograph (Figure 3.3d). The eight states separate into two trajectories involving four states each. State (100) goes to the state (001) via (000), while (110) goes directly to the same point attractor. The other trajectory ends up in a two-state cycle, switching between (101) and (010).

When Kauffman studied the dynamic behavior of much larger randomly constructed networks, the only constraint being on connectivity, $K = 2$, he found a surprising and counterintuitive limitation of dynamic patterns. Even though a network of 1000 genes, for example, has 2^{1000} possible states, the typical result is a limited set of attractors (about 30) that cycle through a small number of states or settle to a stable point (the median cycle length is about fifteen states). On average, N genes resulted in $N^{1/2}$ attractors whose cycle lengths were $\frac{1}{2}N^{1/2}$. This represents a very surprising constraint on the patterns arising from the 2^N states of the network (each of the N genes being either on or off). These networks exhibit an unexpectedly high degree of dynamic order. Many genes settle into stable states that define islands of frozen order, while others repeat moderately long patterns of periodic activity. If we interpret an attractor of the coupled network as a differentiated cell type, then for the estimated 80,000 genes of the human genome the expectation is that there should be of order $(80{,}000)^{1/2} = 283$ different cell types in the human body. Bruce Alberts and his coauthors, in their well-known textbook *The Molecular Biology of the Cell*,[16] estimate 265 cell types. This is a striking correspondence and suggests that there may well be something of considerable value in Kauffman's approach.

The reason for the high degree of localization (the prevalence of relatively few short cycles rather than very long ones) is to be found in a distinctive characteristic of networks with $K = 2$. The key concept is canalization, which describes a property of Boolean functions relevant to logical constraint and localized dynamics. A canalizing Boolean function is one in which a single input, in one of its states (either 0 or 1), determines the output of the function irrespective of the value of the other input. Under this definition, fourteen of the sixteen functions in Figure 3.2 are canalizing functions. For example, looking at the penultimate function in Figure 3.2 with output vector (0, 1, 1, 1), we see that if X takes the value 1, then the output is 1 irrespective of the value of Y. As another example, consider the function with output vector (1, 0, 1, 0) (number 7 in Figure 3.2). Whenever Y is off, the output is on, and whenever Y is on, the output is off. Hence a knowledge of the state of Y is sufficient to determine the output state. All other functions have a similar property except for the functions with output vectors (1, 0, 0, 1) and its complement (0, 1, 1, 0) (8 and 9 in Figure 3.2); for these functions neither input alone determines the output. Hence Boolean functions of two variables have strong constraints on output values relative to inputs, which is an important factor in producing the high degree of localization of dynamic behavior in these networks, irrespective of their size (the value of N).

Networks can be defined with $K = 3$, 4,5,6, etc., so that each gene receives 3, 4, or more inputs. In computer simulations, the typical behavior of these networks diverges rapidly with increasing values of K from that of networks with $K = 2$. Instead of relatively small numbers of attractors (points and short cycles), many long cycles emerge and dominate the dynamic space: there is a divergence of both number of attractors and their length. When $N = K$, so that all genes interact with each other, the number of attractors is of order N/e, where $e = 2.7182\ldots$, the base of natural logarithms. Hence for $N = 80{,}000$ the number of different attractors is around 30,000, very different from the 283 attractors for $K = 2$. Furthermore, the length of these cyclic attractors is on the order of the square root of the number of states, or $2^{40{,}000}$. They are immensely long! Such networks have effectively chaotic dynamics: a very small perturbation in the initial condition of the network is likely to send the system onto a totally different trajectory.

Networks with moderate K values $(3, 4, 5, \ldots)$ show both ordered and chaotic behavior. As might be expected, if there is a predominance of canalizing functions, the dynamics are orderly and localized. But if noncanalizing functions predominate, the behavior is chaotic. With increasing values of K, the proportion of canalizing functions decreases rapidly. For $K = 3$, for example, only 16 of the 256 Boolean functions of three variables are canalizing, and for $K = 4$ the fraction drops below 1%. A useful way of studying the transition from order to chaos in these networks is to vary the proportion p of 0's to 1's in the output column of the Boolean functions, expressed as a fracton less than 1. Figure 3.4 shows three Boolean functions of four variables to illustrate the point. The first one has $p = 0.5$, equal numbers of 0's and 1's in the output column, while the other two have $p = 15/16$, the first with all 0's except one and the second all 1's except one. The latter two are not strictly canalizing functions according to the definition given above, but they have the same effect of constraining the dynamics of the network. Bernard Derrida and Gerard Weisbuch showed that as p increases from 0.5 there is a phase transition in these networks from chaos to order, order emerging for p values near 1 where there is a predominance of canalizing-like functions.

Experimental results on patterns of gene regulation now permit us to examine aspects of this theory of gene activity patterns. Data from

A B C D	E
0 0 0 0	0
0 0 0 1	1
0 0 1 0	0
0 0 1 1	1
0 1 0 0	0
0 1 0 1	1
0 1 1 0	1
0 1 1 1	0
1 0 0 0	1
1 0 0 1	0
1 0 1 0	0
1 0 1 1	1
1 1 0 0	0
1 1 0 1	0
1 1 1 0	1
1 1 1 1	1

a

A B C D	E
0 0 0 0	0
0 0 0 1	0
0 0 1 0	0
0 0 1 1	0
0 1 0 0	0
0 1 0 1	0
0 1 1 0	0
0 1 1 1	0
1 0 0 0	1
1 0 0 1	0
1 0 1 0	0
1 0 1 1	0
1 1 0 0	0
1 1 0 1	0
1 1 1 0	0
1 1 1 1	0

b

A B C D	E
0 0 0 0	1
0 0 0 1	1
0 0 1 0	1
0 0 1 1	0
0 1 0 0	1
0 1 0 1	1
0 1 1 0	1
0 1 1 1	1
1 0 0 0	1
1 0 0 1	1
1 0 1 0	1
1 0 1 1	1
1 1 0 0	1
1 1 0 1	1
1 1 1 0	1
1 1 1 1	1

c

Figure 3.4 Three of the 2^{16} Boolean functions of 4 variables, illustrating how p is defined in terms of the proportions of 0's and 1's. a. $p = 0.5$ b. $p = 15/16$ c. $p = 15/16$

eukaryotic cells identify the number and type of regulatory inputs to different genes (i.e., the connectivity K), and the dynamic patterns of gene activities. Although transcription of messages from genes is not strictly Boolean (off–on), there is often a quasi-discontinuous pattern of gene activity. For example, a gene may show a transcription level of 0.1 of its maximum when control input 1 to that gene is acting alone, 0.15 of the maximum when input 2 is acting alone, and full activity (1.0) when inputs 1 and 2 act together. Such a gene can be approximated by the Boolean AND function (number 5 in Figure 3.2). This behavior is often observed experimentally, though the value of K is not generally restricted to 2.

Can we show experimentally that the fraction of Boolean functions with canalizing functions decreases rapidly as K increases? Stephen Harris and his colleagues[17] have shown in a study of a sample of eukaryotic genes with $K = 3$ and 4 that there is a significant bias toward canalizing functions compared with expectations from random assignments. This implies selection for regulatory gene functions that hold the dynamics in the ordered regime, though not far from the transition to chaos. The result is cells with large numbers of genes having constant activity values (off or on) and groups of genes that cycle through a relatively small set of states. Furthermore, Harris showed in modelling studies that the observed bias for canalizing functions resulted in attractors with considerable overlap in the set of fixed genes. This implies that different cell types share a common core of expressed or silent genes, corresponding to a shared set of basic metabolic and maintenance activities. The fraction of genes that differ between attractors is only a few percent, a result consistent with experimental observations on differentiated states.

These studies reveal generic properties of gene regulation within cells that arise from the occurrence of simple rules of interaction in a complex system. Again we see emergent order, or what Kauffman has called "order for free," in evolution: unexpected constraints on the large-scale dynamic patterns of gene activities that provide the living state with the properties needed for generating flexible, adaptive, and robust behavior. These properties are completely consistent with, and provide an interesting extension of, the dynamic picture of robust metabolism that we saw in the previous section. There we focused on the control of flux rate through metabolic pathways catalyzed by products (enzymes) of the fixed genes common to the attractors of different types of cell. The assumption of constancy of enzyme concentrations in metabolic control analysis is consistent with the "frozen" islands of genes in complex genetic networks, and the analysis of flux helps us understand the robustness of metabolism to genetic change that is revealed by the phenomenon of genetic dominance, a property that comes free with metabolic networks.

The model system of cell growth and division examined by Kaneko and Ko, showing how spontaneous differences of state can arise between initially indentical cells, can be seen as a detailed study of the spontaneously divergent dynamic that is characteristic of the more chaotic aspects of complex dynamics within cells, reflecting sensitivity to initial conditions. The same divergent processes underlie the emergence of a diversity of cell types from a single cell, the fertilized egg, and are consistent with the chaotic component of genetic network dynamics.

It is useful now to extend this inquiry into gene dynamics, differentiation, and morphogenesis by looking at one of the most fundamental morphogenetic processes in vertebrate development: the production of paired somites on either side of the vertebral column. What behavior of genes is involved in this process, how do discrete somites emerge in such an orderly array, and how does this fit into the complex emergent dynamics of the living state?

Spatial Patterns from Temporal Order

Somite formation in amphibia, chick, and mouse has been a subject of detailed embryological study for many years. Figure 3.5 shows a chick embryo at two days of incubation, with 14 of its final total of 50 somites formed. The presomitic mesoderm lies in a band on either side of the central body axis and consists of loosely packed cells that are differentiating toward somite formation. At the tail end is the primitive streak, the site of inward movement of cells from the surface to form the presomitic mesoderm beneath the ectodermal cell layer. For the community of biologists studying the molecular mechanisms of morphogenesis, it had been clear for some time that somite formation could not be accounted for by a spatially periodic standing wave in the concentration of a morphogen, running from head to tail with 50 peaks and troughs, defining a primary pattern for the somites. Such a pattern is virtually impossible to generate reliably, so it lacks the robustness required by morphogenetic processes.

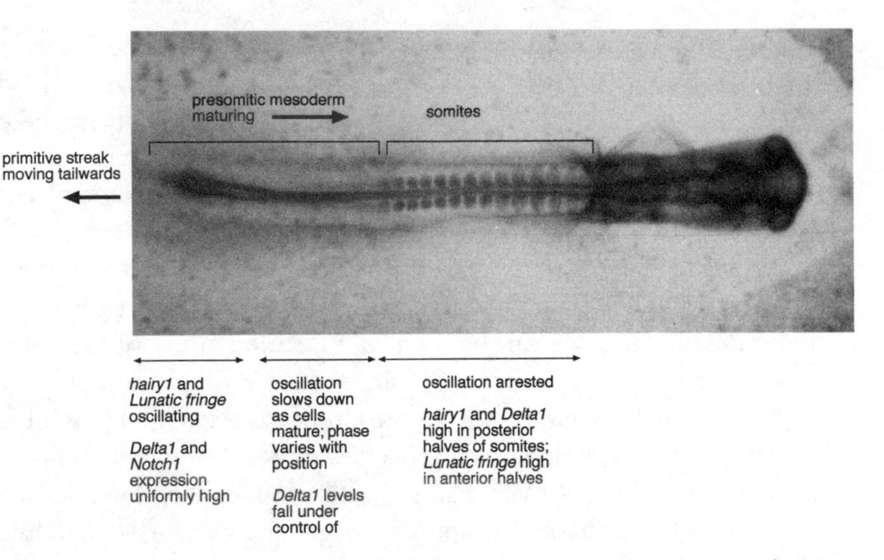

Figure 3.5 Dorsal view of a chick embryo at two days of incubation, showing (from the right) the developing head and neural tube, 14 pairs of somites, and the presomitic mesoderm from which more pairs will form at 90 minute intervals. (From Julian Lewis, with permission.)

Another class of dynamic model under consideration in the early 1970s involved coupling temporal periodicities—oscillations in molecules involved in morphogenesis—to the generation of ordered spatial patterns. One such model was developed precisely to account for somite formation: Jonathan Cooke and Christopher Zeeman's[18] "clock and wave front" model. This consisted of essentially three ingredients: a graded distribution of a morphogen with a maximum at the head end of the presomitic mesoderm and a minimum toward the tail; an oscillation throughout the presomitic mesoderm with a periodicity equal to that of somite formation (90 minutes in the chick); and a discontinuity in the state of cells at a critical stage in their differentiation within the presomitic mesoderm, such that they form a discrete group and detach from the others to form a somite.

The dynamics work as follows. Cells toward the head end of the presomitic mesoderm, where the gradient is highest, are progressing most rapidly toward somite differentiation. The oscillation defines a clock that acts as a gate on somite formation. When the clock in the mesoderm cells reaches a particular phase in its cycle, say a maximum in the oscillating morphogen, all those cells that have passed a critical stage in somite differentiation express a property that results in their forming a tightly adhering group, which detaches from the rest of the presomitic mesoderm to form a discrete somite, one on either side of the body axis. The clock then passes beyond this phase, and the gate closes. When it opens again, another cohort of cells will have reached the stage of preparedness for somite formation, and another pair of somites forms. This discrete event is likely to be controlled by a sudden increase in the adhesiveness of the cells that have become competent for somite formation. In the clock and wave front model, this sudden discontinuity in tissue dynamics is described mathematically as a "catastrophe."[19]

This model has the important qualities of robustness and the capacity to regulate the number of somites according to the length of the individual embryo by simply setting constant boundary values on the head-to-tail gradient. A property that might seem implausible to the experimental biological community, an oscillation throughout the cells of the presomitic mesoderm, is nevertheless perfectly reasonable in terms of the dynamics of cells, which can express oscillations with periods in the range from several minutes to many hours. So a 90-minute cycle is not problematic, even though in the 1990s the experimental community had not encountered evidence of such a mechanism in its studies of gene activities in *Drosophila* or other model organisms.

The value of *Drosophila* is that it allowed the identification of classes of genes involved in morphogenesis, and those connected with somite formation were among the first to be characterized. Among these is a gene called *hairy* because when mutated it causes a disturbance of segment formation such that the little hairs, or denticles, on the ventral surface of the *Drosophila* larva are distributed in a continuous lawn rather in discrete bands within segments, as in normal embryos. This gene was subsequently found to be fundamental to the early stage of segment formation in *Drosophila*, and so it was quite likely that its homologues in other phyla might be involved in segmentation or somitogenesis. The first of these to be studied was *c-hairy1* in the chick. Isabel Palmeirin and colleagues demonstrated in a 1997 paper[20] that this gene is expressed in cyclic waves in the presomitic mesoderm with a 90-minute period equal to that of somite formation. Expression is synchronized throughout the mesoderm, but caudal (posterior) cells are phase advanced relative to more rostral (anterior) cells. There is thus a wavelike advance of gene expression traveling forward in the mesoderm toward the site of somite formation. As this wave progresses, the domain of expression narrows to a band that corresponds to the posterior boundary of the next somite and then stops. This space-time pattern is shown in Figure 3.6, which describes the expression of another segmentation gene, *lunatic fringe*; (see later) closely connected with *c-hairy* and with similar behaviour.

The wave is a result of the timing of oscillations that occur independently in cells; it does not propagate via an inductive signal passing from

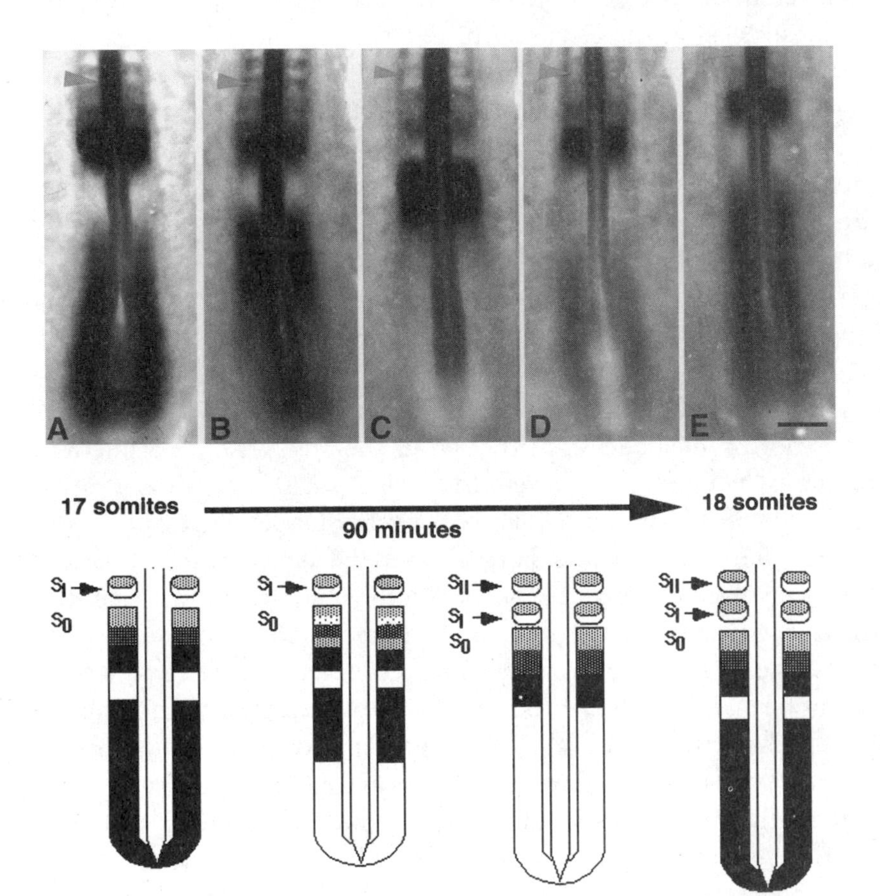

Figure 3.6 Expression pattern of the gene *lunatic Fringe* in the process of somite formation in the chick embryo.
Above: Gene product visualized with a marker, showing the sequence of changes in the expression domain in the presomitic mesoderm. A wave front sweeps from posterior to anterior, narrowing and ending up at a position that defines the boundary of the next somite to form. The arrowhead points to the last somite to be formed.
Below: A diagram of the changes in position of the gene product during the course of the 90 minutes required to form a new somite. (From Palmeirin et al., 1998, with permission.)

cell to cell. When Palmeirin excized a section of presomitic mesoderm to create a gap between two regions, the "wave" continued to appear at the normal time in more anterior tissue. Cells were oscillating autonomously, though their phase relationships must have been organized spatially at an earlier stage. This autonomy was demonstrated also by continuation of the oscillations in explanted sections of presomitic mesoderm.

Palmeirin and her coauthors also showed that the oscillation is not dependent upon protein synthesis, so the mechanism of the clock is not a simple negative feedback of *hairy* protein (a transcription factor) controlling gene transcription. They suggest that *hairy* is probably not, in fact, a part of the clock mechanism involved in somitogenesis, but is controlled by the clock.

The validation of the essential components of a model of somitogenesis some twenty years after its publication is itself of considerable interest. But it is the emergent aspects of this process that we find most significant. With the cellular slime mold discussed in Chapter 1, the study of excitable media as descriptions of essential aspects of space–time order in living (and nonliving) systems provided the conceptual and mathematical tools for understanding the dynamic origins of spiral waves during aggregation and the dynamic instabilities that give rise to stream formation. Similarly, making sense of the experimental observation of a periodic wave of gene expression in somite formation depended on the use of a high-level dynamic model based on the clock and wave front mechanism to simulate the precise spatiotemporal characteristics of the wave. The model closes the causal gap between the level of gene activity within cells and the dynamics of somite formation in space. However, the story is not yet complete. The precise nature of the discontinuity producing discrete somites remains to be discovered. This morphogenetic process is clearly linked to a change in cell–cell adhesiveness, as Cooke and Zeeman originally suggested, but the details of the genes involved have not yet been described. The point to be made is the dialogue that takes place between the description of somitogenesis as an emergent property described by high-level modeling of cell and tissue dynamics and the study of molecular and genetic details. These two levels of description mutually support one another; one is not reduced to the other.

Another gene recently shown to be involved in somitogenesis in the chick is *lunatic Fringe* (lFng), a homologue of *Drosophila Fringe.*

(Biologists clearly have as much fun in naming genes as physicists do in naming fundamental particles such as quarks.) lFng is expressed in the presomitic mesoderm, in a pattern virtually identical to *c-hairy1* except that it has a residual expression in the anterior part of the detached somite, while *hairy* is expressed posteriorly.[21] Also, its expression is blocked by protein inhibition, suggesting that the transcription factor from *c-hairy1* may be controlling this gene and producing the oscillation. lFng is known to be closely connected with two other genes involved in boundary formation: *notch* and *Delta*. It has been suggested that these genes play a crucial role in the separation of the somites, linked to changes in cell–cell adhesion.

These intriguing results on chick somitogenesis have also been extended to the mouse.[22] In this case the gene studied is *Lunatic fringe*, a mouse homologue of the chick gene. Somites form every two hours in the mouse, with an overall spatial and temporal pattern much like that in the chick but with an interesting difference. The wave of transcription initiated caudally in the presomitic mesoderm every two hours takes four hours to complete its apparent journey to the site of somite formation. Hence there are two such propagating waves in the presomitic mesoderm at any moment, one following the other with an exact repetition of pattern. This is a mouse variation on the chick theme. A broader evolutionary question is how many other variations there are on this generic segmentation pattern, and how well it has been conserved over evolutionary time from the first segmented organisms. What we can say at the moment is that this coupling of temporal to spatial order is probably a robust patterning mechanism in morphogenesis, reflecting generic dynamic features of gene activity linked to cell and tissue behavior.

We can also register a word of caution. Homologous genes in different species may be involved in rather different dynamic generators of homologous structures; that is, the way segments are generated in *Drosophila* is no guide to the dynamics of segment formation in other species, invertebrate or vertebrate, despite the involvement of the same gene(s). Changes in the patterns of gene interaction can produce quite different dynamics. Yet the question of generic patterns remains. The expectation from the theoretical study of genetic networks is that cycles of gene activity of the type observed in *c-hairy1* will be found to be the generic pattern in segmentation throughout the animal kingdom, not the highly specialized process seen in *Drosophila*, an atypical insect.

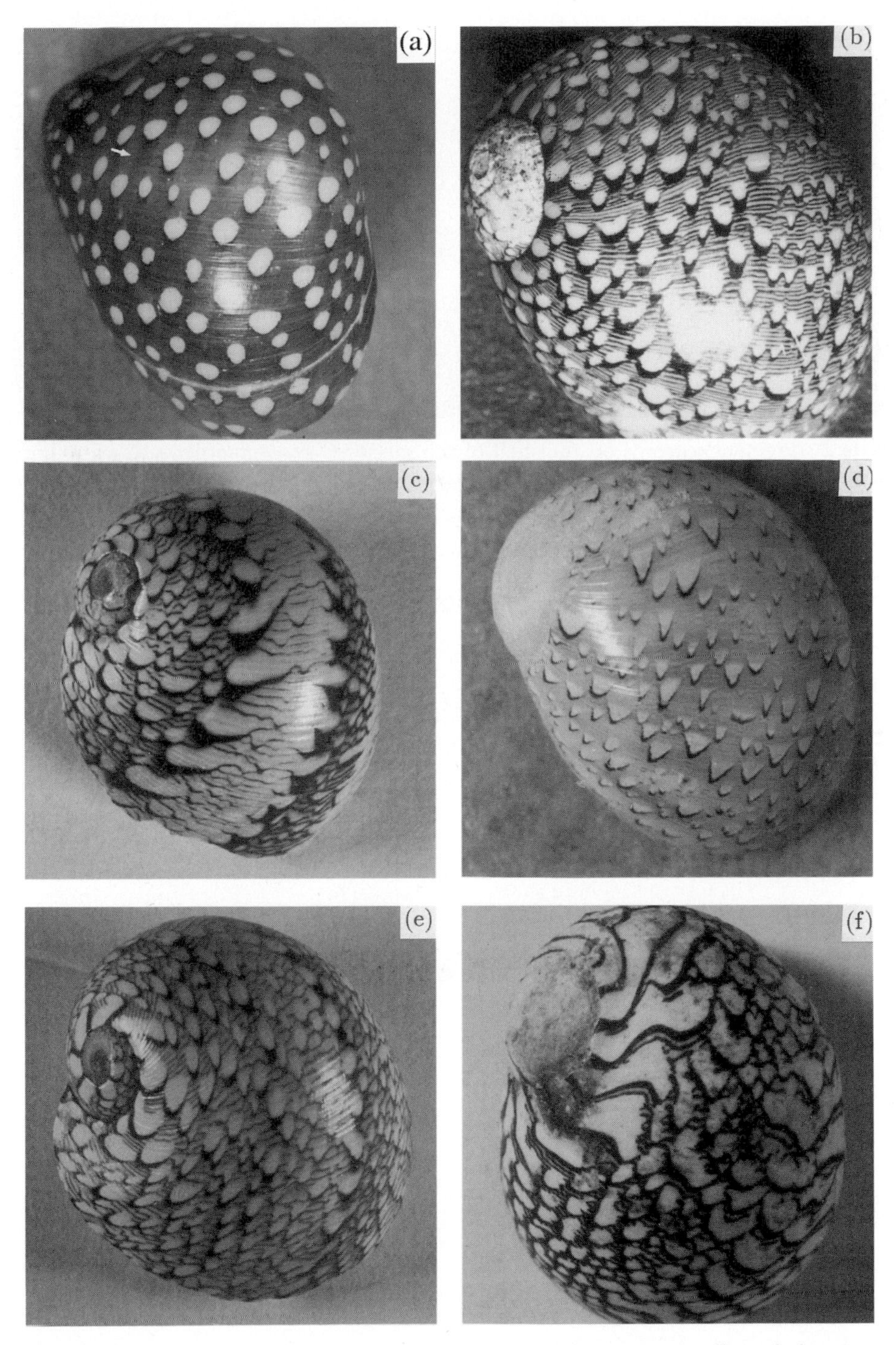

Plate 1 Variety of pigment patterns on the shells of the marine molluscs belonging to the genus *Clithon*. (From Meinhardt, 1995, with permission.)

Plate 2 Pigment patterns on the shell of the species *Cypraea diluculum* showing both sides of the shell, with different modes of pattern formation near the opening (dots) and around the shell (waves). (From Meinhardt, 1995, with permission.)

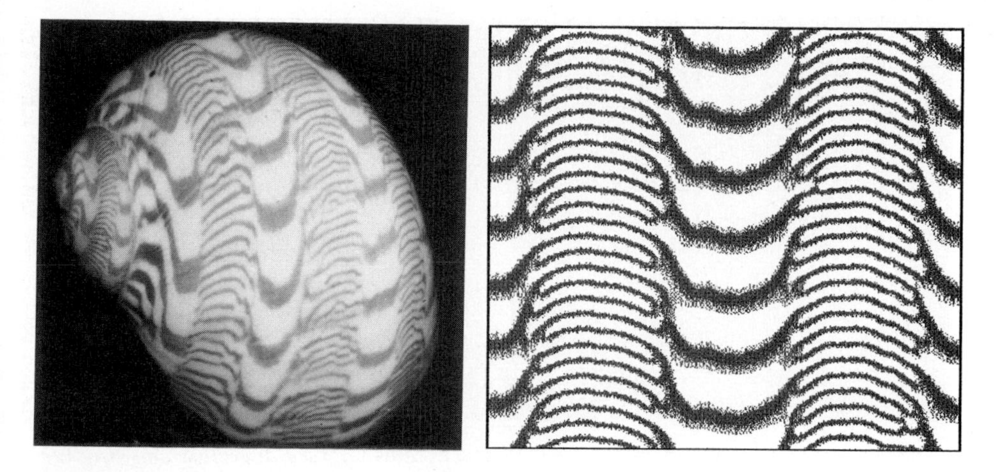

Plate 3 Waves of different frequencies on the shell of *Natica euzona*. (From Meinhardt, 1995, with permission.)

Plate 4 Triangular waves on the shells of species belonging to the genus *Lioconcha*. (From Meinhardt, 1995, with permission.)

The fruit fly is very good for genetic analysis but is misleading for an understanding of generic developmental and morphogenetic dynamics.

Patterns at Play

The study of emergent generic properties of organisms would be greatly facilitated if one could find patterns that are selectively neutral so that functional constraints would be absent and the epigenetic processes could express themselves freely. A likely instance of this is the pattern of arrangement of leaves on the stems of higher plants, called phyllotaxis. It seems that all the patterns observed perform their functions of catching light and exchanging gases effectively, with no significant adaptive differences. The differential abundances of the various patterns (distichous, whorled, spiral, bijugate spiral, and so on) are then accounted for by intrinsic generative dynamic principles related to such characteristics as size and depth of basins of attraction of the morphogenetic dynamic.[23] However, functional constraints could still be playing some role in these patterns, so it would be nice to find a clearer example of the patterning process at play without constraint. Fortunately, there is a wonderful example of this available, which Hans Meinhardt has studied in detail and presented in his beautiful book *The Algorithmic Beauty of Sea Shells*.[24]

Plate 1 shows a striking range of pigmentation patterns on the shells of tropical marine mollusks. These are all closely related species (of the genus *Clithon*), as is evident from the similarity of shell shape, but they show a great variety of patterns. These mollusks live buried in the shallow sediments of the sea floor, where they filter out food particles, and many species are covered with an opaque membrane, the periostracum, which hides the shell decoration. It thus appears that these patterns have no clear function, and there is no significant selection pressure favoring certain forms over others. This creates the opportunity for free expression of whatever intrinsic order there may be in the process of shell decoration.

The pigment patterns are laid down by an ectodermal membrane, which is also responsible for the growth of the shell at the open margin, seen on the right in Plate 2. The pattern is generated either simultaneously with the growth of the shell, new material being laid down along the boundary with pigment deposited along with the calicified material, or an ectodermal protrusion covers an undecorated region of new shell and deposits pigment in a two-dimensional pattern. In the species shown in Plate 2, *Cypraea diluculum*, the wavy stripes were generated by the first mechanism during the growth of the shell, the dots by the second after cessation of growth. The challenge to students of this phenomenon is to account for both the order and variety of the decorations with a plausible developmental pattern-generating process.

Meinhardt[25] has approached the general problem of biological pattern formation using field equations of the type known as reaction–diffusion equations. An example of such equations is shown in Box 2.

A Chemical Pattern Generator

$$\frac{\partial a}{\partial t} = s\left(\frac{a^2}{b} + b_a\right) - r_a a + D_a \frac{\partial^2 a}{\partial x^2}$$

$$\frac{\partial b}{\partial t} = sa^2 - r_b b + b_b + D_b \frac{\partial^2 b}{\partial x^2}$$

Here a and b are two different chemicals produced by cells that interact according to the reaction terms in the equations. They also diffuse, with diffusion constants D_a and D_b, as described by the second partial derivatives of a and b, with x as the spatial variable. The nonlinear reaction terms ensure that for particular ranges of the parameters the spatially uniform solution is unstable, so that distributed patterns in the concentrations of the chemicals arise spontaneously. Stable spatial patterns depend upon particular relations between a and b such that a acts as a short-range (small D_a) activator of itself and b, while b acts as a long-range (large D_b) inhibitor of a. Both substances are actively destroyed at rates proportional to their own concentrations, another source of global stability. The equations also allow for propagating wave solutions to arise; i.e., they define an excitable medium.

The class of equation described in Box 2 was first derived by the great polymath Alan Turing to explain biological pigmentation patterns such as the spots on Dalmatian dogs. Since diffusion eliminates spatial patterns by smoothing out differences, it was utterly counterintuitive that its combination with chemical reactions, which simply describe spatially uniform rates of production and destruction of substances, would produce stable spatial patterns. Turing took the first crucial steps in demonstrating this emergent property of reaction–diffusion equations.[26] Meinhardt, in collaboration with Gierer, reasoned independently that short-range activation together with long-range inhibition would create stable spatial patterns and provide insight into the underlying mechanism. Again we see the importance of working from different levels in the discovery and understanding of emergent properties.

By using equations of this type, with various modifications and extensions, Meinhardt was able to simulate an impressive range of pigment patterns. His book contains several dozen color illustratons of computer-generated shell patterns, all remarkably lifelike. They all share the same generative mechanism, which imposes basic constraints on the patterning process. At the same time they allow for a great diversity of patterns resulting from variations in parameters and from additional mechanisms involving, for instance, secondary inhibitors or hormone-like substances that affect reaction rates. An interesting example is shown in Plate 3. Here, waves of different frequency are simulated by superimposing a stable pattern in the parameter b_b along the growing margin of the shell, with an oscillation whose frequency depends upon the value of this parameter.

Another interesting class of patterns is shown in Plate 4. These shells have triangular patterns with a beautiful combination of order and irregularity. Meinhardt was able to simulate these by assuming that the activator and the inhibitor have antagonistic influences on the pigment-producing process. If the extinguishing effect of the activator dominates, triangles are produced (Fig. 3.7, upper patterns); if the enhancing effect of the inhibitor dominates, branches arise (Fig. 3.7, lower patterns).

The range of pigment patterns that Meinhardt has simulated with reaction–diffusion equations demonstrates that there is a deep generic order to this patterning process. This is what has been anticipated by many who have studied the phenomena, either casually or in depth, and important contributions have pointed clearly in this direction. On the other hand, there are those who take the view that *any* imaginable pattern could be generated under a suitable selection process. This requires that there be no significant intrinsic constraints on the pattern-generating process; it raises the expectation that left to its own devices without selection pressure, the range of patterns would show no coherent order. Meinhardt has explicitly shown that despite their freedom from selective pressure, the extraordinary diversity of patterns

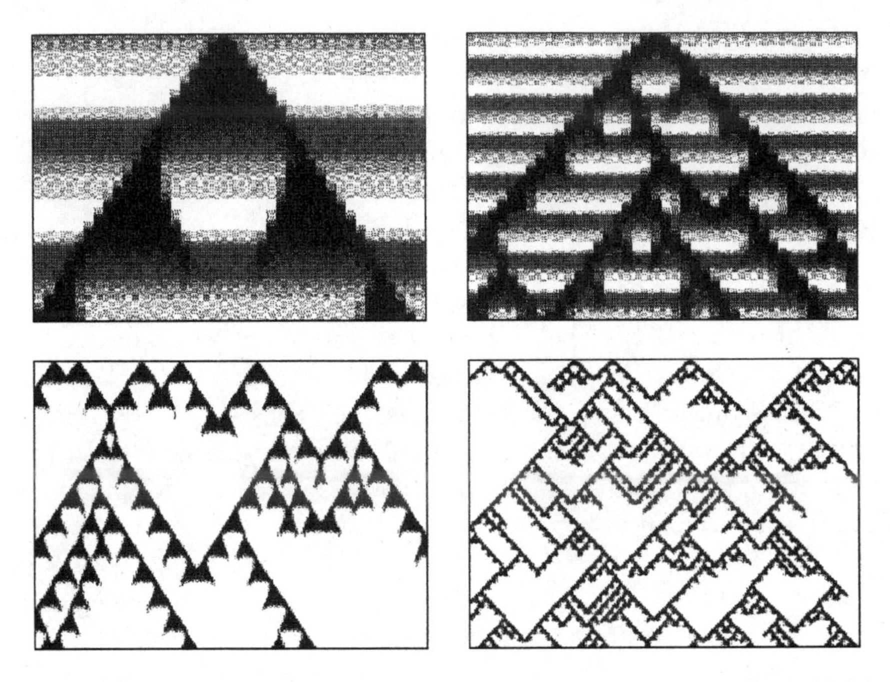

Figure 3.7 Computer model of triangular waves of different types. (From Meinhardt, 1995, with permission.)

nevertheless exhibits a high degree of intrinsic order resulting from a generative mechanism obeying particular mathematical constraints.

Of course, this does not mean that the particular mechanism of reaction–diffusion is necessarily the correct molecular generator of pigment patterns in marine mollusks. Other candidates such as Yukio Gunji's cellular automaton model could do the job equally well.[27] The conclusion from any such model, however, is that there are generative constraints that underlie the generic similarities of the patterns generated; not just anything can happen in evolution. Pigment patterns in mollusks provide one of the most beautiful and convincing demonstrations of constraint arising from intrinsic self-organizing principles of biological pattern formation.

FOUR

Physiology on the Edge of Chaos

. . . . the old controversies about the reducibility of biological facts to physico-chemical facts are now seen to be unnecessary if we realise that we have to deal with regularities which occur at each of these levels. . . .

—J. Needham, *Biochemical Aspects of Form and Growth*

Physiological Dynamics

The modern era of physiology is usually considered to begin with the recognition by nineteenth-century physicians that the human body has mechanisms whereby it regulates its fundamental processes independently of external influences. This maintainenance of internal conditions was called *homeostasis* by the great French physiologist Claude Bernard in his classic text of 1878, *Leçons sur les Phénomènes de la Vie Commun aux Animaux et aux Végétaux*. He recognized that the internal environment of the body, measured by such variables as temperature, blood sugar, salts, and acid–base balance (pH), are maintained at constant values despite changes in the individual's behavior and in external conditions. The body has a remarkable degree of autonomy, sustaining within narrow limits the internal conditions required for life despite long intervals without eating or drinking, and while engaged in the most diverse activities, from sleeping to climbing Mont Blanc.

Working out the details of the interactions between the different body systems that underlie this basic phenomenon occupied physiologists for the next hundred years. The key idea to emerge was that physiological variables interact with one another by feedback. For example, the activity of muscles during exercise increases the concentration of CO_2 in the blood due to conversion of glucose into H_2O and CO_2, with release of energy for muscular contraction. This increased CO_2 level affects the respiratory center in the brain, which stimulates the muscles controlling lung movements to increase their activity, so respiration rate increases and CO_2 levels decrease. This is a case of negative feedback: an *increased* level of CO_2 in the blood results in processes that *decrease* CO_2 concentration. Conversely, a *decreased* level of CO_2 results in a decrease in activity of the resipratory center, so respiration rate decreases and CO_2 level *increases*. Clearly, negative feedback is a source of stability and a major contributor to homeostasis.

Positive feedback works the other way: an increase in a variable activates processes that further increase the variable. This is a source of instability. However, for rapid response within a system this type of amplification can be very effective, provided that it is eventually overridden by a stabilizing process such as negative feedback. The two are often found to work together to produce sensitive responses in the physiological mechanisms of the body.

The principles and language of the preceding two paragraphs obviously invoke mechanical control systems of the type used in the design of anything from central heating to sophisticated missile guidance systems. Cybernetics arose from the recognition of this confluence of principles of control dynamics, whether within organisms or machines. However, first under the stimulus of cybernetic principles and then from the unexpected richness of its own phenomena, the study of physiological dynamics has gone far beyond anything Claude Bernard could have dreamed of. He would certainly have known of the inspiring work in mathematical physics of Henri Poincaré. But he could not have anticipated that the fascinating and puzzling nonlinear dynamics of celestial bodies then being discovered by Poincaré would one day transform physiology from a study of stability and constancy into an exploration of how our bodies are poised on the edge of chaos. The very concept of health, so elusive that it has no place in the lexicon of contemporary physiological concepts, is revealed as a subtle emergent property of the dynamic complexity of living organisms.

To follow the emergence of this idea, we return to the key concept of homeostasis.

Homeostasis

The body maintains the glucose concentration of the blood at a nearly constant value of 5 mmole/liter despite great changes in the rate at which glucose enters or leaves the bloodstream. After a meal or a Mars bar, glucose pours into the blood from the intestine, while during exercise the rate at which muscles withdraw glucose from the blood may increase by a factor of 10. Yet the change in blood glucose concentration is minimal. It has been known for some time that there are two hormones crucially involved in regulating this constancy: insulin and glucagon. Insulin accelerates removal of glucose from the blood and into sites of storage in muscles, liver, and other organs. Glucagon does the opposite: it accelerates the mobilization of glucose from stores and its release into the bloodstream. Working in opposing directions, the two hormones operate on the principle of negative feedback: an *increase* of glucose in the blood above 5 mmole/liter results in release of insulin from the beta cells of the pancreas, which brings about a *decrease* in glucose level; a glucose level below 5 mmole/liter stimulates release of glucagon from alpha cells in the pancreas, which results in an *increase* of glucose in the blood. In principle, either of these would be sufficient to achieve homeosatic regulation. Why have two? This pairing is, in fact, the general rule for hormonal control processes in the body. Is there a dynamic principle that underlies this pairwise counterbalancing action?

Peter Saunders and colleagues[1] investigated this question using insights that came from a quite unexpected quarter: the Daisyworld model developed by Andrew Watson and James Lovelock[2] in 1983 to describe the emergent property of temperature regulation of an imaginary planet based on growth of populations of black and white daisies. In the model, black daisies absorb radiation from the sun and warm the surface, so they thrive better during early planetary evolution when solar radiation is weak. As the sun ages and its radiation increases, conditions begin to favor white daisies, which reflect the sun's rays, cooling themselves and the planet. Both daisy populations have the same optimum temperature for growth; they do not adapt to the prevailing mean temperature during solar evolution. Together they produce a remarkable constancy of planetary temperature over a large range of solar radiation increase.

Lovelock's Gaia theory was originally inspired by a comparison between physiological regulation and the constancy of properties of the earth that are crucial for life, such as temperature, oxygen concentration in the atmosphere, and concentration of salts in the oceans. This gave rise to the metaphor of the earth as a living organism. Saunders and his coauthors, contemplating the emergent temperature regulation of the Daisyworld model, realized that its principles could be applied to physiological regulation. The analogues of black and white daisies are pairs of hormones that have antagonistic effects on the regulated variable. Here is a case of positive feedback within science itself. The theory of glucose regulation that emerged is described in Box 1.

Integral Control of Glucose in the Blood

Denoting concentrations of glucose, glucagon, and insulin by G, A, and B, a general expression for glucose regulation can be written in the form

$$\frac{dG}{dt} = I + \alpha(A, G) - \beta(B, G) - \gamma(R, G). \qquad (1)$$

Here I is the rate of input of glucose into the blood, $\alpha(A, G)$ and $\beta(B, G)$ are functions representing the effects of glucagon and insulin, respectively, on blood glucose concentration, and $\gamma(R, G)$ describes the uptake of glucose by muscles and other tissues. The term R in this last function designates the activity of the organism, which largely determines the rate of glucose utilization. The functions $\alpha(A, G)$, $\beta(B, G)$, and $\gamma(R, G)$ are all taken to be nondecreasing functions of their variables, and $\gamma(R, 0) = 0$.

The hormone concentrations are taken to satisfy the equations

$$\frac{dA}{dt} = A(\phi(G)h_1(A, B) - D_A), \qquad (2)$$

$$\frac{dB}{dt} = B(\psi(G)h_2(A, B) - D_B). \qquad (3)$$

Here $\phi(G)$ is a decreasing function of G and $\psi(G)$ is an increasing function of G, while D_A and D_B are the constant rates of degradation of

A and B. The functions $h_i(A, B)$ represent the effects of the hormones on themselves and on each other, and so must satisfy the relations

$$\frac{\partial h_i}{\partial A} < 0 \quad \text{and} \quad \frac{\partial h_i}{\partial B} < 0.$$

That is, both hormones exert a negative influence on the production of themselves and the other. For the particular case of glucose control we can make the further assumption that $h_1 = h_2$. This implies that both of the hormones, glucagon and insulin, which are produced by different types of cells (the alpha and beta cells of the islets of Langerhans in the pancreas), respond in the same way to the concentrations of the hormones. However, differences in the functions $\phi(G)$ and $\psi(G)$ result in a higher rate of insulin production when blood glucose is above 5 mmole/liter, whereas glucagon is produced at a higher rate when glucose is below that value. Both hormones are produced all the time, but in different proportions under different conditions. This corresponds to physiological observations.

To examine the effectiveness of this glucose control system, we need to determine the steady-state concentration of glucose that is defined by the equations and then see how sensitive it is to different types of perturbation. Equating (2) and (3) to zero, and with the assumption that $h_1 = h_2$, we obtain that

$$\phi(G) = \left(\frac{D_A}{D_B}\right) \psi(G). \tag{4}$$

This can be solved for G, the steady-state value of glucose, which is evidently independent of A and B. With this value of G, equations (2) and (3) can then be used to determine the values of A and B, the steady-state values of glucagon and insulin, for any particular values of I and R, the glucose input and output rates from the bloodstream. What this shows is that for any given $\phi(G)$, $\psi(G), D_A$, and D_B, which can all be taken to be genetically specified in any individual, the glucose concentration in the blood will be constant, and what varies as the input (I) and output (R) change is the values of A and B, the hormones. A simulation of this behavior for the function $h(A, B) = K - A - B$ (K is a constant) in response to a 20-fold change in glucose input is shown in Figure 4.1. The sudden rise in blood glucose beyond $I = 17$ occurs when the steady-state value of glucagon reaches zero, at which point regulation fails.

Saunders and coworkers have generalized this model of physiological regulation involving two antagonistic hormones so that the assumption $h_1 = h_2$ need not hold and a considerably less severe constraint applies to the way the hormones act in concert.

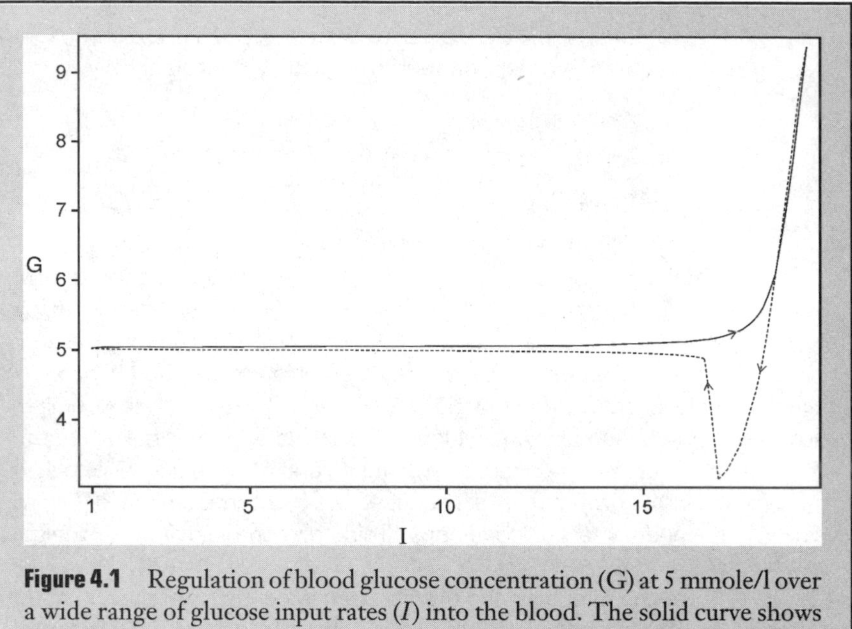

Figure 4.1 Regulation of blood glucose concentration (G) at 5 mmole/l over a wide range of glucose input rates (*I*) into the blood. The solid curve shows the response to increasing rates of glucose input, while the dotted curve is for decreasing rates, as computed from the model of Saunders and coworkers.[2]

The model gives a striking demonstration of the effectiveness of the use of two hormones acting antagonistically above and below the desired set point, rather like the reins of a horse acting on either side of the head to maintain a desired direction. This commonly observed, but previously not well understood, principle of homonal regulation of physiological variables has accordingly been called rein control.

The same authors have extended the model to hormonal regulation of other crucial physiological variables such as blood calcium concentration. These models make specific physiological predictions that are under investigation.

This research provides an interesting example of how an emergent property discovered in the field of Gaia theory or geophysiology, was then generalized mathematically and applied back in the original domain of physiology, which provided the inspiration for Gaia in the first place.

Biological Clocks

One of the first physiological variables to be identified as homeostatic was body temperature. Above 40.6°C (105°F) our brains begin to fry;

below 32°C (90°F) we sink into a coma. Yet if we measure body temperature very precisely throughout a 24-hour period, we find that it does not stay constant but varies with a regular rhythm from about 36.3 to 36.9°C, as shown in Figure 4.2a. There is some variability of amplitude from day to day, as seen in Figure 4.2b, but the times of maximum and minimum body temperature are more precise: close to 6:00 P.M. for the former and 3:00 A.M. for the latter, depending somewhat on the season of the year. How is it that the body has an internal rhythm that maintains such a precise phase relationship to an external periodicity, namely, the rotation of Earth and the day–night cycle? In seeking an answer to this question, biologists have followed a traditional strategy, looking for evidence of similar rhythms in simpler organisms. The evidence turns out to be everywhere: 24-hour cycles are virtually universal in biology, a generic aspect of physiology in organisms ranging from algae to whales.

One of the first single-celled organisms to be systematically studied in the laboratory was a little marine alga called *Gonyaulax polyedra*. This dinoflagellate, which gets its energy from photosynthesis and swims about by means of a whiplike tail, or flagellum, is found in the Caribbean Sea. One of its distinguishing characteristics is its luminescence, which is particularly marked when it is mechanically disturbed, as by wave action; it lights up the surf along the shoreline. Studies in the 1950s and 1960s by Beatrice Sweeney and J. Woodland Hastings on laboratory cultures of these cells showed that there is a daily rhythm of spontaneous flashing luminescence. There is also a rhythm of steady glow, with a peak of luminescent activity in the early morning just before dawn.

The first question is whether this rhythm is intrinsic to the cells or whether it arises in response to the day–night cycle to which the organisms are normally exposed. The way to answer this is to examine the behavior of the cells under constant conditions in the laboratory. Since these cells have to make their own nutrients by photosynthesis, they need light to continue normal physiological functions. First the cultures were grown under a 24-hour light–dark cycle. Under these conditions the cultures exhibited a regular diurnal luminescent rhythm. Then they were exposed to constant dim light, with the results shown in Figure 4.3. The rhythm continues indefinitely, though the amplitude decreases somewhat and the waveform is not so sharp as under periodic light–dark conditions. Clearly, the rhythmic activity of the cells is not dependent on an external day–night cycle.

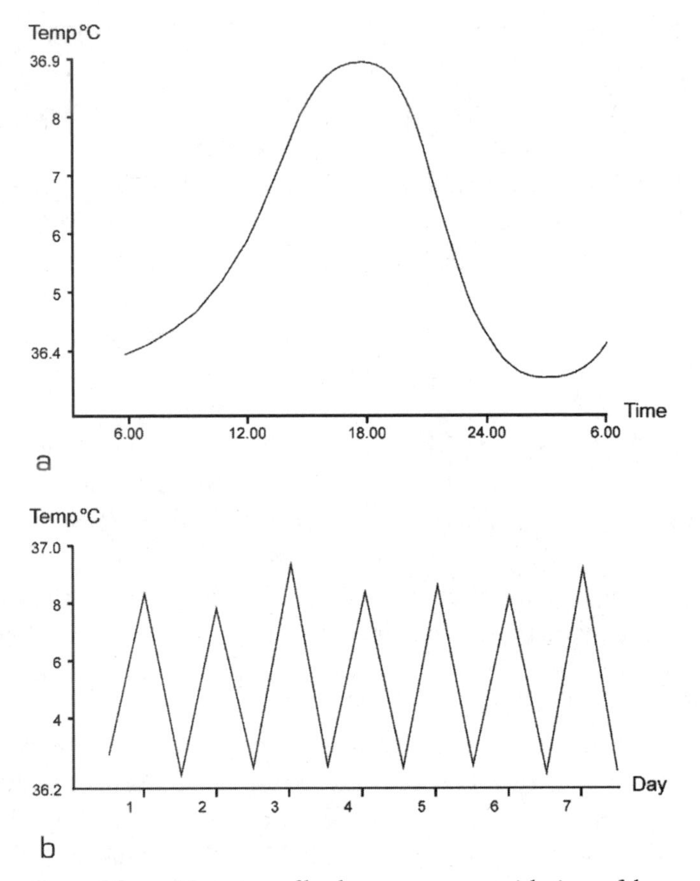

Figure 4.2 a. Variation of body temperature with time of day. b. Maximum and minimum values of body temperature over several days.

As observations always do, these results raise a number of further questions. First, how is it that a culture of individual cells, moving freely relative to one another, maintains a collective rhythm? If there were no interactions between cells, one would expect that small differences of intrinsic period would result, over time, in a complete loss of collective rhythm. Evidently, these cells are interacting in some way, either via light signals or chemicals released in association with their rhythmic activity. This takes us back to Isabel Ko's observations on bacterial cultures and the modeling results of Kaneko and Yomo showing that interactions between cells can maintain conditions of local synchrony. That the whole culture of *Gonyaulax* maintains synchrony implies that there are significant periodic interactions between the cells. They reveal

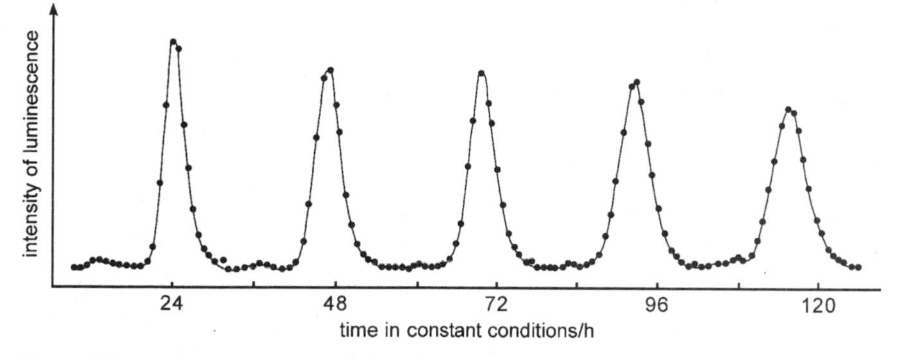

Figure 4.3 Variations in luminescent glow in cultures of the unicellular dinoflagellate *Gonyaulax polyedra* under conditions of constant dim light.

the well-known phenomenon of synchronous locking of interacting nonlinear oscillators.

Another question arising from Figure 4.3 relates to the fact that the period of the rhythm is no longer exactly 24 hours; it has decreased to about 23 hours. This must be the natural mean period of the cells under these conditions, called their free-running rhythm. Evidently, the daily light–dark cycle forces this rhythm into a 24-hour period, another well-known nonlinear phenomenon called *entrainment.* This can occur only if the free-running period of the cells is sufficiently close to the entraining cycle. The cells' natural rhythm of 23 hours clearly satisfies this condition. Because this is close to 24 hours, the intrinsic cycle of the cell is referred to as a circadian (*circa diem*, about one day) rhythm or a circadian clock.

Biochemical processes in cells are sensitive to temperature. What if *Gonyaulax* cells are grown at different temperatures? Will the circadian period change? The period of 23 hours was recorded for cultures at 19°C. When the temperature was increased to 21°C, the period did indeed change, but to 24.5 hours. Biochemical reactions generally run faster at higher temperatures, but the clock runs slower at 21° than at 19°. Temperatures in the Caribbean can change throughout the year by more than 10°C. As a general rule, reaction speeds increase by a factor of 2 with a 10° rise in temperature. If this happened to the circadian clock, the period would fall outside the range of entrainment and cells would fail to synchronize with the rising and setting of the sun. It is in their interest to do so because their lives depend upon organizing their metabolic activities so that they fit the diurnal cycle of the planet. They prepare for the coming of dawn by synthesizing molecules required for photosynthesis, and they carry out light-sensitive reactions during the dark hours. So it appears that the circadian clock has developed the property of *temperature compensation* to ensure that its natural period does not change significantly despite changes in external temperature. The fact that the *Gonyaulax* clock runs more slowly as temperature increases suggests a bit of overcompensation, but the period remains well within the entrainment range throughout the temperature variations that the cells experience in the Caribbean. How can we account for this intriguing property? Is this a natural emergent property of nonlinear systems, or is it a special adaptive feature of circadian rhythms? At this point we need a mathematical model of the circadian clock, based on plausible intracellular processes.

Modeling the Circadian Clock

As noted, negative feedback is a source of stability in control systems. This stability, however, need not be the constancy implied by the term homeostasis. It could equally well be a stable oscillation—a limit cycle—since negative feedback can produce this form of activity under certain conditions. Gene activity is involved, because synthesis of mRNA and proteins is necessary for clock function. This was known by the 1960s, a period of intense activity in molecular biology and also of very active, pioneering studies of biological clocks. Negative feedback control processes had been identified both in metabolic pathways and in the control of gene activity, wherein the final product of a reaction sequence inhibits the reactions leading to it. It was therefore natural to propose that the basic mechanism of the biological clock might be a feedback control loop from gene activity to metabolic product and back to the gene. Goodwin[3] showed that this type of negative feedback system could oscillate under certain conditions. The class of equations involved is described in Box 2.

Oscillations from Feedback Control

$$\frac{dX}{dt} = \frac{k_1}{(Z^n + 1)} - k_4 X,$$
$$\frac{dY}{dt} = k_2 X - k_5 Y, \qquad (5)$$
$$\frac{dZ}{dt} = k_3 Y - k_6 Z.$$

Here X represents the concentration of mRNA produced by a gene, Y is the concentration of the protein (an enzyme) produced by the mRNA, and Z is a metabolite resulting from catalytic activity of Y. The k_i's are rate constants for the different reactions. All reactions are simple first-order kinetics except for the feedback term in which Z inhibits the activity of the gene with stoichiometry n: n molecules of the metabolite form a complex that inhibits transcription of mRNA from the gene according to normal mechanisms for inhibition of activity, giving the first term with the expression $(Z^n + 1)$ in the denominator. For $n > 8$, these equations result in oscillations of the type shown in Figure 4.4. These are clearly nonlinear, and they behave as limit cycles: for given values of the parameters there is a stable cycle to which the system returns after perturbation.

There are different interpretations possible for the variables X, Y, and Z in these equations. The variable X can be seen as nuclear mRNA that is processed to give the active cytoplasmic form Y for translation into protein, in which case Z is a protein product rather than a metabolite. As such it plays the role of a transcription factor that feeds back to control gene activity. The large value of n required for oscillatory activity is unrealistic in relation to known transcription complexes that control gene activity, though $n = 4$ is not unusual. To get oscillations with $n < 8$ it is necessary either to add further linear equations to the sequence, representing more steps in the production of active transcription complex, or to make some of the steps nonlinear. Equations (5) are simply a convenient set for describing the qualitative properties of this class of molecular control process.

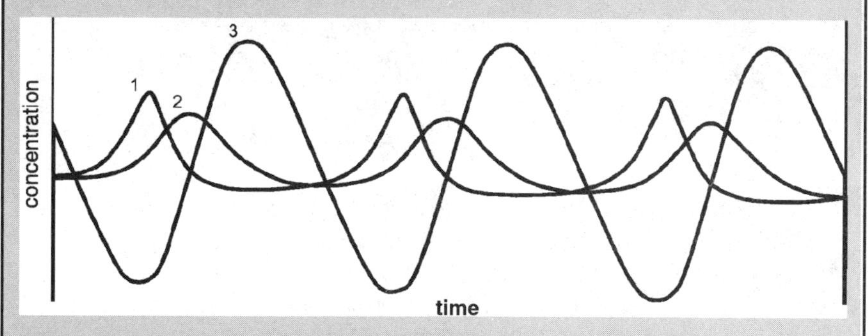

Figure 4.4 Oscillatory behavior of the components of a feedback circuit to a gene modelled by Goodwin.[4] 1. Concentration of messenger RNA. 2. Concentration of protein. 3. Concentration of the feedback control molecule.

It was the properties of the Goodwin model, together with experimental observations showing that the model conformed to basic molecular mechanisms of the biological clock, that led Peter Ruoff and Ludger Rensing[4] to propose how temperature compensation could arise in a remarkably simple way from the collective behavior of the whole circuit. Ruoff called the basic idea *antagonistic balance*. If the effects of temperature on the amplifying processes (the steps with rate constants k_1, k_2, and k_3) and on the inhibiting or stabilizing processes (with rate constants k_4, k_5, and k_6; see Box 2) controlling clock dynamics are balanced, then the period of the oscillator will be temperature compensated. The experimental observations indicating that the model has the right general structure came from genetic studies of clock mutants in two of the classic organisms used in the study of circadian rhythms: the bread mold *Neurospora crassa*[5] and the fruit fly *Drosophila melanogaster*.[6] These showed that the protein products of genes basically implicated in clock function (*frequency* in *Neurospora* and *period* in *Drosophila*) feed back to inhibit transcriptional activity of the gene. Furthermore, it was shown[7] that mutations in *frequency* also affect temperature compensation, implying that the feedback mechanism and temperature compensation are linked. So Ruoff and Lensing proceeded to explore how these might be connected in the model, as described in Box 3.

How Antagonistic Balance Emerges from Feedback

Temperature influence on chemical reactions is decribed by the Arrhenius equation, which links rate constants to temperature as follows:

$$k_i = A_i \exp\left\{\frac{-E_i}{RT}\right\}.$$

Here A_i is a constant, E_i is the activation energy of the reaction (both independent of temperature), R is a universal constant, and T is the temperature. It is not possible to describe precisely the effect of temperature on the period of oscillations in a sytem such as (5). However, Ruoff and Lensing derived an approximate expression by introducing empirical parameters that are found by calculating the partial derivatives $\alpha_i = \partial \log P_{num}/\partial \log k_i$ from numerically determined circadian period lengths. These are then used in an expression for approximate period length P_{appr} as a function of temperature. Temperature compensation implies that $\partial P_{appr}/\partial T = 0$. This leads to the expression $\sum \alpha_i E_i = 0$ for temperature compensation. This relation can be satisfied because while the E_i's are always positive, the α_i's are positive for the reactions in (5) with rate constants k_1, k_2, and k_3 and negative for reactions with rate constants k_4, k_5, and k_6. Making reasonable assumptions about the values of activation energies, Ruoff and Lensing showed how oscillation periods of equations (5) can quite naturally have temperature compensation. Not all chemical oscillation schemes have this property, and they conclude that this feature of what are called the Goodwin equations strengthens their candidacy for the circadian clock. Clearly, they are a highly simplified version of the real processes involved, but they appear to capture essential aspects of clock behavior, including their unexpected property of relative insensitivity to temperature change.

Temperature compensation emerges from this analysis as another example of order for free: negative feedback circuits of the type involved in circadian clocks are naturally insensitive to temperature variation so this property does not have to be selected in evolution. How general is this property of temperature compensation in the molecular activities of cells?

One of the most fundamental processes of cells is growth and division. *Gonyaulax* cells tend to divide at a particular time of day, early in the morning. However, the mean growth rate of a culture is dependent on temperature: the higher the temperature, the greater the growth rate and the shorter the interval between cell divisions in any one cell, up to an optimum temperature above which the growth rate decreases. So these processes are not temperature compensated. Nevertheless, their timing is controlled by the circadian clock: a cell that happens to be ready to divide in the evening has to wait until morning to enter the division process. The circadian clock and cell doubling are not synchronized with each other. Cell growth proceeds at its own rate, which varies directly with temperature; but there is a window of time every day, specified by the circadian clock, when cells can divide if they are ready to do so. Hence the overall economy and organization of a cell are a result of different types of process, with different temperature dependencies, whose complex order arises from a loose confederacy of dynamic interactions with subtle relationships.

In warm-blooded animals such as humans, temperature regulation has been taken over by control processes acting on the whole body. These include rates of metabolism and heat production through shivering, and heat conservation and loss through contraction and dilation of blood vessels, panting, and sweating. The circadian rhythm that modulates our relatively constant temperature also involves high-level control processes dependent on neural and hormonal activities. Prominent among these is the pineal gland, which secretes the hormone melatonin in a rhythmic pattern. Its concentration in the blood rises in the evening and puts us to sleep, then falls during the day. The activity of the pineal is affected by stimuli from the eyes, keeping its activity tuned in to the cycle of day and night. After jet travel across time zones, you can reset your body clock by taking a pill containing melatonin half an hour before going to bed. Together with altered eating times and the pattern of light stimuli from the eyes in the new time zone, this helps to shift the whole physiology into a rhythm attuned to the new context.

Despite this higher-level regulation of the body clock, our cells and organs still have their own circadian clocks, and these, too, need to be shifted into the new day–night rhythm. Our bodies are more like multilevel clock shops than single-level control systems. Nowhere is this dynamic complexity more evident than in the regulation of the basic reproductive process in mammals: the menstrual cycle.

The Dynamic Control of the Menstrual Cycle

Reproduction in higher organisms is rather more complex than in single cells. In primates, including humans, the basic cycle of egg maturation and release, and preparation of the uterus for receiving a fertilized egg, involves a set of component activities among which we can recognize the basic constituents of homeostatic mechanisms. But they are organized in a manner that is remarkable for the robustness that emerges from complex dynamic interactions. What follows is a somewhat simplified account of this process from which the essential dynamic principles emerge. The basic physiological components of the 28-day menstrual cycle are shown schematically in Figure 4.5. The cycle is divided into two phases, punctuated by ovulation and menstruation, which are separated by about 14 days. Hormonal signals (gonadotropins) from the pituitary gland in the brain are released into the bloodstream and stimulate the growth of a follicle within the ovary. The follicle consists of a single haploid germ cell, together with a surrounding sheath of diploid cells that assist the growth and maturation of the germ cell into an oocyte. The developing follicle releases a hormone (oestradiol) that stimulates pituitary secretions, constituting a positive feedback loop. As a result, pituitary activity increases progressively. This stimulates growth of the endometrium, the soft tissue on the inner surface of the uterus where implantation will occur if the oocyte is fertilized. Gonadotropins also stimulate maturation of the oocyte and its release from the follicle into the space outside the ovary, that is, ovulation.

The empty follicle, called a corpus luteum, remains within the ovary and changes production from oestradiol to progesterone, which

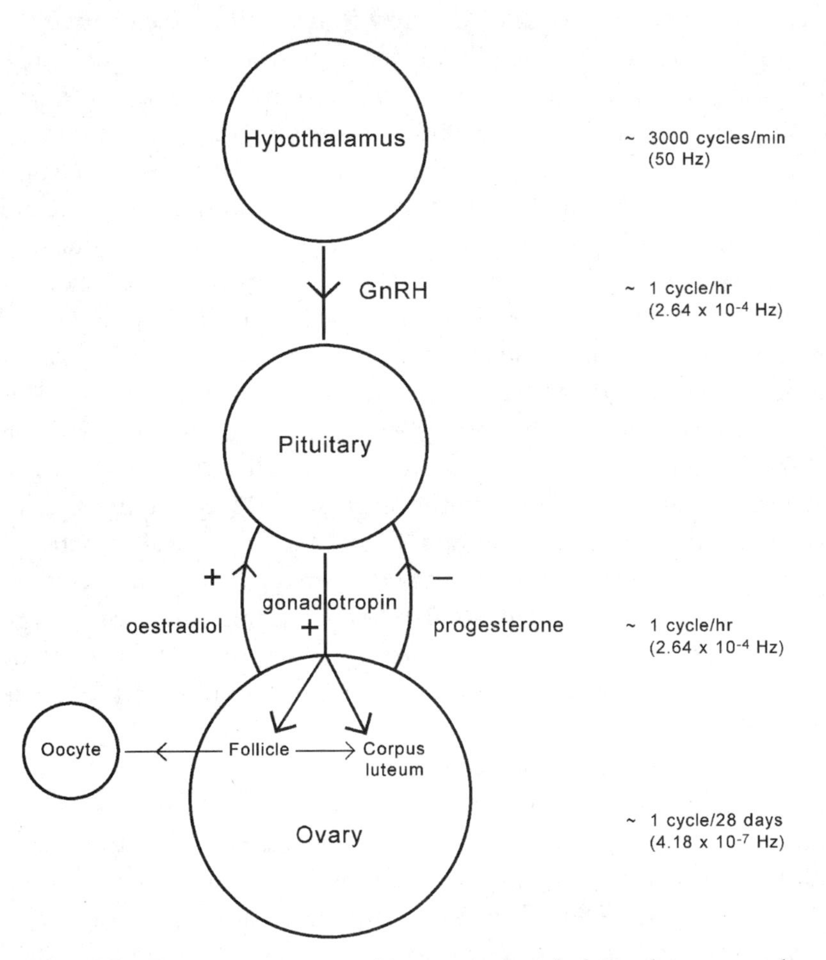

Figure 4.5 Basic components and interactions underlying the menstrual cycle, showing the positive feedback loop during follicular growth and the negative feedback loop during the luteal phase of the cycle. See text for details.

inhibits pituitary activity: what was a positive feedback loop turns into negative feedback, and pituitary activity subsides. The oocyte enters a fallopian tube and travels toward the uterus. Sperm can fertilize the oocyte during this passage, and if this occurs, then implantation can follow, with maintenance of the pregnancy by hormones produced by the developing embryo. If there is no fertilization or implantation, the follicle degenerates and progestrone production falls away. The uterine endometrium degenerates, and menstruation occurs around the fourteenth day after ovulation. The cycle then starts again with release of the pituitary from inhibitory influence and the beginning of another phase of positive feedback from a new follicle within the ovary.

At first sight the principles of this cycle appear to manifest physiological control principles of the most simple and straightforward kind: positive and negative feedback mechanisms alternate to produce a regular rhythm whose normal period of 28 days is a result of the slow processes of follicle growth, oocyte maturation, and follicle degeneration. But there is another player in the cycle: the hypothalamus. Located deep in the brainstem, this part of the brain secretes a hormone called gonadotropin releasing hormone (GnRH). This stimulates the pituitary to produce gonadotropins, as the name of the hormone indicates. One might have expected that the pattern of GnRH synthesis would also wax and wane on a 28-day cycle, reinforcing the rhythm of the pituitary secretions. However, this is where the dynamics take an interesting turn. The hypothalamus introduces a quite different frequency into the control system. There is an electrical activity cycle with a period of around 1 hour, with neurons changing their frequency of firing from about 1,000 to 3,000 spikes per minute, as shown in Figure 4.6 (the data are from rhesus monkeys).[8] This correlates directly with hourly pulses of gonadotropin release from the pituitary (Figure 4.6, upper trace).

This hourly rhythm of pituitary activity became evident only when the gonadotropin concentration in the blood was assayed every 10 minutes or so. How does this high-frequency rhythm vary over the 28-day menstrual cycle? Normally menstruating women volunteered to have blood samples taken every 10 minutes throughout a 24-hour period at different stages of the menstrual cycle. Results are shown in Figure 4.7 for early, mid, and late follicular phases in the first 14 days during which the follicle is developing and the oocyte maturing.[9] Periods of sleep are indicated by shaded bars. Here we see a complex

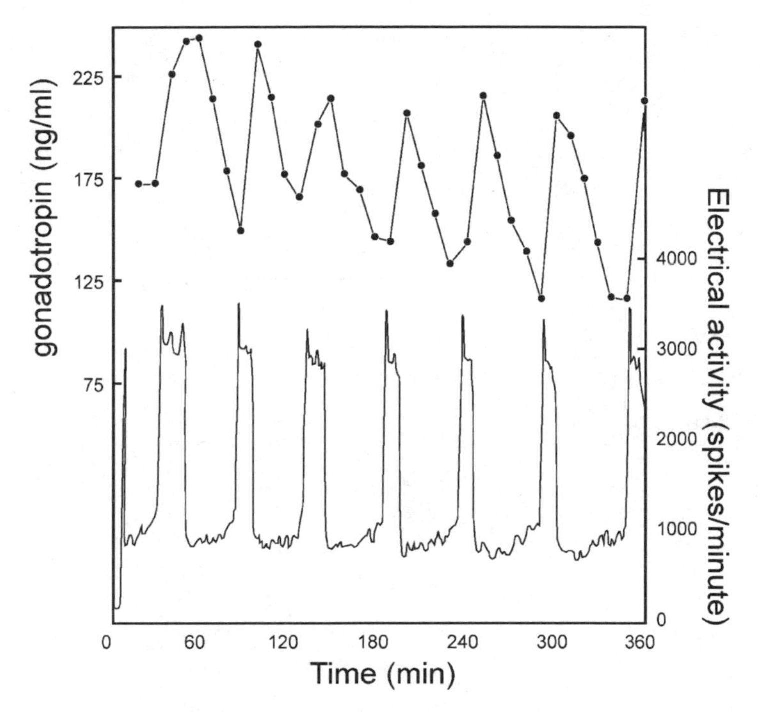

Figure 4.6 Variations in blood gonadotropin concentration (upper curve) and electrical activity in the hypothalamus (lower curve) in *rhesus* monkeys.

dynamic pattern in which the amplitude of the hourly gonadotropin rhythm increases. In the early phase, sleep interrupts this rhythm: the circadian cycle exerts an influence. Later in the cycle, however, the hourly pulsing continues throughout sleep. What had been previously taken to be a continuous increase in gonadotropin levels during the first 14 days of the menstrual cycle turned out to be an increasing amplitude of an hourly pulse. During the subsequent 14 days, the luteal phase of the cycle, there is diminution of amplitude as well as frequency. What is the significance of this pattern?

Ernst Knobil's experiments on rhesus monkeys in the late 1970s showed that ovulation occurred only if the GnRH pulses were within a relatively narrow frequency range centered on 1 per hour. If the period is too long (>1.5 hours) or too short (<0.5 hours), the follicle does not mature, and no ovulation occurs. This led to the realization that various types of ovulatory disturbances in women might have their origins in dynamic disorders of GnRH secretion. Normal ovulation could be restored in many types of disturbed ovulation by hourly

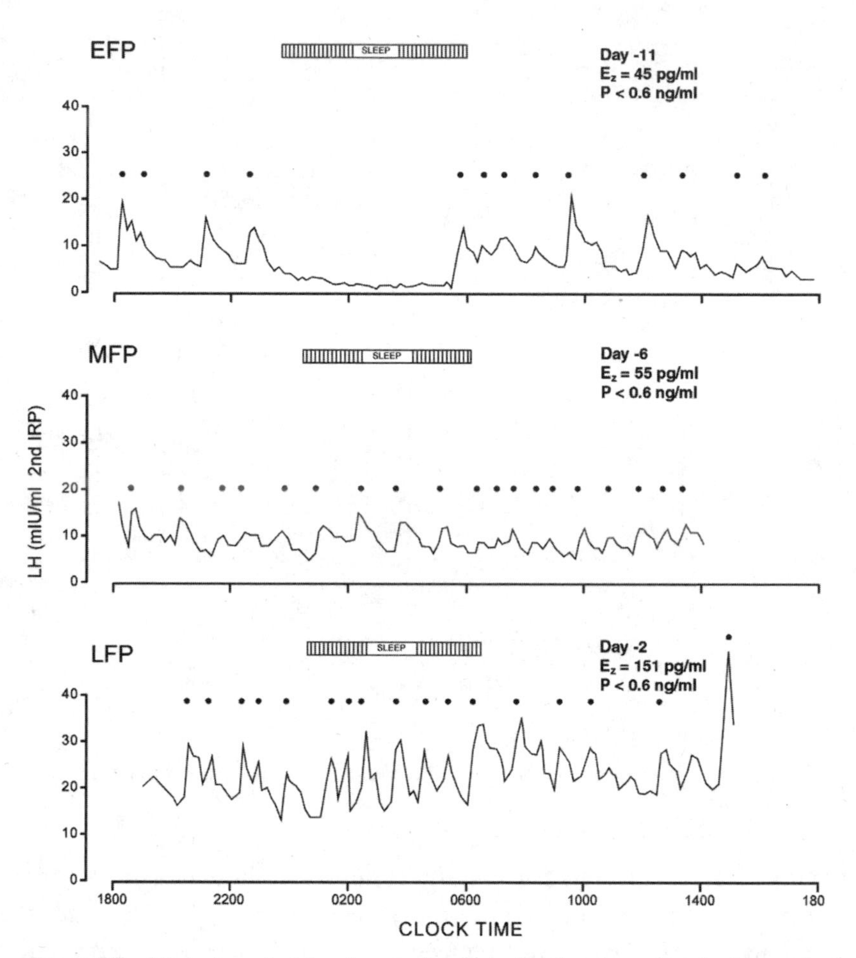

Figure 4.7 Variations in blood gonadotropin levels of women during early (upper curve), middle (middle curve), and late (lower curve) follicular phases of the menstrual cycle.

pulses of this hormone, whereas daily injections that maintain the same average hormone levels are ineffective. It thus emerged that the menstrual cycle is crucially dependent upon an oscillation of unexpectedly high frequency relative to the overall 28-day cycle. The follicle, tuned to a narrow frequency range of the pulsatile gonadotropin signal, fails to grow and mature if this range is violated. Frequency control is widely recognized as a reliable signaling process in electronic devices, with receivers tuned to resonate at particular frequency ranges. Extending this idea to the biological domain, researchers such as Paul Rapp[10] and Albert Goldbeter[11] have demonstrated that frequency coding of physiological signals is more reliable than amplitude control. The dynamics of menstrual cycle control strongly support this view.

It is particularly striking that control of ovulation is so robust despite the complexity of the dynamics. There is a great deal of irregularity in the patterns of gonadotropin shown in Figure 4.7, an interesting mixture of order and disorder that nevertheless produces a reliable physiological result. For systems as complex as living organisms this is perhaps not surprising, since a great many different processes, operating on different time scales, are all interacting. An extraordinary range of frequencies are integrated into a single coherent cycle: hypothalamic firing patterns of 3,000 cycles per minute (50 Hz) alternating with 17 Hz firing frequency at a rhythm of 1 cycle per hour (2.64×10^{-4} Hz); influence of the circadian cycle with a frequency of 1.1×10^{-5} Hz; and an overall frequency of 1 cycle per 28 days (4.14×10^{-7} Hz). This is eight orders of magnitude in the frequency range! And physiological processes other than those described in the simplified picture of Figure 4.5 also exert their influence, such as signals connected with emotional and nutritional states that can disturb the menstrual cycle. A dynamic complexity and sophistication of control processes has entered physiology and led to two new concepts that are changing the understanding of health and disease: dynamical disease and scaling laws.

Dynamical Disease

Ovulatory disorders need not arise from any deficiency in the quantity of gonadotropin but solely in what Marco Filicori[9] calls the quality of the signal, such as the amplitude or the frequency of the periodic pulses of GnRH. The recognition that many physiological signals are intrinsically periodic means that the temporal dimension of phase and frequency relations is essential to normal physiological function, and disturbances to these relations may underlie many medical conditions. The importance of this realization of course lies in treatment: simply increasing the amount of a hormone or a drug may be ineffective in treating the disorder. Phase or frequency can be crucial. This is now widely recognized, and a growing research community is actively engaged in the investigation of dynamic disorders and their treatment.[12] The concept of dynamical disease was introduced by Michael Mackey and Leon Glass in 1977.[13]

Sudden cardiac arrest is one of the most striking examples of a particularly acute dynamical disease. Many individuals who die of heart failure reveal in autopsy no evidence of any previous damage to the heart. Their heart failure appears to have arisen from the sudden onset of the rapid, dysfunctional beating that characterizes ventricular fibrillation: disorganized, autonomous contractions of the ventricles running at five times the normal rhythm of the cardiac pacemaker, with consequent failure to pump the blood to the body. What could be the origin of such a condition?

The study of excitable media was what gave the clue to this phenomenon.[14] Adult heart muscle is not itself capable of generating an autonomous oscillation of the type that recurs in the sinoatrial node, the region of specialized pacemaker cells that initiates the regular propagating contraction wave of the normal heartbeat. Experimental studies of propagating patterns in heart tissue had suggested that fibrillation may be caused by an excitation wave that travels in a closed loop in heart tissue. This is a perfectly natural state for an excitable medium. It is one of the attractors we encountered in Chapter 1 in connection with the generator of spiral waves in the cellular slime mold. This type of reentrant wave, circus movement, or rotor was shown to occur in a square section of healthy sheep heart when two successive electrical stimuli were applied to adjacent edges of the tissue, initiating a wave that propagated in a closed loop around the tissue. The periodicity of this wave was about 200 msec, which corresponds to 300 beats per minute, a frequency that is within the range observed in a fibrillating heart. This suggests that fibrillation can arise if heart tissue receives an additional electical stimulus (from autonomic nerve stimulation of the heart, say) or if there is an island of damaged tissue that is unable to conduct the propagating wave, which therefore travels around it and can develop into a closed propagating loop around the infarct. In the former case heart tissue would be perfectly normal and we would be dealing with an imbalance of electrical stimulation that initiates an abnormal rotor in either atrial or ventricular tissue. This is an instance of dynamical disease that could be treated by, say, a beta blocker that modulates autonomic impulses to the heart or by relaxation therapy that reduces the intensity of autonomic activity. Once again, understanding the dynamic origins of a physiological condition can lead to effective treatment by temporal balancing and harmonizing of the process.

As with the menstrual cycle, what is striking about heart activity is the diversity of the influences that act upon it and the robustness of its performance. The heart, one might say, lives within an utterly unpredictable context, with demands coming from change of muscular activity, respiration, and hormonal and electrical signals reflecting changing emotional and physiological conditions. Homeostasis would suggest that the heart tends to maintain a constant rate of beating for any particular state of the body (lying, sitting, walking, running). However, the evidence is that strict constancy is what the healthy heart avoids, and if it occurs, it is a clear sign of danger.

Living with Uncertainty

Cardiologists have known for some time that the electrocardiograms of children have a very noticeable irregularity: the intervals between heartbeats vary unpredictably even if the child is sitting quite still. Whereas the mean heart rate has a very well defined and repeatable value whenever the child is sitting or lying quietly, the interbeat interval shows significant spontaneous variation. This variability is also characteristic of a healthy adult, though it is less marked than in children; and the older the individual, the less the variation, as a general rule. Does this variability provide some insight into the dynamic strategy of a healthy heart in an unpredictable body?

During the past ten years or so, this question has been addressed by a number of different groups investigating physiological dynamics with the new tools available from nonlinear analysis and complexity theory. One such group is at the Beth Israel Hospital in Boston, under the direction of Ary Goldberger. This interdisciplinary team has used a variety of analytical methods to investigate the characteristics of the interbeat interval in both healthy individuals and those with various types of cardiac arrhythmias. The analysis starts with a plot of instantaneous heart rate as a function of time, determined by measuring the interval between heartbeats (called the R-R interval), as shown in Figure 4.8. The instantaneous heart rate is proportional to the inverse of the interbeat interval and is presented in Figure 4.9. Here we see the pattern of variation in a healthy subject compared with two individuals with cardiac disease. Each has a well-defined mean value around which the rate varies, but whereas the normal pattern is extremely variable, the two diseased hearts are much more orderly, one showing a clear

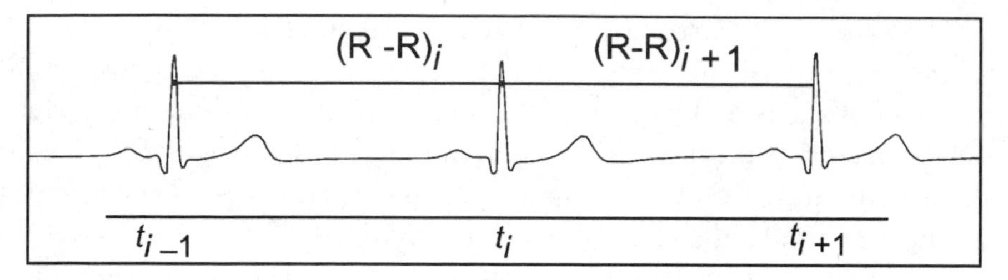

Figure 4.8 The interbeat intervals of successive heartbeats are measured by the time intervals between the so-called R component of the electrical signal, the most prominent feature of the electrocardiogram.

oscillation, while the other is almost constant. The latter is the pattern predicted for a homeostatic process that keeps the heart rate constant. Yet the person with this pattern is very ill and in danger of cardiac arrest, as is the individual with the periodic variation in heart rate. Here, too much order is a sign of danger. Although we all know that this maxim holds for people's behavior in life generally, it is at first

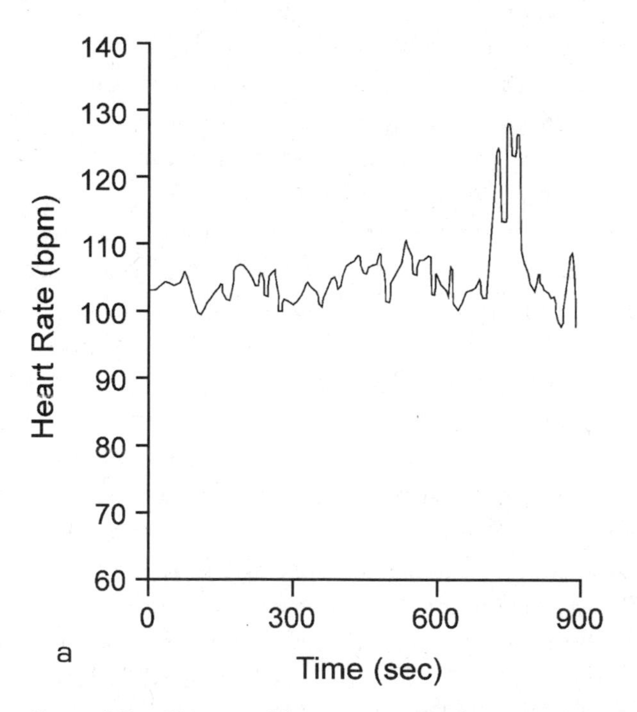

Figure 4.9 Change of heart rate, which is inversely proportional to the interbeat interval, in (a) a healthy individual, (b) and (c) individuals with heart conditions.

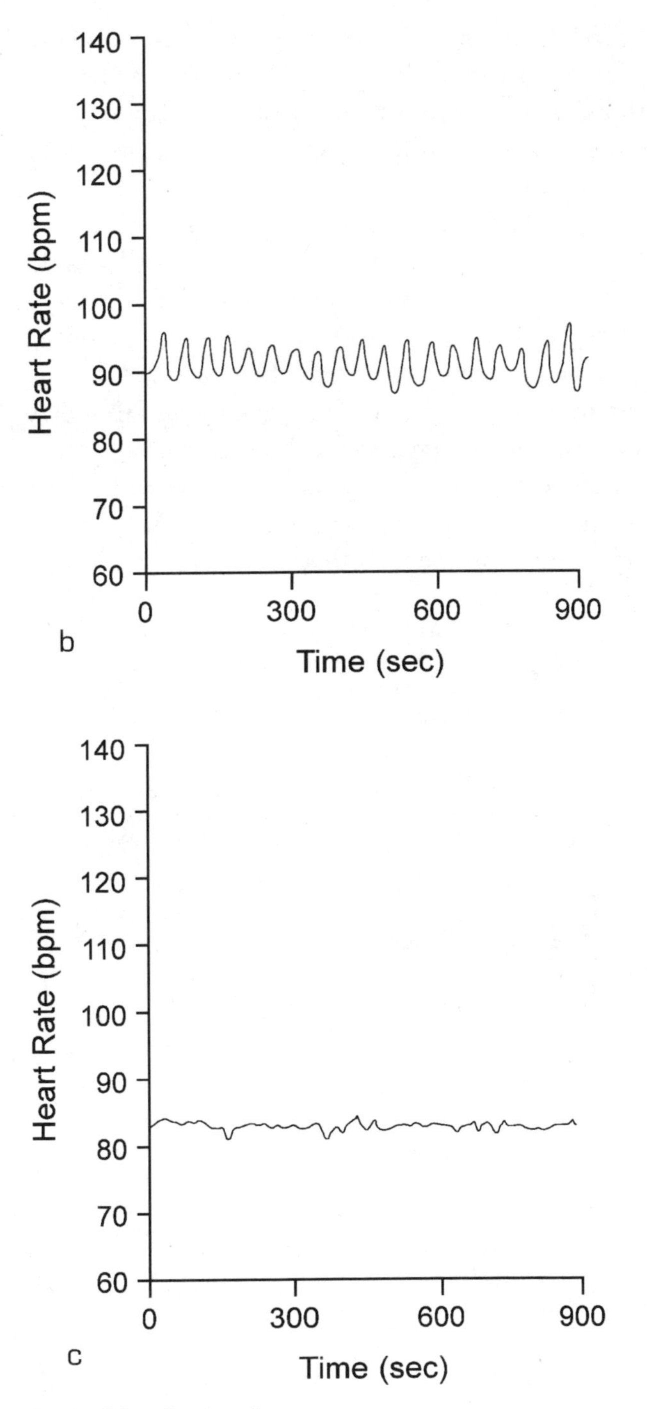

Figure 4.9 *Continued*

surprising to encounter it in physiology. A question that now arises is whether the healthy individual also shows some order, but perhaps of a much more subtle kind than the pathological cases. This is the issue that Goldberger and his colleagues have pursued, with very interesting consequences and insights.

A first approach to this issue involved examining the data in Figure 4.9 by applying some conventional mathematical techniques such as Fourier analysis. This procedure identifies the frequency components of time series, which are shown in Figure 4.10 as Fourier spectra. It is clear from visual examination that there is a major frequency in the periodic pattern of subject 2, in Figure 4.9b, and this shows up in the Fourier spectrum as a strong peak at the corresponding frequency (Figure 4.10b). Subject 3, with nearly constant interbeat intervals, has very weak frequency components. However, the healthy subject has a broad range of frequencies in the data that are strongest for low frequencies

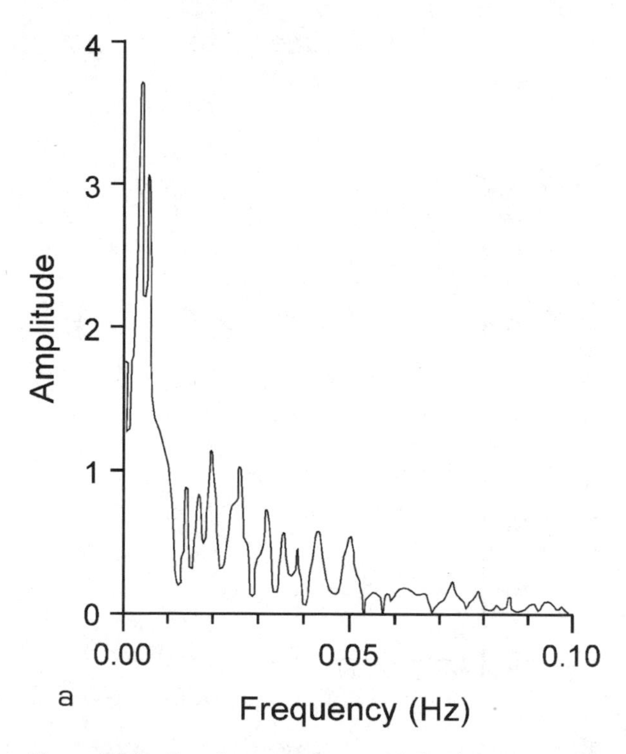

Figure 4.10 Fourier transforms of the heart rates in Figure 4.9, showing the frequency composition of healthy (a) and abnormal (b, c) heart rates.

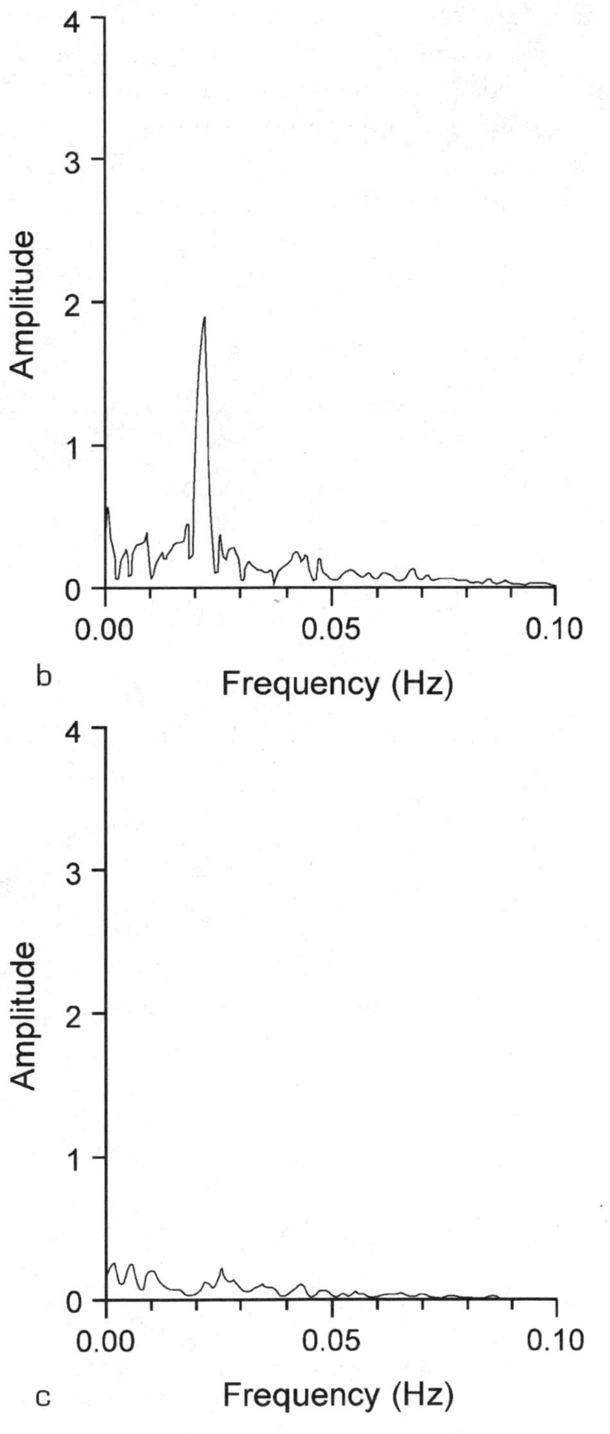

Figure 4.10 *Continued*

and drop off rapidly at higher frequencies. Such a pattern is consistent with, but not diagnostic of, a dynamical system with self-similarity and long-range order, as occurs in deterministic chaos. Goldberger identified self-similarity by examining the time series on different time scales and observing similar variation, as shown in Figure 4.11. Here the heart rate was examined on different time scales: a 30 minute section of a 300 minute record was expanded, revealing a pattern of fluctuations similar to the original, and also for a 3 minute section expanded again by

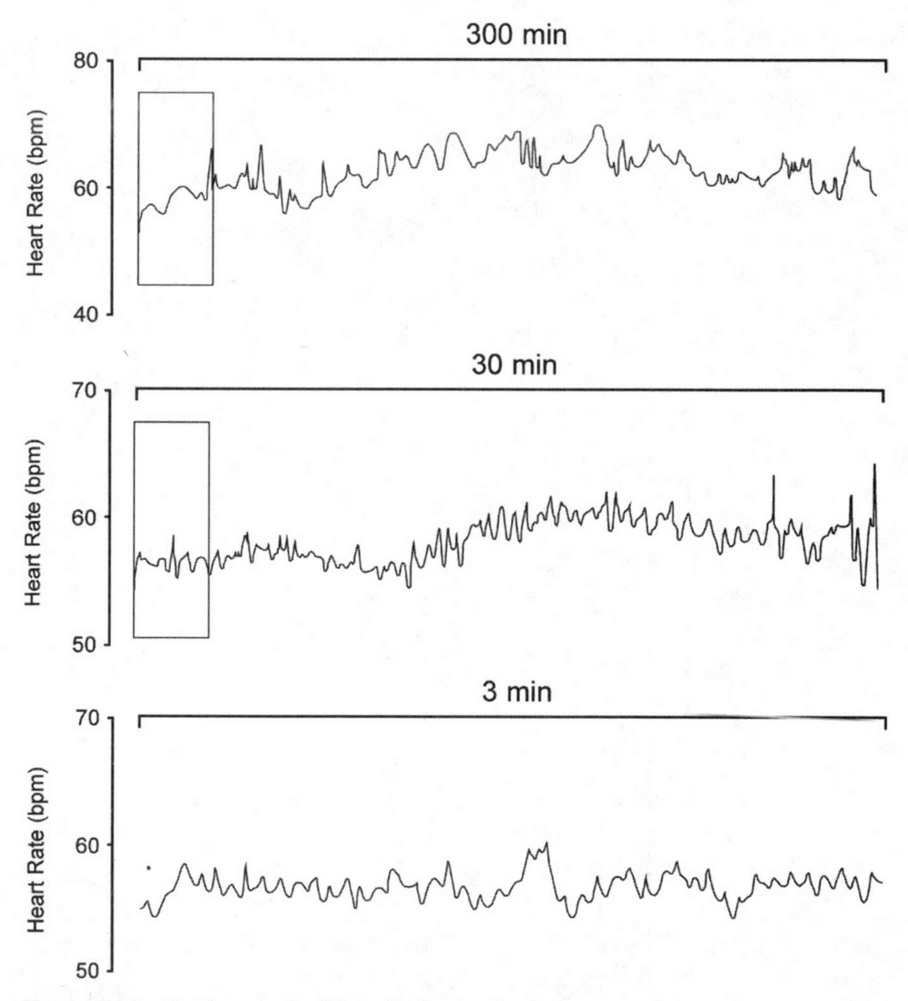

Figure 4.11 Evidence of self-similarity in the heart rate time series: expanding small sections of the time series by the same factor of 10 reveals progressively finer detail without any characteristic time scale.

the same factor of 10. A subsequent electrocardiogram study revealed that the behavior of the healthy heart is indeed characterized by chaotic dynamics, while individuals with congestive heart failure had a more random pattern of variation with a weaker expression of chaos.[15] Can we further characterize the property of health? Is there some identifiable feature of heart dynamics that is a signature of health? This question requires a more sophisticated analysis of the data that involves a study of properties using the renormalization procedure described in Chapter 2.

A time series such as that of instantaneous heart rate shown in Figure 4.9 is not stationary; that is, it does not have constant statistical properties. To identify a generic, or typical, property of such a time series, one must use analytical techniques that do not require stationarity. The Boston team[16] reported on the result of applying such an analytical technique to a comparison of heart rate data between two sets of individuals: a healthy group and one suffering from obstructive sleep apnea, a condition of intermittently interrupted breathing during sleep. Electrocardiograms from 18 healthy persons and 16 apnea subjects were transformed into time series by measuring the R-R intervals and plotting these as a function of time. These data were then studied using what is called the wavelet transform, also called the "mathematical microscope" because it allows one to study a time series on any scale one considers significant. For the heart data, the focus chosen was 8 beats, so that very high frequency variations were filtered out and dynamic correlations could be examined in the range of 30 seconds to 1 minute, considered to be the range of most physiological significance. Beat to beat variations of each individual were then described by construction of probability amplitudes for these variations. Each individual, whether healthy or with sleep apnea, had a distinct probability function of these amplitudes.

The next step was to see if there is some property of these functions that is shared by the healthy group but not by the apnea group. The researchers found that the probability functions of the healthy group all belonged to the same class: they could be transformed into one another by use of a scaling function with a single adjustable parameter. The apnea group did not have this property. This shows that despite individual differences, all the healthy subjects share the same type of self-similarity in the pattern of the interbeat intervals. As discussed in Chapter 2, this occurs for different types of physical systems near a critical point, such as the transition of a gas to a liquid, the appearance of magnetization in a ferromagnet, or superfluidity in a liquid at critical temperatures. These phase transitions all show similar behavior with respect to such features as the correlation length, or the sizes of clusters with collective properties, at the critical point. These are scale invariant, with the property of self-similarity over different spatial scales and obeying a characteristic power law distribution.

Despite this striking phenomenon of scaling that reveals a basic similarity between phase transitions in physical systems and the dynamics of a healthy heart, there is also an important difference. Phase transitions occur in physical systems only under very specific conditions such as those of temperature and pressure; they are not the typical state of these systems. On the other hand, health *is* the typical or natural condition of an organism; it is the dynamic attractor to which the self-healing organism tends to return spontaneously. The apnea subjects have been disturbed from this state, and their heartbeat dynamics do not belong to the same class as the healthy group. The expectation is that they could undergo therapy of some kind that would help their dynamic state return to the healthy condition.

What, then, is the physiological significance of the scaling behavior revealed by this study of the heartbeat? The analysis reveals the presence of long-range correlations in heart dynamics that appear to be an emergent property of complex physiology. The resulting balance or coherence is subtle, but it means that the heart avoids getting locked into any dominant frequency that might prevail under particular patterns of the individual's behavior. The different influences that act on the heart operate over many different time scales: millisecond intervals from neural impulses, which include emotional changes; seconds from respiratory demands; minutes from hormonal signals; hours from behavior patterns such as sleeping or sitting or walking; daily rhythms as we have seen; monthly hormonal patterns; and annual changes of season and habit. The heart rate can get locked into any one of these that happens to dominate, or it can fall into dynamic attractors characteristic of its own condition as an excitable medium, such as high-frequency fibrillation or other cardiac arrhythmias. In sleep apnea, it appears that the heart becomes locked into a mode arising from periodic breathing dynamics, thus disturbing the typical distribution of interbeat intervals that is described by a scaling law. This distribution is not simply the sum of the diverse influences that act on the heart from its unpredictable physiological context; rather, the characteristic dynamics of health are a result of nonlinear influences that describe the dynamic coherence of the whole organism as a single unified system, an emergent entity with distinctive properties of subtle dynamic order. The state of health is the normal biological attractor that combines both order and chaos.

One of the most striking implications of this conclusion concerns the relationship between part and whole. By examining the dynamics of a part of the organism, in this case the heart, one can draw inferences about pathologies of other parts of the whole, in this case the respiratory system. This provides a potentially powerful diagnostic tool for identifying dynamic imbalances of physiological function. There is nothing new about this procedure. Good health practitioners, whatever their training and tradition, diagnose conditions of health and disease from examination of parts to reveal the condition of the whole. Pulse rate, body temperature, blood cell counts, and many other variables are used in allopathic diagnosis, but a good diagnostician also pays attention to complexion, tone of voice, posture, and other qualitative features of the person to reach a conclusion about health or disease. A practitioner of traditional Chinese medicine is trained to read the condition of the whole body and imbalance in any part by pulse diagnosis alone, reading it both quantitatively and qualitatively. Physiological dynamics may now be reaching the point of convergence between these different traditions through the recognition of health as an emergent dynamic property of the whole organism.

FIVE

Brain Dynamics

We are barely beginning to address the fact that interactions among many non-contiguous brain regions probably yield highly complex biological states that are vastly more than the sum of its parts.

—Antonio Damasio

Into the Black Box

The human brain is the most astonishing and mysterious of all known complex systems. Inside this mass of billions of neurons, information flows in ways that we are only starting to understand. The memories of a summer day on the beach when we were kids; imagination; our dreams of impossible worlds. Consciousness. Our surprising capacity for mathematical generalization and understanding of deep, sometimes counterintuitive questions about the universe. Our brains are capable of this and much more. How? We don't know: the mind is a daunting problem for science.

When a brain fails, its failure can be as puzzling as normal function. Sometimes mental disorders are comical, as with poor Don Quixote, who, after many nights of no sleep and too much reading of old books, started to confuse reality with tales about dragons and chivalry (Figure 5.1). Sometimes the injured brain acts in ways that are simply bizarre. The essays of Oliver Sacks give us an impressive list of examples.[1] One of Sacks's patients was a musician and teacher who began to have problems recognizing the faces of his students. He could recognize, for example, familiar voices, but faces became strange

Figure 5.1 Don Quixote's imagination, inflamed by romances of chivalry.

objects. Walking through the streets he sometimes stopped to pat the heads of parking meters, thinking them the heads of children. Sometimes, to his embarrassment, "he would amiably address carved knobs on the furniture, and he was astounded when they did not reply." Sacks discovered that our spontaneous ability to recognize faces as whole entities (and not a set of independent objects like the nose or the mouth) was lost in this patient. He could not identify people by looking at them: faces had lost their meaning as coherent structures.

In many cases, brain damage in some area tells us what has failed. The destruction of part of the cortex can lead to loss of hearing or visual capacities, or to motor disorders. These observations led neurologists to see the brain as a compartmentalized system whose specific capacities were localized in well-defined regions. This is shown in Figure 5.2, a nineteenth-century drawing of the brain areas. In recent years, the development of brain imaging techniques has led to some renewal of this view. Brain tomography, nuclear resonance, and other techniques are based on the detection of metabolic brain activity levels. It is assumed that for a given mental task those regions more involved will require a larger energy consumption and thus a higher activity level. When we see these color-coded pictures of the brain in many newspapers and journals, the captions sometimes suggest that we are looking at thoughts themselves. The neurophysiologist Steven Rose

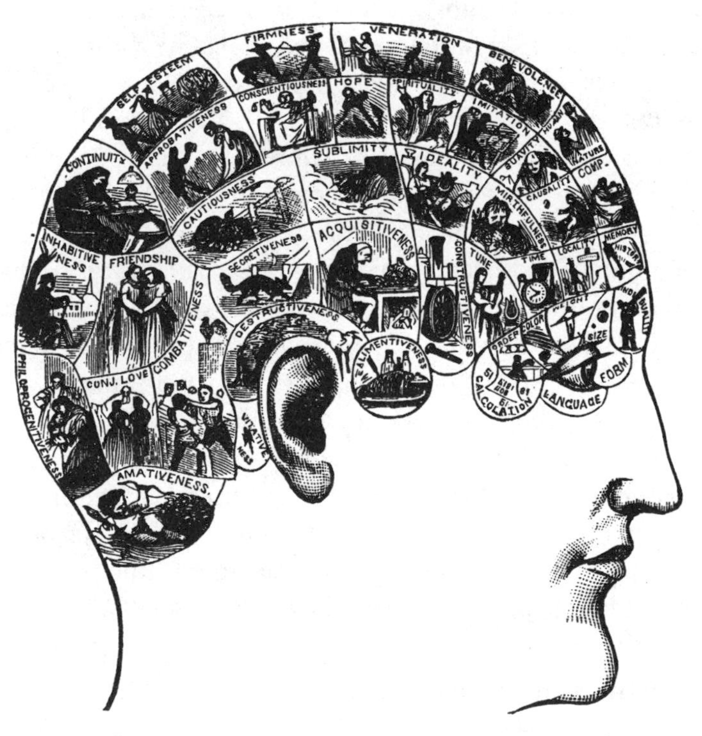

Figure 5.2 An old fashioned view of the compartmentalization of brain functions.

clearly points out the need for care about the meaning of these spatial patterns of activity:

> It is hard not to feel when one first sees them that such techniques will surely answer all the questions we might have about brain function and its relation to mental processes. But it isn't quite so simple. Showing that a brain region is active when a person is learning or remembering is not the same as showing that the memory "resides" in that part of the brain. The "store" might be somewhere quite different, somewhere that doesn't need a great flurry of glucose utilization to activate it.[2]

The history of brain research is a succession of metaphors. To some, the mind and the brain are different entities, and as a consequence the first cannot be explained through an understanding of the second. To others, the human brain should be analyzed like a rat in a maze: certain inputs lead to certain outputs. Some physiologists consider the brain as a black box, claiming that by applying appropriate sets of stimuli and looking at the resulting responses they can decipher some part of the internal mechanisms. Unfortunately, this is rather like trying to know the details of a movie by watching people leave the theater.

Is it possible to model the brain? Some authors have argued against the possibility of understanding the brain (and the mind), supporting this position with theories ranging from Gödel's incompleteness theorem to quantum mechanics. To others, one of the most appealing metaphors is that the brain is a computer. It is very tempting to think that since our brains perform computations and store information, brains must be much like computers. John von Neumann tried to develop this analogy, but he failed to obtain an appropriate description of brain function.[3] All of these metaphors ultimately failed—as we will see in this chapter, the brain operates very differently from an electronic computer—but all of them were in some way very helpful. The brain is not a computer, but computers are very useful in helping us learn about the brain. Early, both Von Neumann and Alan Turing had the important insight that an appropriate modeling of the brain required networks of simple, perhaps randomly wired, elements.

From Neurons to Networks

The basic units of brain tissue are the neurons. A view of the brain cortex under the microscope reveals a dense matrix of neural cells with many interconnections. In Figure 5.3 we show a famous drawing of the great Spanish neurologist Santiago Ramón y Cajal, showing an area of the human cerebral cortex that controls voluntary movement. Cajal received the Nobel Prize in 1904 for his remarkable series of publications on the circuitry of many regions of the brain. He was the first scientist to suggest that neurons are the fundamental units of brain tissue and that they communicate by contact, instead of being part of a continuous reticulum like the arteries and veins of the circulatory system. This conjecture was strongly supported by later research and fully demonstrated when electron microscopy became available.

The basic structure of a neuron is shown in Figure 5.4. Other neurons send inputs across the synapses, the gaps between the ends of the dendrites and the membrane of the adjoining cell. Eventually, after some threshold is reached, the neuron is able to generate a signal that propagates to other neighboring neurons. Nerve cells, however, are far from simple firing devices. A very important part of neuroscience research over the past few decades has been dedicated to understanding the transmission of signals from one neuron to the next. These studies have shown that most synaptic connections are chemical (instead of electrical, as we might expect). Neural transmission is thus a complex interplay of chemical and electrical processes.

One of the main problems of early neuroscience was to explain the origins of neural activity.[4] If we look at a single neuron and measure its membrane potential—the difference in charge between the inside and the outside of the cell—we find that it has a value of −65 mV: the inside of the membrane carries a negative charge. This is because certain membrane proteins (called potassium channels) actively transfer negative ions into the cell. This membrane potential is a characteristic of the neuron at rest and so is called the resting potential. Signals from other neurons can raise or lower it. If the membrane potential crosses a given threshold, the membrane rapidly depolarizes and reaches a peak (of about +40 mV). Afterwards, it drops again to low negative levels and recovers the resting value. The depolarization propagates along the membrane in a coherent wave that moves down the axon toward the next synaptic terminals. In this way a nerve impulse can spread through the brain.

The complexity of neurons is strongly reduced in models of many interconnected neurons. These models, called *neural networks*, became very popular after the work of the physicist John Hopfield. In 1982

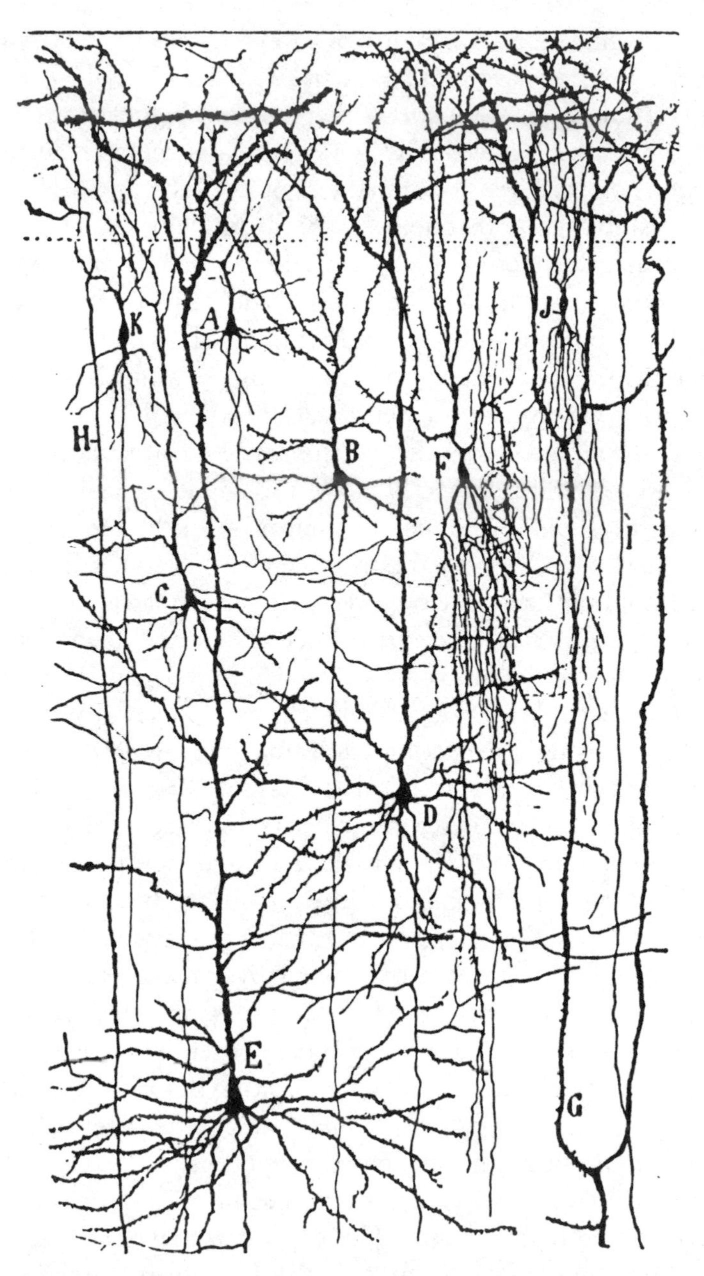

Figure 5.3 A drawing by Ramon y Cajal of neurons in the cerebral cortex.

Hopfield published a paper in which he introduced an oversimplified neural network, comprising a set of fully connected binary units, as a metaphor of neural computation.[5] The most remarkable feature of this

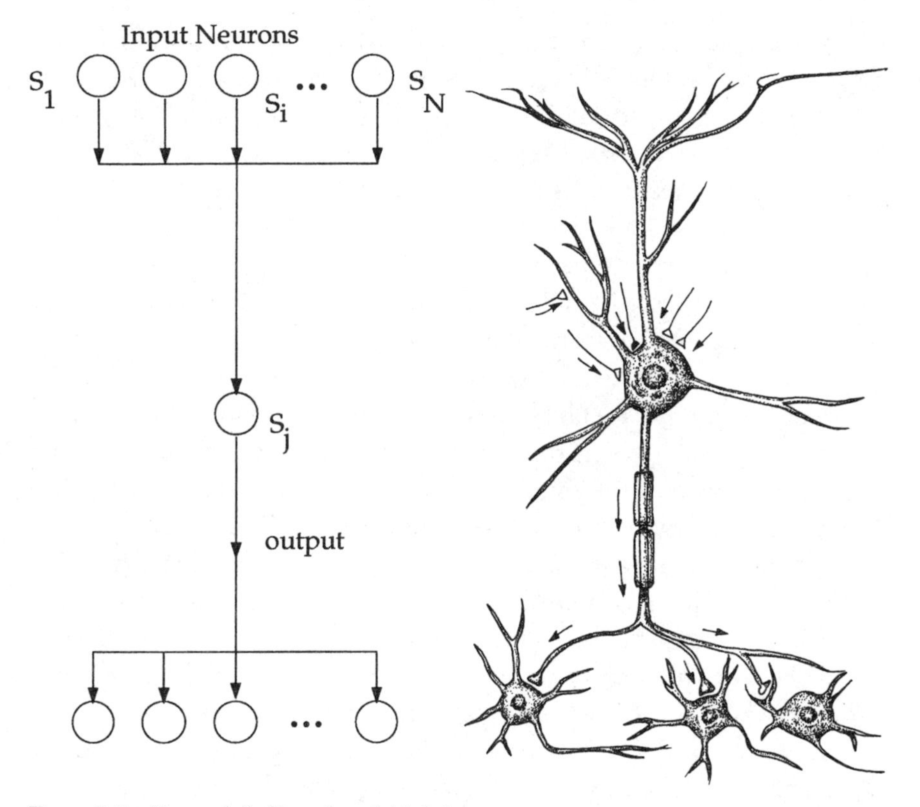

Figure 5.4 Formal (left) and real (right) neurons.

model was that it could learn by association and was quite insensitive to noise. A Hopfield net is able to recognize previously learned patterns that are not complete or have been corrupted by noise. This capacity for recognition under incomplete information is something we should find very familiar. Take a cartoon of a familiar face. We identify the face even though the picture is just a schematic drawing and a two-dimensional object. Even if some part of the cartoon is covered by a piece of paper we can still (up to some limit) recognize the face. This is a fascinating property of brains, and it seems that it should be difficult to reproduce it from a simple model. But this is precisely what the Hopfield model does.

This network uses formal neurons (Figure 5.4, left) as basic units. They can take only two values: either at rest (-1) or firing ($+1$). All neurons are connected with all the others, and connections can be of two types: excitatory (positive) or inhibitory (negative). The massive parallelism is a good picture of some areas of the brain cortex, where each neuron receives several thousand inputs. Hopfield also assumed some biologically unrealistic properties, such as the symmetry of connections: the strength and type of input received by neuron A from neuron B are the same as those neuron B receives from A. Neurons change state by means of a simple threshold function (see Box 1). If the sum of the inputs is positive, the neuron will switch to (or stay in) the +1 state. Otherwise, it takes the −1 value.

These rules describe only the basic structure and dynamics. What about learning? Where are memories stored? This requires an additional rule, known as *Hebb's rule* because it was proposed by the neurologist Donald Hebb to describe how external stimuli can be translated into permanent changes in synaptic connections. When two neurons receive the same type of stimulus, their connections are reinforced. Otherwise, they are weakened. In the Hopfield description, the so-called learning phase involves a process in which the network is shown a number of patterns once. Each of the training patterns generates changes in the connections by means of the Hebb rule. At the end of this phase, the set of memories has been stored *in a collective way* into the matrix of connections. Now, the pattern are, in fact attractors of the dynamics, and the net is able to recognize them even when a significant part of their properties has been randomly perturbed.

Figure 5.5 shows an example of the dynamics followed by the Hopfield network with $N = 25$, where N is the number of neurons. During the training phase, the network has stored a set of patterns corresponding to letters of the alphabet. One of them is the letter A, which is shown to the net with a number of switched units (left top, a). The network evolves by modifying the state of different units following the previous rules. After some sequence of changes (b–d) the network is able to fully reconstruct the complete pattern (e). If the amount of noise introduced at the beginning is too large, then the network cannot identify the letter correctly.

The previous dynamics can be shown to be understandable in terms of a decrease in energy when an appropriate energy function is used. In physical systems, it is often possible to define some kind of function that is minimized through the dynamics. This function (defined in Box 1) is, in fact, a multidimensional hypersurface, since the number of variables is N and the total number of possible network states is

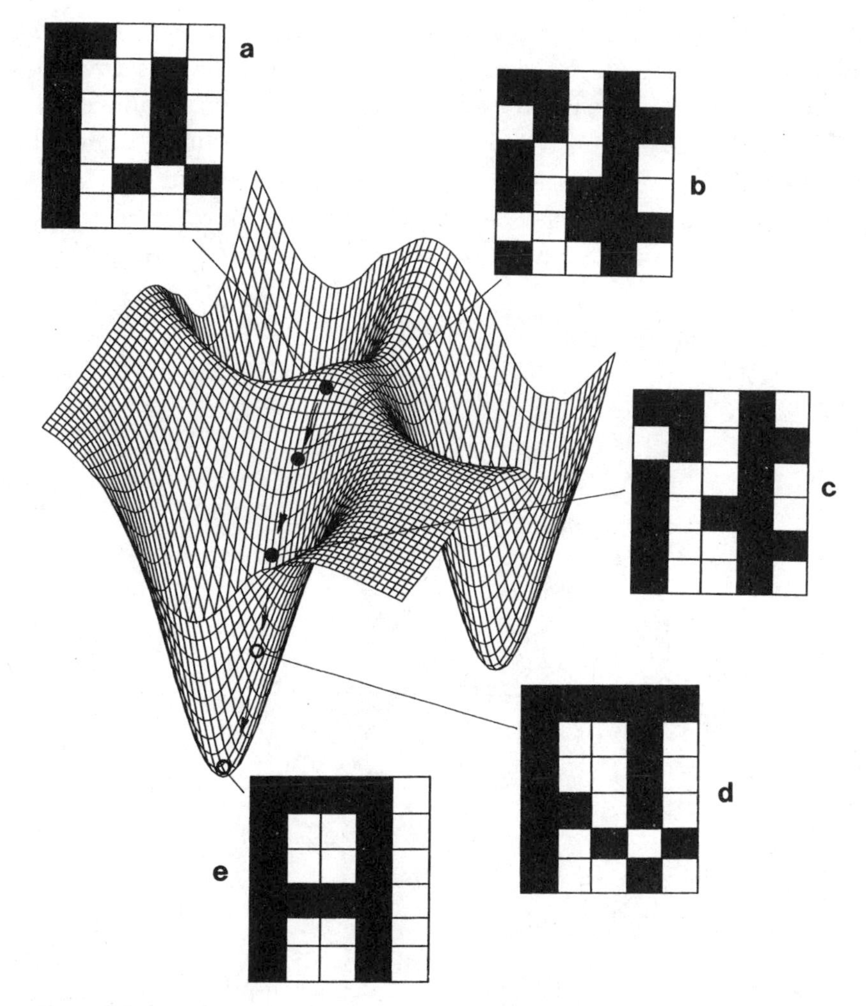

Figure 5.5 A schematic representation of the energy landscape for the Hopfield model (the real landscape is a discrete, multidimensional object). Here the initial state is indicated by a circle, located at some point on the surface. This state is in fact a corrupted memory state. Through the dynamics, the system evolves in time until the right memory state is recovered (here the letter "A").

2^N. But we can get an intuitive idea of it by plotting a simple two-dimensional surface such as the one shown in Figure 5.5. The bottom of each valley corresponds to a memory state (i.e., to a previously learned pattern). Starting from a given initial condition, we can visualize the process of pattern recognition as a downhill movement to the bottom of energy valleys. Each of these valleys defines a *basin of attraction*, and the initial conditions (corrupted patterns) that evolve toward a given memory state define the valley and its size. Too strong a noise level in one initial pattern can place the initial condition under the influence of a different basin of attraction and thus lead to a wrong pattern recognition, which is precisely what occurs when previously learned information is too noisy or incomplete: we can fail to recognize it properly.

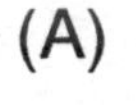

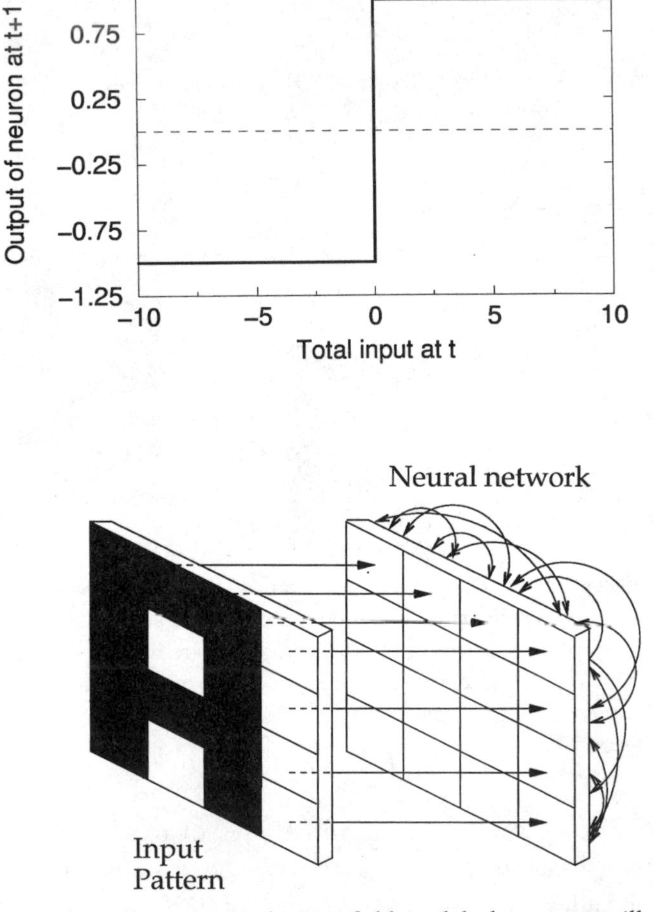

Figure 5.6 (a) Response of neurons in the Hopfield model: the neuron will be at the −1 state if the input is negative and at the +1 state if positive; (b) presenting a pattern to the network: here the neurons have been arranged in a two-dimensional "retina."

The Hopfield Model

Let $S_i(t) \epsilon \{-1, +1\}$ be the state of a given neuron (here $i = 1, 2, \ldots, N$). Neurons are connected though a set of couplings J_{ij}, where J_{ij} gives the strength and sign of the input from the ith to the jth neuron. Symmetry ($J_{ij} = J_{ji}$ for all pairs of neurons) and no self-interaction ($J_{ii} = 0$ for all neurons) are assumed. The temporal dynamics are described by a step function (Figure 5.6.a, thick line). The state of the ith neuron at step $t+1$ will be

$$S_i(t+1) = \Phi\left[\sum_{j=1}^{N} J_{ij} S_j(t)\right],$$

where the function $\Phi(x)$ is equal to $+1$ when $x > 0$ and to -1 when $x \leq 0$. This dynamic rule is applied in the following way: at each step we pick up a unit at random and update it according to the step function. During this process, the states of other neurons are not modified.

The network must be trained in the first place in order to store the desired set of patterns. A set of input strings of size N is used as the set of input vectors for the net. These are the patterns that have to be stored and later retrieved in a robust way. If these vectors are spatially ordered as two-dimensional matrices, they can be viewed as images (such as letters or numbers). The values $+1$ and -1 can be associated to black and white pixels, respectively (Figure 5.6.b). The input patterns will be indicated as $\xi_\mu = \left(\xi_\mu^1, \ldots, \xi_\mu^N\right)$, where ξ_μ^j is the state of the jth pixel corresponding to the μth pattern. A set of P patterns is used in the training phase.

The training phase involves the application of Hebb's rule (see text) to each input pattern. If two neurons receive the inputs ξ_μ^i and ξ_μ^j, their mutual connection J_{ij} will be modified to the new value

$$J_{ij}^{\text{new}} = J_{ij} + \frac{1}{N} \xi_\mu^i \xi_\mu^j.$$

We can see that since $\xi_\mu^i \epsilon \{-1, +1\}$, the product $\xi_\mu^i \xi_\mu^j$ will be positive if both neurons are in the same state (i.e., correlated) and negative otherwise. This is a basic implementation of Hebb's rule, and we can also see that the maximum increase per update is $1/N$. This rule is applied for each pair of neurons and for each of the P patterns. Once the training phase is finished, it is not difficult to see that the connections are given by

$$J_{ij} = \frac{1}{N} \sum_{\mu=1}^{P} \xi_\mu^i \xi_\mu^j,$$

which formally describes Hebb's rule.

If the total number of input patterns is such that the net capacity $\alpha = P/N$ satisfies $\alpha < \alpha_c = 0.15$, then this model is able to work as an associative memory. If the capacity exceeds the critical value, a phase transition occurs where the network fails to operate as an associative memory, and new *spurious* states emerge (these are linear combinations of the set of patterns stored during the learning phase). This transition is illustrated in Figure 5.7, where the fraction of erroneous recognitions is plotted in relation to the network's capacity. We can see that the network becomes random when the critical value is reached.

It can be shown that the function

$$H(t) = \frac{1}{2} \sum_{i=1}^{N} \sum_{j=1}^{N} J_{ij} S_i(t) S_j(t)$$

is an appropiate energy function for the Hopfield model (such as the surface shown in Figure 5.5, but on a high-dimensional space). In other words, the minimum of H corresponds (for $\alpha < \alpha_C$) to the stored patterns. To see this, let us consider a given neuron with state $S_i(t)$. The input to this neuron will be

$$h_i(t) = \sum_j J_{ij} S_j(t).$$

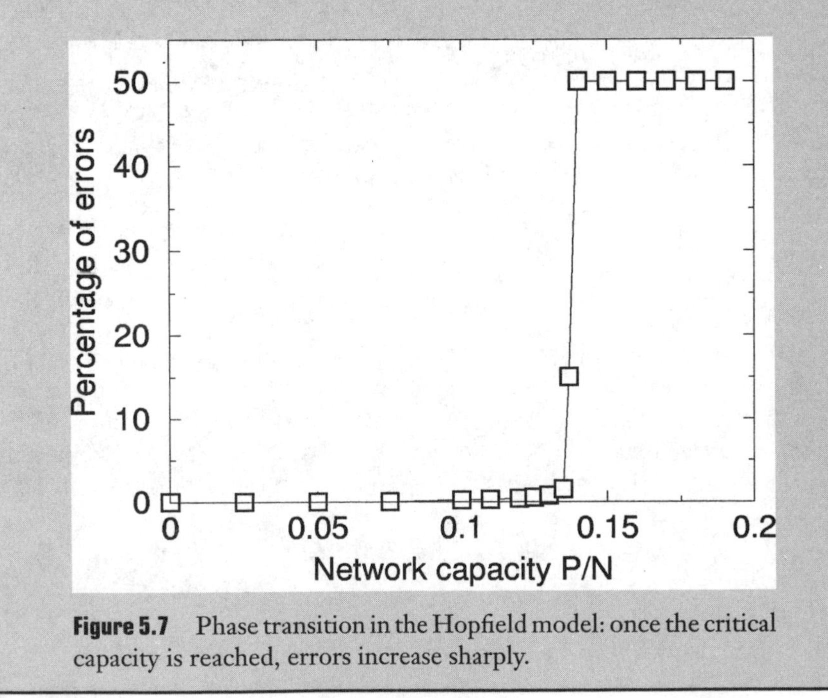

Figure 5.7 Phase transition in the Hopfield model: once the critical capacity is reached, errors increase sharply.

The state of the neuron will change if $S_i(t)h_i(t) < 0$. In other words, only if some difference exists between the signs of the input field and the neuron state will a change occur (otherwise, the energy remains the same). Let us see how in this case the energy decreases. The increase in energy, $\Delta H = H(t+1) - H(t)$ is given by

$$\Delta H(t) = -\frac{1}{2}\sum_{i=1}^{N}\sum_{j=1}^{N} J_{ij}[S_i(t+1)S_j(t+1) - S_i(t)S_j(t)].$$

Since only neuron S_i changes, we can write the previous difference as

$$\Delta H(t) = -\frac{1}{2}[S_i(t+1)h_i(t+1) - S_i(t)h_i(t)],$$

but in fact, we have $S_i(t)h_i(t) < 0$ and $S_i(t+1)h_i(t+1) > 0$, and thus $\Delta H < 0$.

The Hopfield model started an explosion in the theory and application of neural networks.[6] The outcome of the model is remarkable. It shows that memory can be stored as a collective pattern of connections, and that correct recognition is possible from incomplete information. The introduction of damage into the connections has no effect until a significant number of connections have been removed. Again, this accords with the observation that brain function is highly robust against neural loss (up to some limits). But as a metaphor for real neural structures, the Hopfield model has some important limitations. For instance, the dynamics are restricted to steady states: memories are attractors, and once they are reached, the net remains frozen. But the brain and other neural structures are highly dynamic, far-from-equilibrium entities. They are an example of *excitable medium*. An element of an excitable medium is defined by the capacity for bursts of activity initiated by an external perturbation that forces the unit beyond a given threshold value, after which the unit returns to its initial state. This behavior typically results in propagation of a traveling excitation pulse.[7] Such details as protein channels and the type of ions pumped through the membrane are of little relevance at this larger scale: waves are inevitable once the universal conditions are met. Interestingly, models with neurons able to display a threshold response to local excitation generate several types of wave patterns, from spiral waves to spatiotemporal chaos, much like what we see when looking not at a neuron but at the whole brain.

Neural Chaos

In previous chapters we have seen complex dynamics in a wide variety of systems. The brain, another example of nonlinearity at work, exhibits many different types of rhythms that are not strictly periodic. The electric dynamics of the brain can be recorded relatively simply with an electroencephalogram (EEG), a measurement that enables us to glimpse the collective activity of the cerebral cortex. This technique was first discovered by the English physiologist Richard Caton in 1879 and later used in human brains by the Austrian psychiatrist Hans Berger in 1929. Today, the EEG is recorded from different locations on the skull, and the result is a parallel time series of fluctuating potentials. One of these records is shown in Figure 5.8, where we can see some

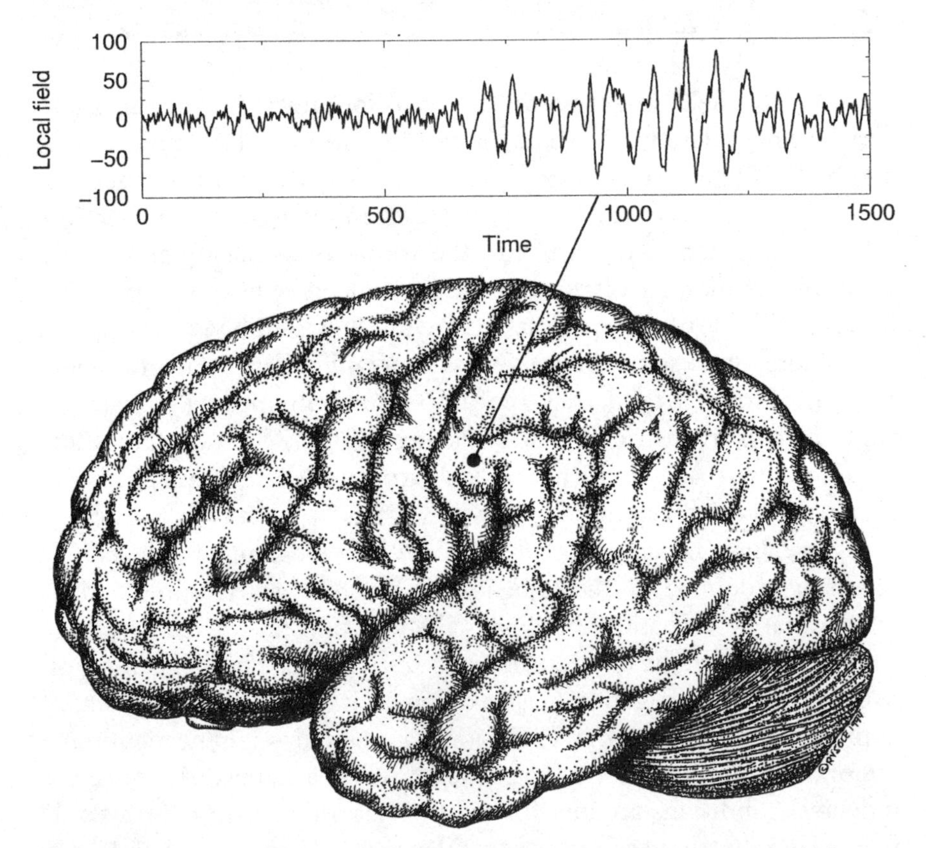

Figure 5.8 Brain waves: here the local brain activity recorded from a specific area is shown. Note the transition from two different dynamical patterns, due to the start of a seizure.

change in the local activity as a consequence of some pathological state (the EEG is mostly used for diagnosis of neural conditions such as epilepsy).

Looking at the EEG records, we perceive some degree of order (the presence of some oscillating components) and disorder (the oscillations vary in amplitude and frequency through time). Since the collective dynamics of the brain are caused by the highly nonlinear interaction of millions of neurons, we might wonder whether EEGs in fact reveal the fingerprint of chaotic dynamics.[8] Simple mathematical models can help: it is known that the basic oscillatory behavior of neural assemblies comes from the interaction between excitatory and inhibitory sets of neurons, and when such sets are modeled, they easily yield oscillations and chaos (Box 2).

Oscillations and Chaos in Neural Nets

The temporal fluctuations present in neural systems are mainly caused by the presence of groups of neurons in interaction. These groups include both excitatory and inhibitory neurons. Excitatory neurons send inputs to other neurons, which trigger their excitation and bursting. Inhibitory neurons act in the opposite way: they tend to suppress the activity of other neurons. Let us consider a neural system composed of two groups of excitatory and inhibitory neurons (Figure 5.9a). The states of these neurons (as measured in terms of firing rates) are given by

$$\{e_k(t)\}, k = 1, \ldots, N_e,$$

for the excitatory neurons (small circles, Figure 5.9a, left), and by

$$\{i_l(t)\}, l = 1, \ldots, N_i,$$

for the inhibitory ones (small squares, Figure 5.9a, right). It can be shown that their dynamics can be described by means of the following dynamical system:[9]

$$\frac{de_k}{dt} = -e_k + \Phi\left[a_e\left(\frac{1}{N_e}\sum_{l=1}^{N_e} u_{kl}e_l - \frac{1}{N_i}\sum_{l=1}^{N_i} v_{kl}i_l - \theta_k^e + p_k\right)\right],$$

$$k = 1, 2, \ldots, N_e,$$

$$\frac{di_k}{dt} = -i_k + \Phi\left[a_i\left(\frac{1}{N_e}\sum_{l=1}^{N_e} w_{kl}e_l - \frac{1}{N_i}\sum_{l=1}^{N_i} z_{kl}i_l - \theta_k^i\right)\right],$$

$$k = 1, 2, \ldots, N_i,$$

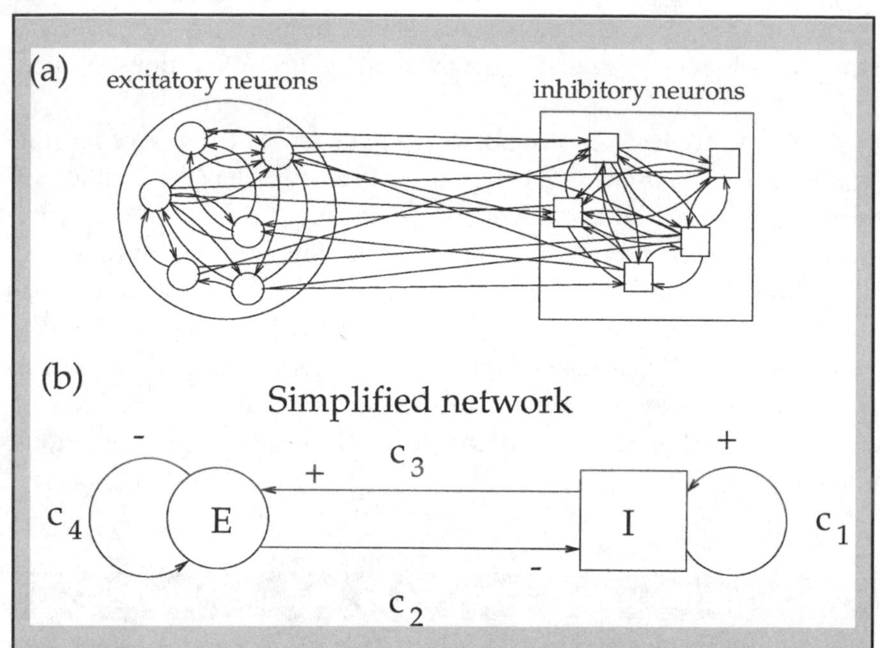

Figure 5.9 Neural network model of oscillatory cortex. The model involves two populations of neurons (a) of two types (see Box II). The model can be strongly simplified (b) assuming that each group of neurons can be replaced by an average neuron.

defining a network, as shown in Figure 5.9a. Several types of connections are defined, involving self-interactions and cross-interactions. The terms p_k introduce external inputs into the excitatory neurons. A sigmoidal function $\Phi(x) = \left[1 + e^{-x}\right]^{-1}$ is used. The sigmoidal function is similar to the previously defined step function for large values (both negative and positive) of the argument x. But close to zero it grows linearly instead of showing a sharp jump from -1 to $+1$. This model is rather complicated, but Schuster and Wagner, at the University of Kiel, showed that the average activity of these sets can be quite well described in terms of a simple two-compartment set of equations. This is a so called mean field model. We define the average activity of each group as

$$E(t) = \frac{1}{N_e} \sum_j e_j(t) \text{and} I(t) = \frac{1}{N_i} \sum_l i_l(t).$$

These variables have dynamics described by the two-dimensional system

$$\frac{dE}{dt} = -E + \Phi\left[a_e\left(c_1 E - c_2 I - \Theta^e + P\right)\right],$$
$$\frac{dI}{dt} = -I + \Phi\left[a_i\left(c_3 E - c_4 I - \Theta^i\right)\right]$$

where $\Theta^e = 1/N_e \Theta_k^e$, $\Theta^i = 1/N_i \Theta_k^i$ and $P = p_k/N_e$. The basic topology corresponding to the mean field model is shown in Figure 5.9b. This model exhibits oscillations in activity, but it is not difficult to obtain more complex patterns by coupling several of these modules (as occurs in the cerebral cortex).

The first study of chaos in brain function was performed by Agnes Babloyantz, at the Free University in Brussels.[10] She and her team used a particular type of EEG pattern: the one corresponding to a class of epileptic seizure known as petit mal. In these seizures the patient does not suffer convulsions but experiences a transient loss of consciousness. Externally, the individual looks absent, and for this reason these are also called *absence seizures*. Because epilepsies involve the synchronization of large areas of the brain cortex, an epileptic EEG appears much more ordered than a normal one (such as the second part of the EEG shown in Figure 5.8). Babloyantz used nonlinear dynamics to explore these data sets and made an astonishing discovery: brain dynamics were shown to be low-dimensional. The behavior of those billions of neurons in synchrony could be described by means of a chaotic dynamical system with about three degrees of freedom, as measured from the fractal dimension of the attractor. Further studies,[8] some of which are summarized in the table below, confirmed Babloyantz's results.

State/System	*Fractal Dimension*
dream sleep (2)	$D = 5.03$
dream sleep (4)	$D = 4.0$ to 4.4
awake (alpha rhythm)	$D = 6.1$
closed eyes	$D = 2.4$ to 2.6
Creutzfeldt–Jakob disease	$D = 3.7$ to 5.4
epilepsy	$D = 2.05$

These studies showed that several regimes of brain function are associated with low-dimensional chaotic dynamics. What do these patterns mean? To answer this question we require the kind of well-defined experimental system that is, for obvious reasons, not possible in humans. But some neural structures in animals have been a great source of information for neuroscientists. One of them, the olfactory bulb, the simplest among the sensory systems, has been especially illuminating.

Research on nonlinear dynamics in the olfactory bulb has been carried out over more than thirty years at the University of California at Berkeley by the group led by the neurobiologist Walter Freeman.[11] Their results give strong support to the view that perception cannot be understood solely by examining properties of individual neurons but depends on the cooperative activity of millions of cortical neurons. To explore the workings of perception and pattern recognition, Freeman used the olfactory cortex of rabbits, a brain regime whose basic structure is shown in Figure 5.10a. When an individual sniffs a given odorant, molecules carrying the scent are captured by a few of the receptor neurons in the nasal passages; these receptors are are on cells specialized in the kinds of odorants to which they respond. Cells that become excited fire action potentials, which propagate to the olfactory bulb. The number of activated receptors indicates the intensity of the stimulus, and their location in the nose relates to the nature of the scent.

Freeman summarized the further steps this way:

> The bulb analyzes each input pattern and then synthesizes its own message, which it transmits via axons to another part of the olfactory system, the olfactory cortex. From there, new signals are sent to many parts of the brain—not the least of which is an area called the entorhinal cortex, where the signals are combined with those from other sensory systems. The result is a meaning-laden perception, a gestalt, that is unique to each individual. For a dog, the recognition of the scent of a fox may carry the memory of food and expectation of a meal. For a rabbit, the same scent may arouse memories of chase and fear of attack.

Freeman recorded the electrical activity of the olfactory neurons using a fixed array of microelectrodes to obtain a detailed spatiotemporal pattern of neural activity. From the large amount of data he obtained on the behavior of different cells involved and their wiring connections, Freeman was able to build up an accurate mathematical model of the olfactory cortex (the basic architecture of that model is shown in Figure 5.10b). Using the mathematical model he could also follow the patterns of activity of both single cells and global dynamics. In particular, it was possible to analyze the attractors involved in the dynamics of different parts of the cortex. Both the recorded activity patterns and those obtained from the simulation model were completely consistent: chaotic dynamics were at work in the olfactory bulb.

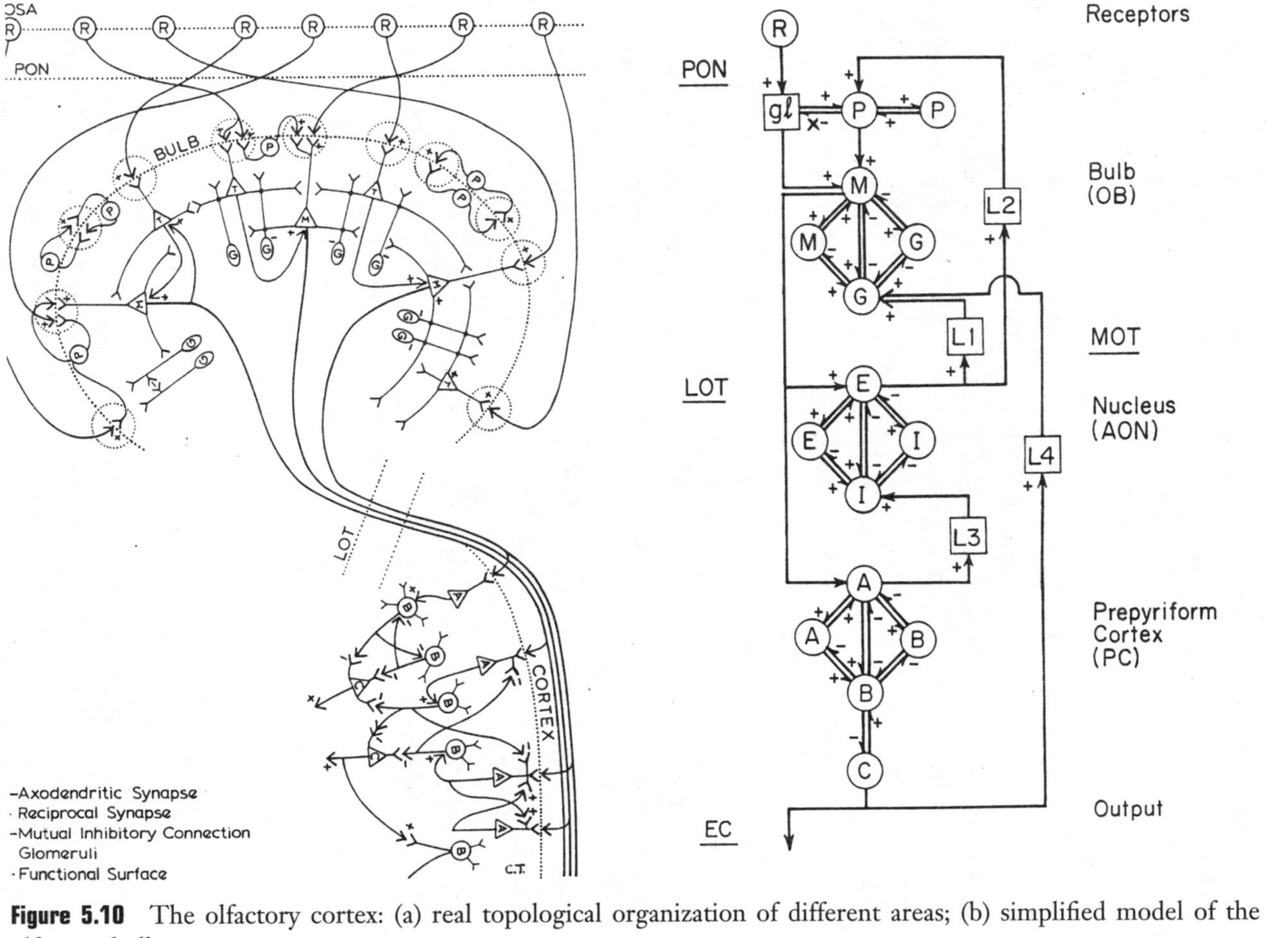

Figure 5.10 The olfactory cortex: (a) real topological organization of different areas; (b) simplified model of the olfactory bulb.

The signals recorded by the electrodes revealed that in fact, chaotic dynamics (as shown by the observed strange attractors) represented the normal state when the animal was attentive, in the absence of a stimulus. However, these attractors underwent dramatic changes when some familiar odor was introduced, and the behavior of the individual displayed recognition of a previously stored memory. The pattern of activity changed both in space and time (since measurements in different parts of the bulb were recorded, it was possible to observe that different odors generated different spatial patterns of activity). The new dynamical pattern shown by the fluctuations of neural assemblies was much more ordered (very much like a limit cycle; see Chapter 1), and the spatial pattern exhibited a well-defined stable structure, which was, in fact, characteristic for the specific odor used. In other words, different learned stimuli were stored as a spatiotemporal pattern of neural activity, and the strange attractor characteristic of the attention state was replaced by a new, much more ordered attractor related to the recognition process. Each (strange) attractor was thus shown to be linked to the behavior the system settles into when it is under the influence of a particular familiar input odorant. Freeman suggests that "an act of perception consists of an explosive leap of the dynamic system from the basin of one chaotic attractor to another." These results indicate that the olfactory bulb and cortex maintain many chaotic attractors, one for each odorant an animal or human being can discriminate. Whenever an odorant becomes meaningful in some way, changes in the synaptic connections between neurons in different parts of the olfactory cortex take place. Just as in the Hopfield model and other neural networks, these changes are able to create another attractor, and all the others undergo slight modication.

Freeman and Christine Skarda published some of these results in a very influential paper entitled "How brains make chaos in order to make sense the world."[12] This paper played a key role in convincing neuroscientists that chaos, as an emergent property of intrinsically unstable neural masses, is very important to brain dynamics. One profound advantage chaos may confer on the brain is that chaotic systems continually produce novel activity patterns. Since novelty is always present, it provides a source of flexibility to the individual. But since chaos is an ordered state, such flexibility is under control. Freeman also proposed that such patterns are crucial to the development of nerve cell assemblies.

The observation that brain function is chaotic is puzzling. One might imagine that ordered dynamics would be the state of "normal" brain dynamics, but in fact, as we saw with the heart, a shift to more ordered dynamics is typically associated with pathological states. Most current models of neural networks perform remarkably well in a wide range of areas, but are clearly different from the far-from-equilibrium behavior of real neural assemblies. Understanding the origins of neural complexity requires new model approaches to exploit the still unknown laws of brain dynamics. Why is the brain chaotic? Perhaps because the computational constraints involved in memory dynamics and perception require a large amount of flexibility. Chaos provides dynamics that are at once ordered and innovative. Some experiments suggest that in fact, our brains might be operating on the edge of instability.

Phase Transitions in the Human Brain

We have seen that neural systems display large-scale coherent patterns of activity in both space and time. In many cases these can be very well simulated by simple models displaying the same coherent, although chaotic, patterns of behavior. Many situations, such as epileptic seizures or Freeman's experiments with pattern recognition in the olfactory cortex, involve shifts from one such chaotic pattern to another. Inspired by these observations, the German physicist Hermann Haken has strongly advocated a study of pattern recognition in terms of far-from-equilibrium dynamics.[13] The idea is that these complex, highly nonlinear systems made of a very large number of basic interacting units exhibit phase transitions and other collective phenomena characteristic of some analogous physical systems. Switches between different dynamical regimes occur, for example, in Bénard's fluid instability (see Chapter 1). In this case, as some control parameter (the temperature) is tuned, the system makes a transition from a spatially homogeneous pattern to a convective, heterogeneous one. This transition occurs by means of a symmetry-breaking instability. As we discussed in Chapter 2, symmetry-breaking involves the choice between two symmetric final states. Consider a ball perched precisely on the peak of a steep ridge: the tiniest chance fluctuation in the forces acting on it (such as a stray gust of wind) may send it on one of two widely divergent paths (Figure 5.11).

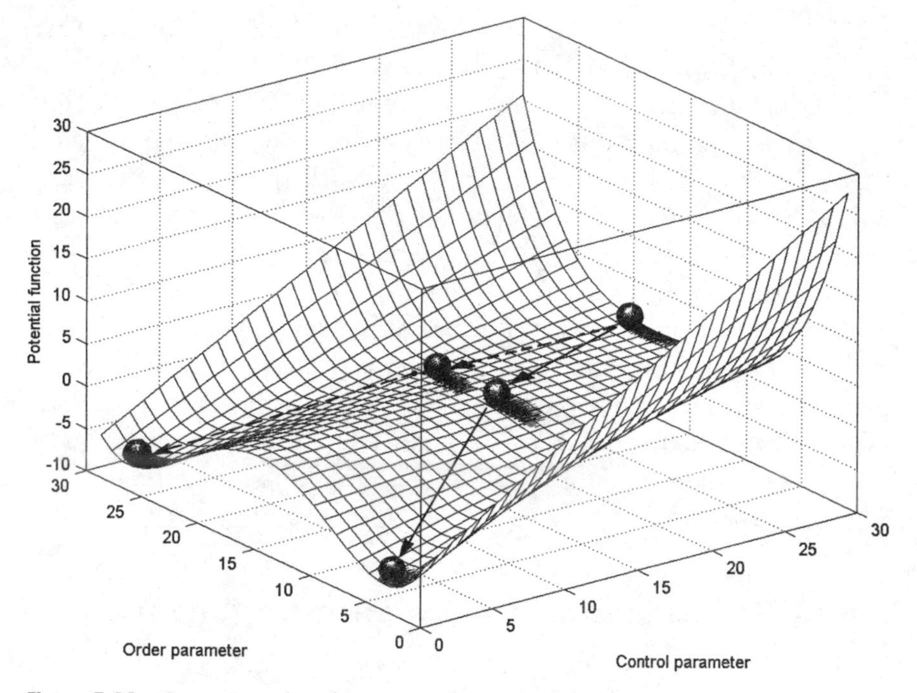

Figure 5.11 Symmetry-breaking: a mechanical analog.

Given that complex systems exhibit well-known universal properties in the vicinity of instability points, Haken suggested that brains, as far-from-equilibrium complex systems, might also exhibit these properties. If brains operate close to such instability points we should be able to measure some characteristic features of criticality. One such feature is well known in physics: the phenomenon of *hysteresis*. This is illustrated by Haken through a simple experiment involving visual perception. Figure 5.12 shows a sequence of similar figures. Let us start at the top left and view each figure in turn from left to right. The first picture is clearly a face, and the last a woman. When we follow the sequence, at some point we switch from the preception of a face to the perception of a woman. But as we follow the reverse sequence, the transition occurs at some different point. Our brain has experienced a sudden jump at two different points, depending on the inmediate past. In other words, the actual state of the system depends on the history. In this case one speaks of hysteresis. This is a very common phenomenon in far-from-equilibrium phase transitions.

Brain dynamics are nonstationary. Although several studies have shown that low-dimensional chaos is at work in different situations

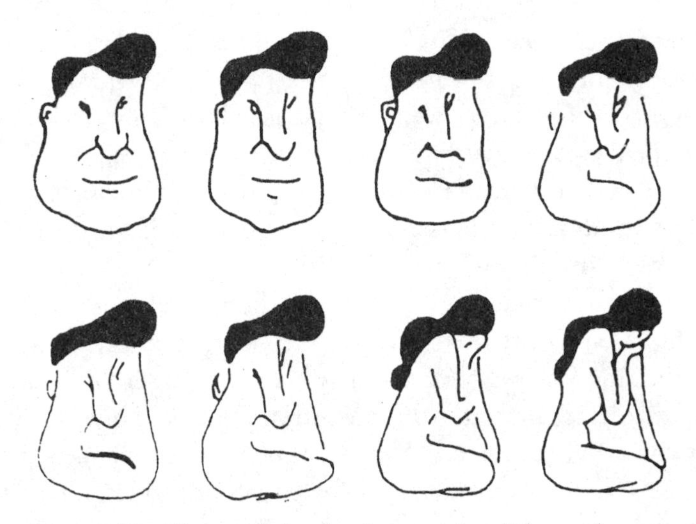

Figure 5.12 Hysteresis in visual perception. The reader can perform the experiment by looking at the different figures starting from the left in the top row and then at the second row (again from left to right) and doing the same in reverse order. At some point we switch our perception from a face to a woman. But this occurs at different pictures depending on the sequence followed.

(pathological or not), there is some amount of variability in the results of the analysis of EEG patterns obtained by different researchers. In particular, the fractal dimension measured over short periods of time shows good evidence of low-dimensional dynamics, but it increases in some cases as the measurements are extended over larger intervals. This variability strongly suggests that in spite of the rather ordered behavior of neural systems in some circumstances, transitions between different dynamical patterns (attractors) are at work all the time. This assertion is not easy to test. Physicists analyze phase transitions using well-defined experimental conditions where control parameters (such as temperature) are appropriately tuned, and they can make careful, precise measurements. Imposing similar control over conditions within a human brain seems an impossible task. Yet such experiments are possible: Scott Kelso, a neurobiologist at Florida Atlantic University, and his team were able to see phase transitions at work in the human brain.[14]

The experimental setup requires a simple definition of the transition phenomenon and an appropriate measurement system. In a classic experiment, Kelso showed that a phase transition occurred when an experimental subject's two index fingers, moving in parallel, shifted to antiparallel motion as the frequency of movement increased beyond a threshold. The subjects were either asked verbally or forced by a metronome to move their fingers more quickly in parallel (as in Figure 5.13a). At a certain critical frequency, a sudden and involuntary change of the finger movement pattern occurs, and you will start to move the two fingers in antiphase (Figure 5.13b).

Kelso and coworkers developed a mathematical model of this phenomenon based on the theory of phase transitions. Following Haken's synergetics, the frequency of finger movement can easily be identified as the control parameter for the transition. The next obvious step is to identify the order parameter. As the frequency of movement increases, the system crosses a critical point leading to a sudden change in the qualitative behavior. These authors used the relative phase Φ between both fingers as an appropriate order parameter. When fingers move in phase (i.e., in parallel), the relative phase is zero. Once the bifurcation takes place, the phase becomes nonzero up to a maximum. By placing electrodes on the appropriate muscles to measure their electromyographic (EMG) activity, Kelso could clearly measure the sudden shift from one pattern to the other. And not only this: all the measured properties were consistent with a critical phenomenon. For instance, the size of the fluctuations in neural activity dramatically increases close to the critical point (as occurs with other physical systems; see Chapter 2).

Changes in behavioral patterns such as finger movement must be correlated with changes in global brain activity. Kelso and his team used a highly sensitive experimental system to measure the spatiotemporal patterns displayed by the brain on large scales. This system is a SQUID (superconducting quantum interference device), an array of sensors placed around the head of the subject (Figure 5.14), whose purpose is to detect magnetic fields arising from electrical activity in the brain.

The underlying idea in Kelso's studies was that the brain is a self-organizing, pattern-forming system that operates close to instability points, thereby allowing it to switch flexibly and spontaneously from one coherent state to another. Behavioral and spatiotemporal dynamical patterns of brain activity should be linked in some clear way, and Kelso's work strongly suggests that behavioral changes relate to phase transitions.

Kelso's SQUID experiment was similar to the one described above. But in this case, the subject was exposed to acoustic periodic stimuli,

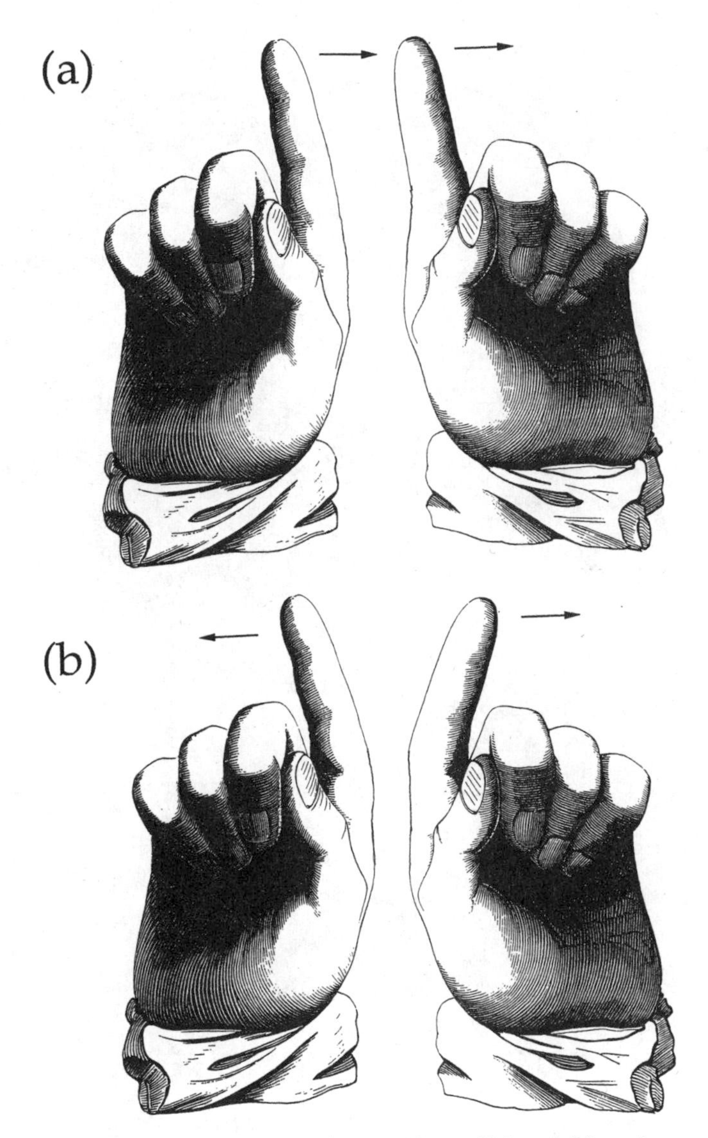

Figure 5.13 Kelso's experiment: (a) parallel and (b) antiparallel motion.

and the task was to press a button in between two consecutive tones. The stimulus frequency increases by a fixed amount after ten stimulus repetitions. The two types of movements are (1) a synchronous response of finger movement to the external periodic auditory signal and (2) an out-of-phase movement, where the signal and the finger are in asynchrony. As the frequency of the auditory signal increases, the movement

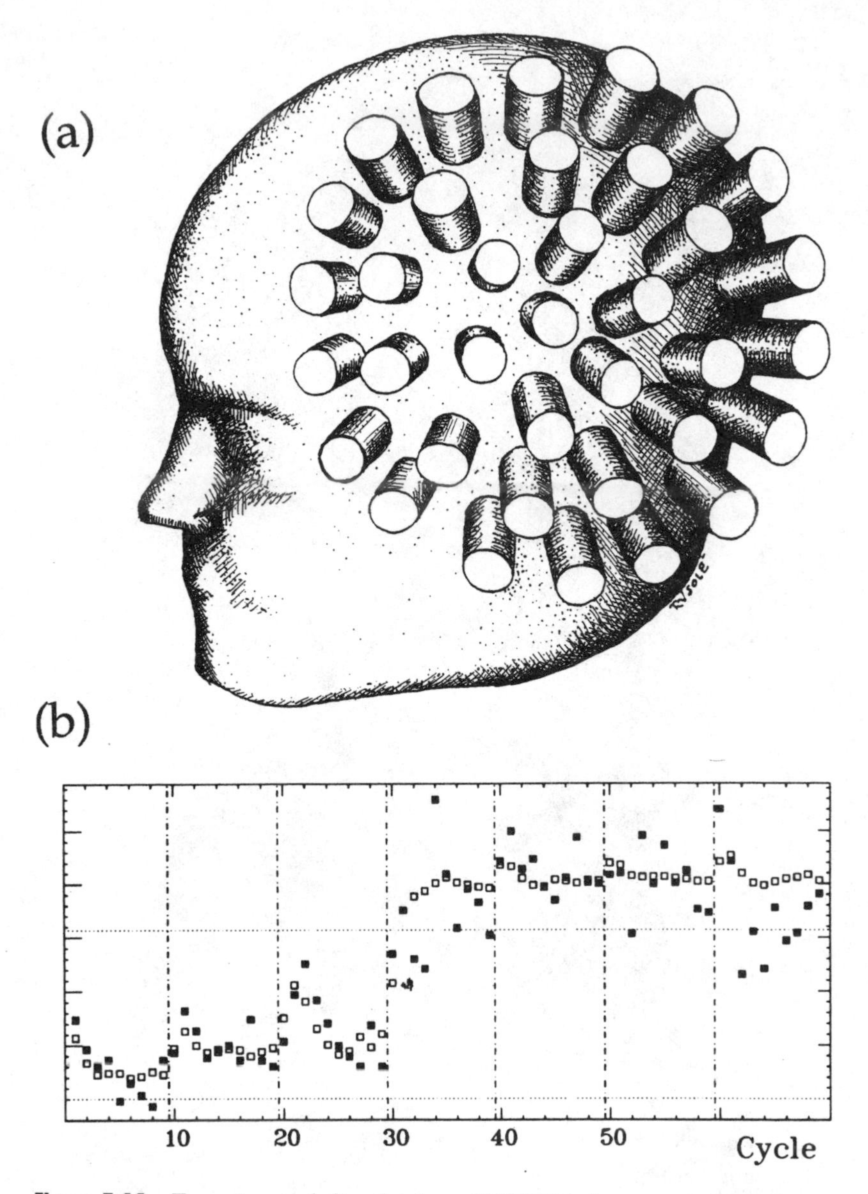

Figure 5.14 Experimental distribution of SQUID detectors in Kelso's experiment.

eventually shifts from in-phase to anti-phase. Specifically, at a critical frequency of around 2 Hz, the subject was no longer able to syncopate and switched to a synchronized behavior. During the experiment, the neuromagnetic field data were recorded over the left cortex using the SQUID device. At the transition point between syncopated and synchronized movement, the global dynamics of brain waves displayed a strong qualitative change: the spatiotemporal activity of the subject's cortex experienced a well-defined shift from one attractor to another, and the changes were accompanied by the characteristic features of phase transitions.

Kelso then went further, analyzing the spatial information available from the array of SQUID sensors. The whole spatial pattern of activity can be decomposed in a number of modes (like the harmonics in a wave), and the relevance of each component to the whole pattern can be measured and used as a quantitative characterization of brain dynamics. This method, known as the K-L decomposition, reveals information that is not evident in the full signal. In particular, Kelso's group obtained evidence for the existence of different types of attractors before and after the transition (Figure 5.15, upper plots). A simple mathematical model allowed them to reproduce these experimental observations in a qualitatively accurate way. These results gave further support to Freeman's studies, suggesting that brain dynamics might involve multiple attractors and that a coherent and flexible information processing system requires both order and disorder to operate.

The neurophysiologist Charles S. Sherrington famously described the human brain as "an enchanted loom, where millions of flashing shuttles weave a dissolving pattern." After decades of intensive research, we know only a little about its function, yet our knowledge has increased exponentially. Over the last few years, neuroscientists such as Francis Crick and Chris Koch have suggested that even consciousness, the most surprising and puzzling emergent phenomenon of all, should be a matter of scientific (and not only philosophic) exploration. As Crick points out, for many years consciousness was a taboo concept in psychology, and until recently most cognitive scientists ignored it as a research problem.[15] But this attitude ressembles that of physicists a generation or two ago toward cosmological theories. It seemed to them that the origin of the universe was a purely abstract, philosophical question. Yet cosmology is now a well-established scientific discipline involving thousands of theorists and experimentalists.

The depths of the human mind are no less impressive. Instead of billions of distant galaxies we have billions of interacting neurons. It is these interactions that create our thoughts and trigger our imagination. Chaotic and strange, they can dream of alternative worlds that do not exist except in the mind.

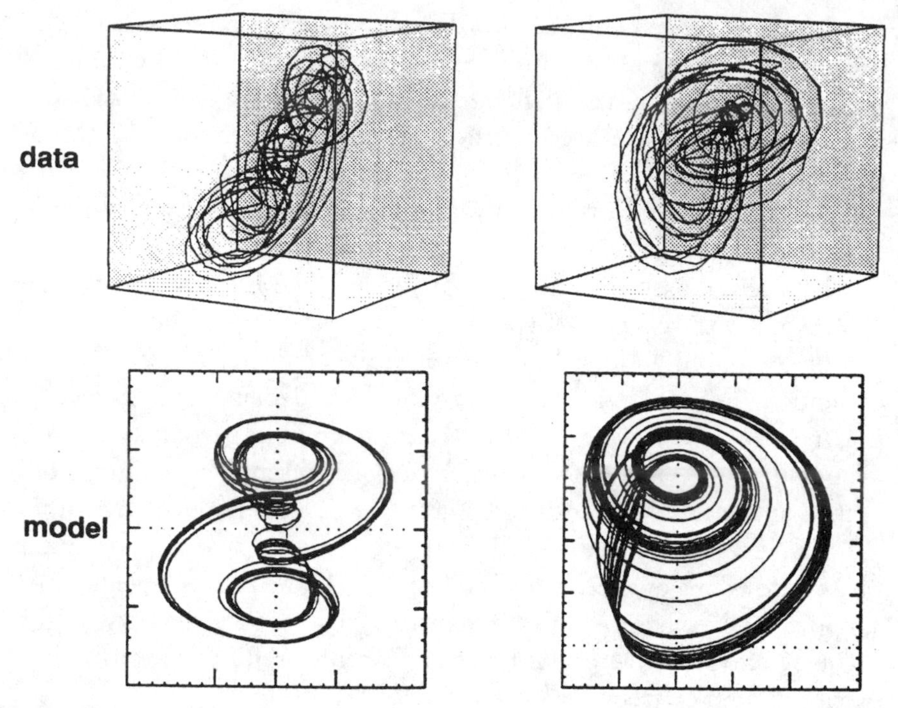

Figure 5.15 Transition between different attractors in Kelso's experiments. The upper row shows the different attractors previous to and after the transition occurs. The bottom row shows an example of the simulated attractors before and after the bifurcation.

SIX

Ants, Brains, and Chaos

Achilles: Familiar to me? What do you mean? I have never looked at an ant colony on anything but the ant level.

Anteater: Maybe not, but ant colonies are no different from brains in many respects . . .

—Douglas Hofstadter, *Gödel, Escher, Bach*

The Superorganism

Social insects display some of the best examples of what we call emergent behavior. It is difficult not to become fascinated by the abundance of patterns shown by the work of ants, termites, bees, and social wasps. The huge nests of termites and raid patterns of army ants traveling through the rain forest are just two examples. We are fascinated by their collective behavior, but also by their ecological success: the dry weight of ants and termites in some rainforests is about four times that of all the other land animals (Figure 6.1). In some ecosystems ants compete successfully with rodents and other vertebrates. We find them all around the world, from deserts to the jungle, and they are strong competitors. Some authors even propose that this strong competitive ability leads to a well-defined partition of habitats, with ants and termites playing a central role and solitary insects having much less ecological relevance.[1]

But while colonies of social insects behave in complex ways, the capacities of individuals are relatively limited. The brain size of a single ant varies from species to species. In numbers of neurons it ranges over several orders of magnitude. Generally speaking, single ants behave in a simple way. As Holldobler and Wilson have written, "In the course

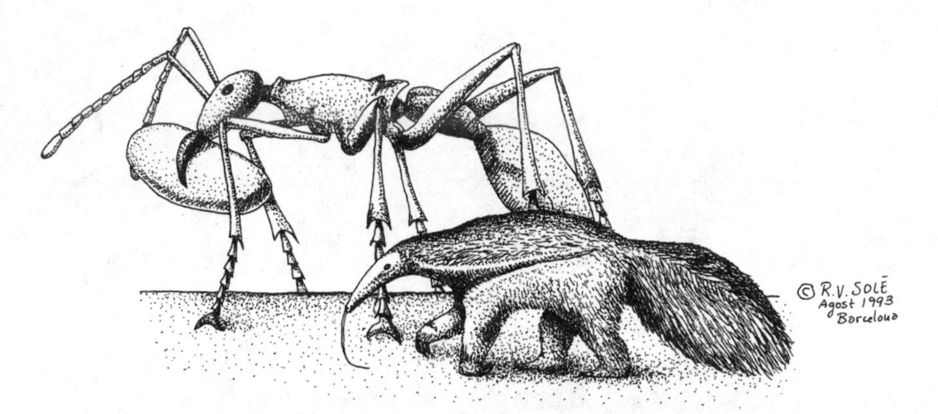

Figure 6.1 The ecological importance of ants: in some rainforests, the dry weight of social insects is about four times that of all other land animals.

of evolution the brain capacity of individual ants has been pushed close to the limit . . . to watch a single ant apart from the rest of the colony is to see at most a huntress in the field or a small creature of ordinary demeanor digging a hole in the ground. One ant alone is a disappointment; it is really not ant at all."[2] But then, how do social insects reach such remarkable goals? The answer, as discussed here, comes to a large extent from self-organization: insect societies share basic dynamic properties with other complex systems, such as brains (Figure 6.2).

In both ant colonies and brains, individual units (ants or neurons) do not gather, store, and process information by themselves. Instead, they interact with each other in such a way that information is manipulated

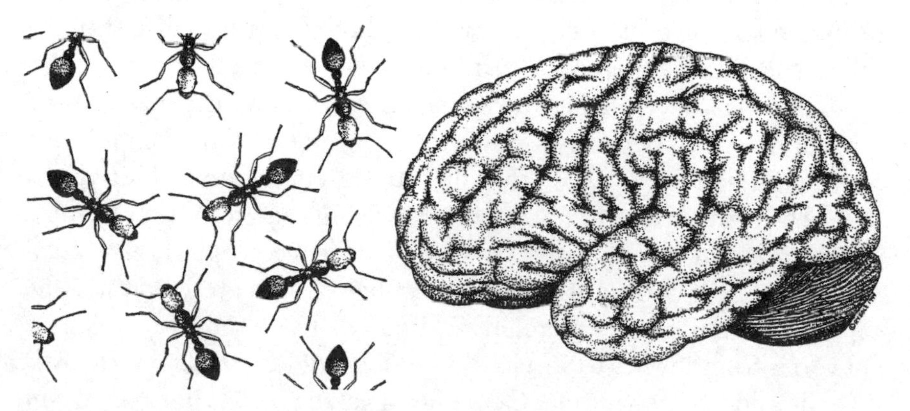

Figure 6.2 Ant colonies and brains: two related systems.

by the collective. The whole colony is the organism, the basic entity that we must understand. And like the brain, the colony is formed by many individuals in interaction; individual units can switch from one type of activity to others; they can fail or be removed without any harm to the collective (i.e., the collective is robust), and they have only a limited behavioral repertoire, gathering information in a local way. The basic similarities are summarized in the Table that follows.

	Ant Colonies	*Neural Networks*
Number of units	high	high
Robustness	high	high
Connectivity	local	local
Memory	short-term	short/long-term
Stability of individual connections	weak	high
Global spatial pattern of activity	trails	brain waves
Complex dynamics (1/f)	Observed	Common

It is not surprising that the main differences arise from connectivity: direct contact between individual ants is a transient phenomenon, whereas synaptic connections among neurons usually have a long lifetime. A direct consequence is that memory in ant colonies will typically be short. This is partially compensated (particularly in large colonies) by the use of chemicals, which can create spatial structures that clearly involve long-term memory effects. Deborah Gordon, of Stanford University, has noted the deep analogies between ant colonies and other complex systems,[3] including the molecular interactions within a living cell, the unfolding pattern of cells and tissues in an embryo, and the activity of the neurons that produce the mind. Not surprisingly, Gordon, working with Brian Goodwin and Lynn Trainor, has used a Hopfield-like neural network model to explain many of her most interesting field observations.[4]

In spite of the local character of ant communication (a given ant gathers information only from its nearest nestmates), ant colonies obviously form long-range structures without relying on the centralized, hierarchical control used in human organizations. Yet the colony is able to respond to its collective needs by means of a constant monitoring of the environment. A dramatic example is the army ant raids. Army ant colonies comprise hundreds of thousands of individuals and have no permanent nest structure. When they are ready to move, they form a so-called bivouac, a conglomerate of the entire colony from which the raiding column emerges and spreads in a well-defined pattern. There are different types of basic swarm patterns. Some are more linear and others are highly dendritic, with fractal-like features. In all cases, the swarm behaves as a single entity, searching and expanding over space as though guided by some kind of intelligence. The colony, however, is blind and responds only to local concentrations of pheromones laid down by its individual members. There is no central control or individually complex behavior. This emerges from the interactions between ants.

Most earlier theories of the origins of social behavior in insects have been based in the neodarwinian view of evolution and have emphasized caste and natural selection at the gene level. While some of these ideas have been been successful, gene-based interpretations of colony behavior have raised some problems. Since most collective phenomena observable in social insects arise from the interactions among individuals and must be described in terms of higher-level dynamics, we may wonder how selection processes, acting at the level of genotypes might ever create these phenomena. The interactions among ants in the colony involve two different levels (figure 6.3): first, the individual level, where each ant responds to, say, the local chemical field released by other ants; and second, the global chemical field itself. These levels are strongly connected, and in fact they cannot really be separated as two entities: individual ants respond to the global field and change it. This is what Hermann Haken calls "circular causality".

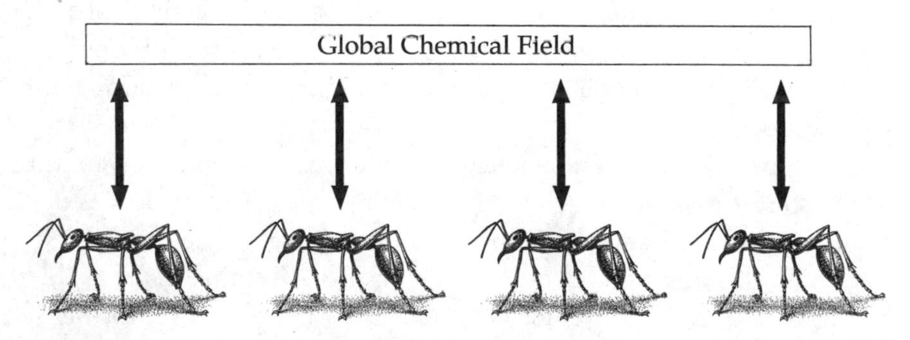

Figure 6.3 The two levels of emergence: individuals create chemical fields used in communication, but the chemical field itself is a source of behavioral change.

The importance of self-organization to the emergence of insect social behavior has been deeply analysed by Eric Bonabeau at the Santa Fe Institute, Guy Theraulaz at the Paul Sabotier University and Jean-Louis Deneubourg at the Free University of Brussels. These authors start by giving an accurate definition of self-organization (SO): *a set of dynamical mechanisms whereby structures appear at the global level of a system from interactions among its lower-level components*. In this context, the interactions between constituent units are determined entirely by local information, without reference to the global pattern, which is an emergent property of the system rather than a property imposed upon the system by an external ordering influence. An example is foraging in ants: the emerging structures include spatially and temporally organized networks of pheromone trails. But how do such structures emerge?

The characteristic signatures of SO would include include:[5]

The creation of spatiotemporal structures in an initially homogeneous medium.

The possible existence of several stable states (multistability): because structures emerge by amplification of random fluctuations, any such fluctuation can be amplified, and the system converges to one among several possible stable states, depending on initial conditions.

The existence of bifurcations when some parameters are varied: the behaviour of a self-organized system changes dramatically at bifurcations. For example pillars built by termites can emerge only if there is a critical density of termites. The system undergoes a bifurcation at this critical number: no pillar emerges below it, but pillars can emerge above it.

From these characteristics, we can arrive at a collective-level description of social insects in terms of nonlinear dynamics and emergent phenomena. In this context, the specific details of single individuals will be rather irrelevant. The following examples will give us a good perspective on how order out of individual chaos emerges through simple interactions between units.

Order by Fluctuations

One of the first examples of emergent phenomena in social insects was proposed by Jean-Louis Deneubourg in 1977. In this remarkable paper,[6] Deneubourg used the ideas of Turing's pattern formation theory (chapter 3) as well as those of Ilya Prigogine and Gregoire Nicolis, who extended Turing's ideas into a general framework. In social insects, particularly in termites, there may be several orders of magnitude difference between the size of an individual and the size of a nest built by the colony. Some nests are like clay cathedrals. The termite genus *Macrotermes* build nests that can reach 6 or 7 meters high, about 600 times the length of a worker.

To build their nests, termite workers use soil pellets joined with a cement whose pheromones attract other workers. The first stages of nest building involve the construction of some strips and pillars with these pellets. Eventually, arches are formed between the pillars, and finally, the space between pillars is filled to make walls. Deneubourg created a simple mathematical model to describe these first stages of termite nest building. An essential component of the model is *stigmergy*, a term coined by the French naturalist Paul Grassé. Grassé showed that the coordination and regulation of building activities do not depend on the workers themselves but are mainly achieved by nest structure. A given spatial configuration (some local, early distribution of soil pellets) triggers the response of a termite worker (Figure 6.4, A) which modifies the configuration (adding new soil pellets, Figure 6.4, A_1). As the configuration changes, so do the behavioral patterns of the individuals. There is thus constant feedback between the emerging structure and the spatial distribution of activity and worker activity.

Deneubourg's model involves three basic ingredients of termite nest worker behavior: the density of termites, the amount of cement at each point, and the presence of the pheromone that the workers add to the cement. The basic dynamics are simple: it is assumed that at the beginning termites move randomly on a given surface and deposit small amounts of pellets together with some amount of pheromone. As new termite workers reach the same area, positive feedback starts to operate: those random places where the local concentration of pheromone is higher will be more attractive to the workers carrying cement, and they will leave their pellets at those positions. Since these positions also receive more pheromone, they attract still more workers, with a further increase in their size and attractiveness. This is not far from the theoretical models of pattern formation in development discussed in Chapter 3. The building of a termite nest is close to a developmental process.

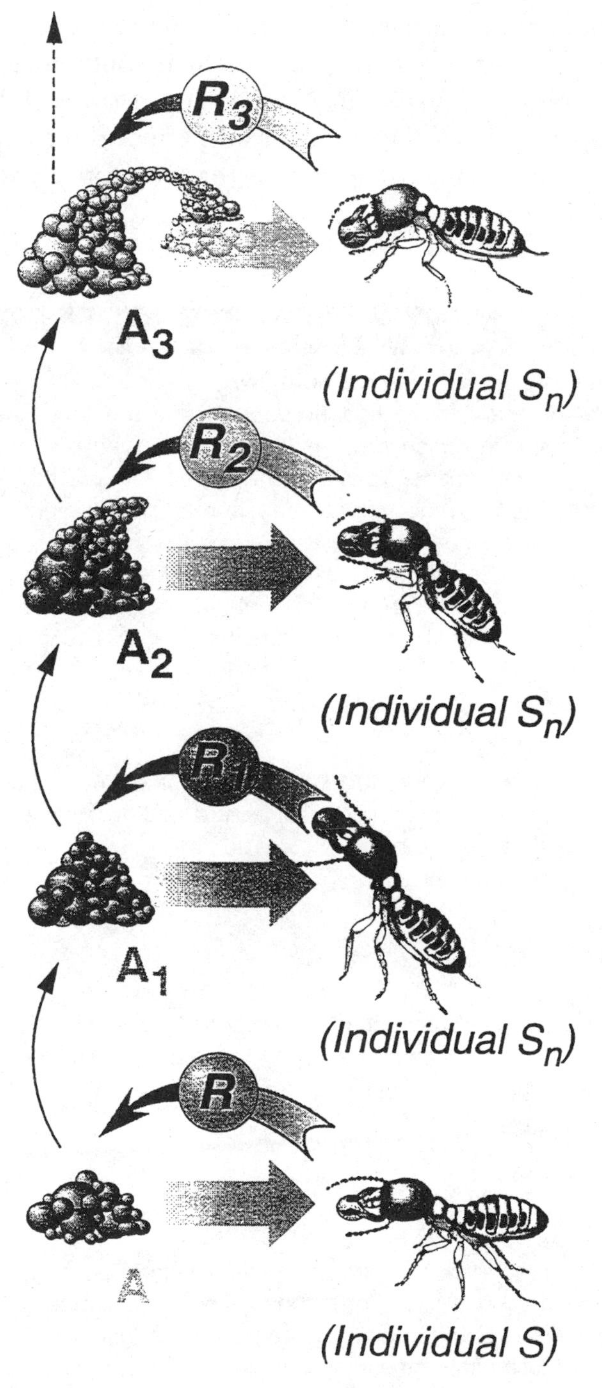

Figure 6.4 Stigmergy: as individuals create structures, the emerging spatial configurations (A, A1, A2, etc) change the behavioral patterns of the individuals.

Deneubourg's model (see Box 1) treats the density of termites and the local amount of cement or pheromone as continuous fields. This is a common approximation in physics, and it successfuly describes many complex systems. The model leads, under some conditions, to a set of well-defined regular structures that qualitatively resemble the formation of pillars in termite nests.

Building Pillars in Termite Nests

The model presented by Deneubourg involves three basic variables. Let us indicate by r and t the spatial location and the time step, respectively. Then the local concentrations of marked cement (i.e., cement with pheromone), pheromone, and termites are indicated by $P(r,t)$, $H(r,t)$, and $C(r,t)$, respectively. It is assumed that the pheromone can diffuse freely through space and that a constant flow of individuals, Φ, is present. Termites carrying soil pellets will be attracted to the local positions with higher pheromone concentrations. The pheromone is also degraded proportionally to its concentration. The first basic equation is

$$\frac{\partial P(r,t)}{\partial t} = k_1 C - P(r,t),$$

which means that the deposited cement with pheromone increases with the presence of termites carrying soil pellets and decreases as a consequence of the pheromone decay.

The second equation,

$$\frac{\partial H(r,t)}{\partial t} = k_2 P - k_4 H + D_h \nabla^2 H(r,t),$$

tells us that the local amount of pheromone grows with the amount of deposited pellets and decays at a rate $-k_4 H$. Additionally, there is passive diffusion as indicated by the last term.

The final equation,

$$\frac{\partial C(r,t)}{\partial t} = \Phi - k_1 P + D_c C(r,t) + \gamma \frac{\partial}{\partial r}\left(C \frac{\partial H(r,t)}{\partial r} \right),$$

indicates that termites carrying soil pellets appear at a given rate, but become inactive once they find areas with higher amounts of marked cement. There is another diffusion term corresponding to the random movement of termites, and a final term that involves the active movement of individuals toward increasing concentrations of pheromone. In fact, the last two terms will compete: the increasing creation of pillars will modify the random movements of ants.

In Figure 6.5a we can see an example of the starting initial condition, with a random distribution of small amounts of marked material. These initial fluctuations are amplified, as shown in the intermediate state in Figure 6.5b, and the final stationary state is a number of regularly spaced peaks of material (Figure 6.5c). Although the average distance between pillars is a characteristic value (once the parameters are specified), their specific location is highly dependent on the random initial conditions. As the pillars start to grow, they also compete for termites. The specific final configuration is thus path-dependent (i.e., multiple attractors are possible) just like most emergent patterns in social systems. It is very interesting to see how simple is the answer to our initial problem: individuals dealing only with local, noisy information are able to generate ordered, large-scale structures through the amplification of initial perturbations. These individuals are unaware of the progressive emergence of higher-order structures, although it is they who create them. There is an underlying dialogue between the individuals and the structures they create.

Prigogine and the chemist and philosopher Isabelle Stengers have discussed the importance of order through fluctuations. They mention

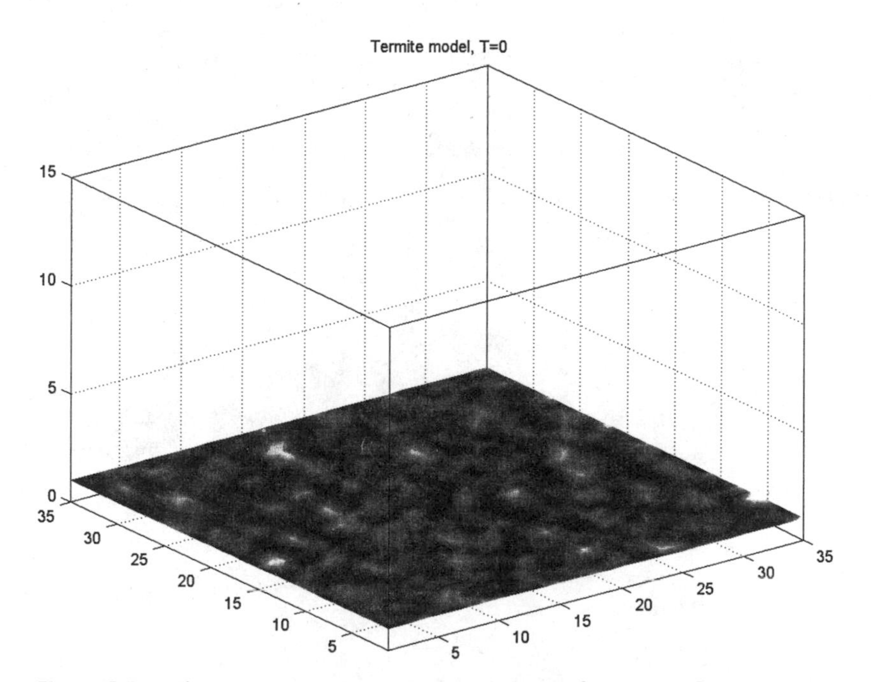

Figure 6.5 Three steps in the early termite nest formation from Deneubourg's model at three different times in the simulation.

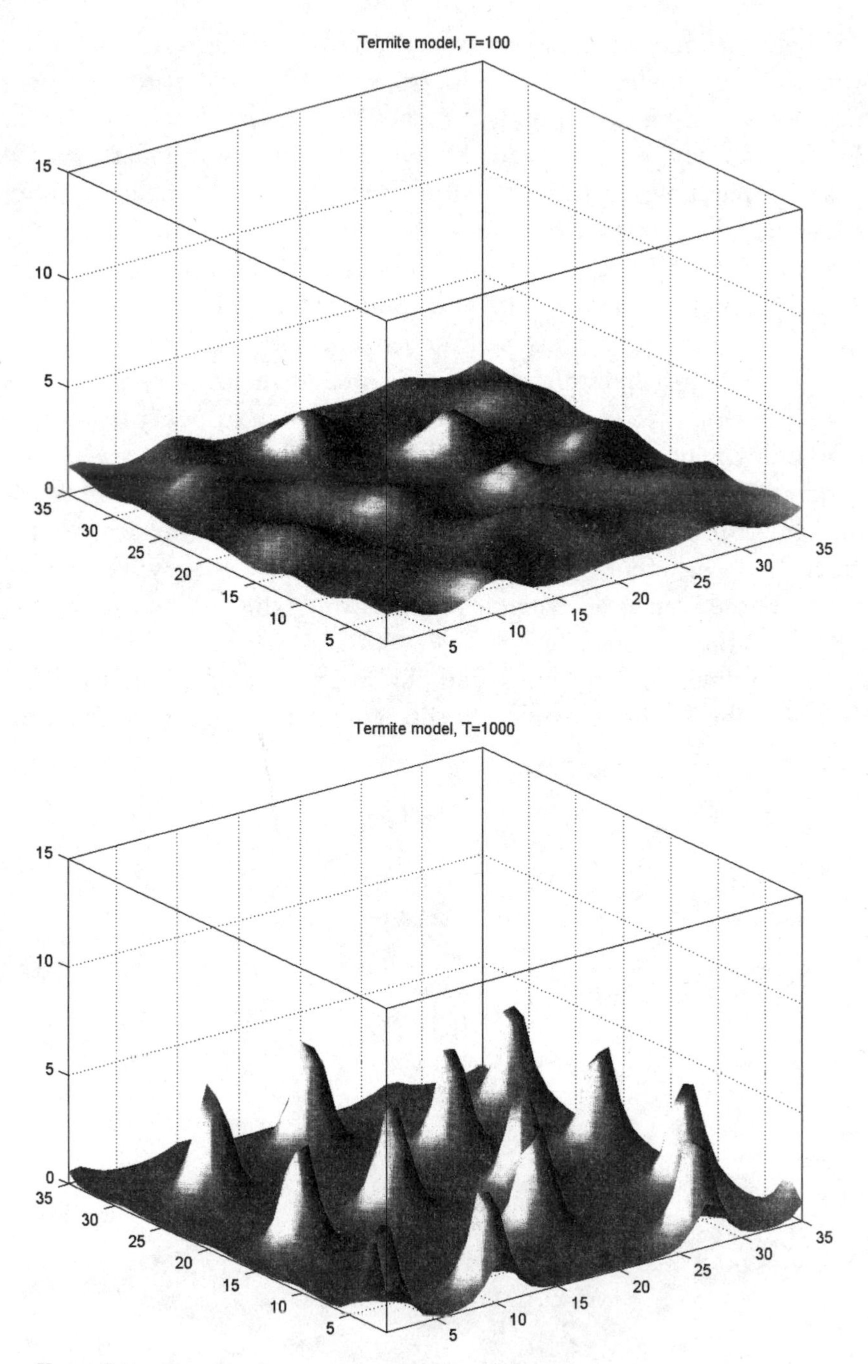

Figure 6.5 *Continued*

a very important aspect: as the complexity of a system increases, the number of types of fluctuations that threaten its stability also increases. To avoid instability, complex systems employ communication (in our example, through pheromone diffusion). There is, in fact, "competition between stabilization through communication and instability through fluctuations." The competition between stabilizing and destabilizing forces often finds a compromise at the edge of stability, as we will see in the following examples.

Oscillations and Chaos in Ant Colonies

Let us now consider a very interesting emergent phenomenon that has been observed among ants of the genus *Leptothorax*, whose colonies typically consist of fifty to a few hundred individuals. These colonies are easily maintained in the laboratory, since they require only the space provided by one microscope slide slightly elevated over another. Such nests were studied independently by Nigel Franks,[7] at the University of Bath, England, and Blaine Cole,[8] working in Houston, Texas. By monitoring the activity of ants inside their nests, Franks and Cole discovered a pulsatile activity with a short periodicity (Figure 6.6). The number of ants active at a given moment exhibited a regular fluctuation with 3 to 4 peaks an hour. Sometimes all ants were stationary, and sometimes all were moving, performing such tasks as nest maintenance or tending the brood and the queen. But this is rather counterintuitive: why should ants show synchronized rhythms instead of a steady level of activity?

Synchronized global behavior could be the result of coupling together a set of already periodic elements. This is typical of oscillators in nature: each unit is a regular clock. When they interact, they can synchronize. However, Cole's experimental results are not consistent with this idea.[9] In monitoring the activity of single ants, Cole found that individuals are not regular at all: they are chaotic! A careful study revealed that the temporal fluctuations of single ants are describable through a low-dimensional chaotic attractor. This implies that the global colony behavior must be an emergent phenomenon: the regular dynamics at the colony level must arise from the interactions between chaotic individuals. Cole also analyzed the behavior of groups of different numbers of ants and found that as the density of ants increased, the oscillation became more and more clear. Somehow, an increase in

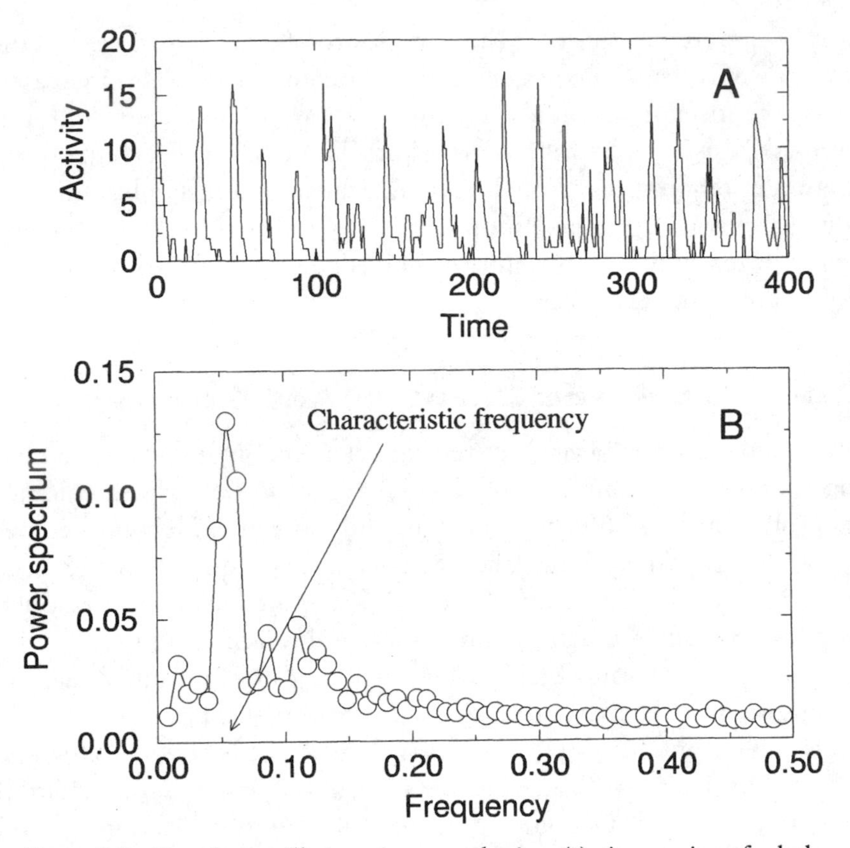

Figure 6.6 Regular oscillations in ant colonies: (a) time series of whole colony oscillations; (b) Fourier spectrum of the previous data set showing the presence of a characteristic peak at a given frequency (data kindly provided by Octavio Miramontes).

the number of elements was able to generate the emergent dynamics. How does this happen?

Inactive ants can become active in two ways: either an active ant touches the inactive one, or an inactive ant can spontaneously become active with some small probability of activation. Active ants remain active over a variable time scale but become inactive in a spontaneous way at low densities of active ants. Again, two basic components are at work: some intrinsic noise leading to spontaneous activation, and the interactions among nearby ants. In collaboration with Octavio Miramontes, now at the Universidad Nacional Autonoma, in Mexico (UNAM), we developed a model of *Leptothorax* dynamics based on the previous analogies between ant colonies and neural networks.[10,11] Active and inactive ants are like firing and nonfiring neurons. In our model these ants, if active, move on a two-dimensional square grid. Inactive ants remain frozen at their sites. When a given ant becomes active, it moves at random to adjacent positions (if not occupied by other ants). This random movement makes possible interactions between nearest ants and thus the propagation of activity.

This model was called a *fluid neural network* because our basic units, the virtual ants, behave like standard model neurons except that they can move through space. The explicit rules of our model are described in Box 2. The model reproduces the basic set of observations for a suitable range of parameters. In Figure 6.7 we can see several runs of this model as the density of ants is increased (here the density is defined as the number of ants divided by the total number of nodes in the lattice).

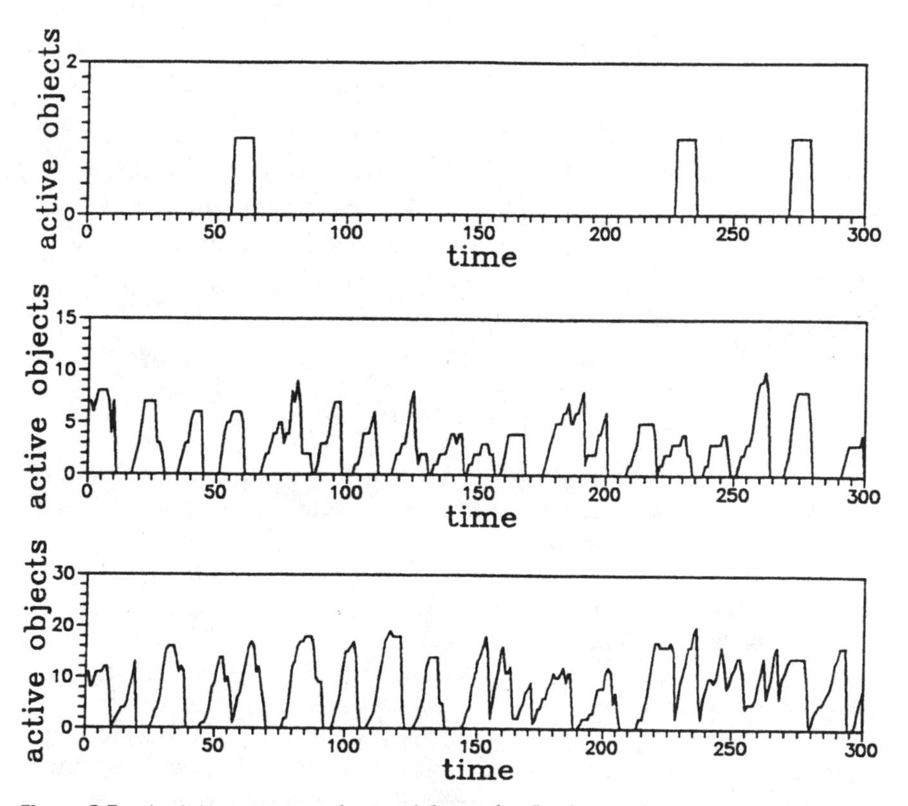

Figure 6.7 Activity patterns obtained from the fluid neural network model: from top to bottom, the number of active ants are displayed through time for increasing densities. We can see the emergence of a regular pattern as the number of individuals, and thus of interactions, increases.

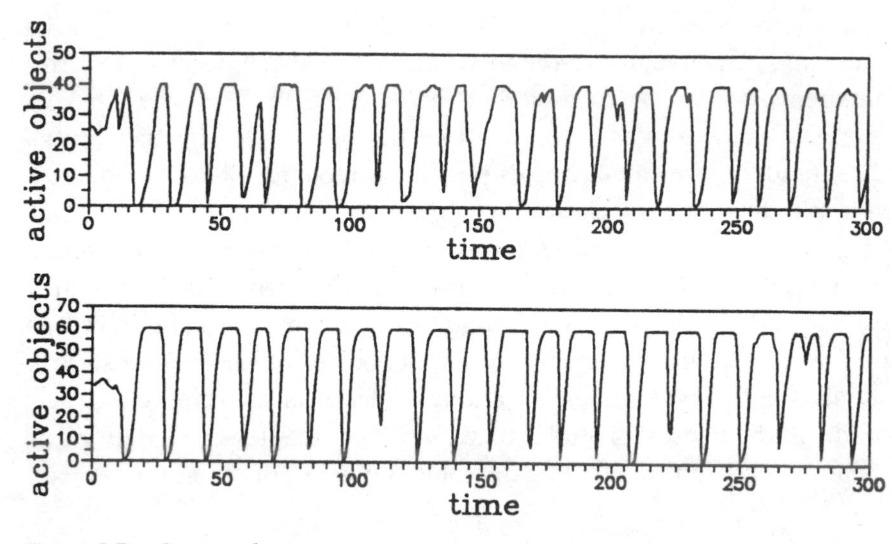

Figure 6.7 *Continued*

From top to bottom we have increasing densities of ants $\rho = 0.01$ (a single ant), $\rho = 0.1$, 0.2, 0.4, and $\rho = 0.6$. Single ants activate at random,* and thus they do not show any kind of regular pattern. But as more and more ants are added, a pattern starts to emerge. At high densities this pattern is highly regular. Yet real ant colonies never show such a regular pattern; their oscillations are neither totally disordered nor totally ordered. Are ants using an intermediate density between order and disorder to maintain themselves at the edge of chaos?

Fluid Neural Networks

In our model each ant is described as a formal neuron, with an internal state given by $S_i(t)$, where $i = 1, 2, \ldots, N_{\text{ants}}$. Here t is the time step. Ants are distributed on a two-dimensional lattice (Figure 6.8a). Two basic states are defined: inactive ants (white circles) and active ants (black circles). Active elements can move to the available nearest sites for this given unit (indicated by arrows). Active and inactive states are defined through the condition $S_i(t) > \theta$ and $S_i(t) < \theta$, respectively. Here θ is a positive parameter that acts as a threshold. Inactive ants will remain

*More realistically, chaotic dynamics can be introduced at the individual level, but the results are very similar, provided that the frequency of activation events is the same.

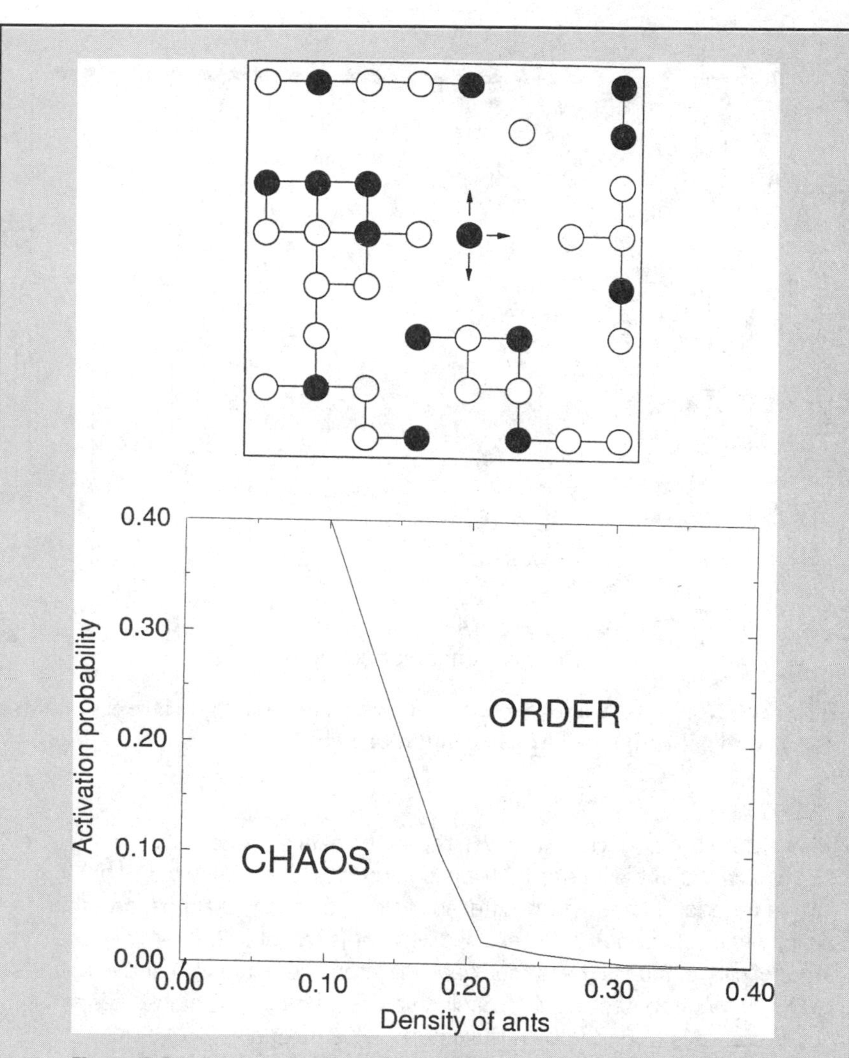

Figure 6.8 Upper plot: the fluid neural network involves formal neurons moving on a lattice if they are active (black circles) and becoming frozen if not (white circles). Bottom: phase space of the model, showing two phases. The real ant colonies appear to be located at the phase transition boundary.

frozen at their spatial location. Each lattice point can be occupied by a single ant. Interactions among ants occur only between individuals placed at nearest positions, and the state of each ant is updated by

$$S_i(t+1) = \tanh\left[g \sum_{\text{nearest}} S_j(t)\right],$$

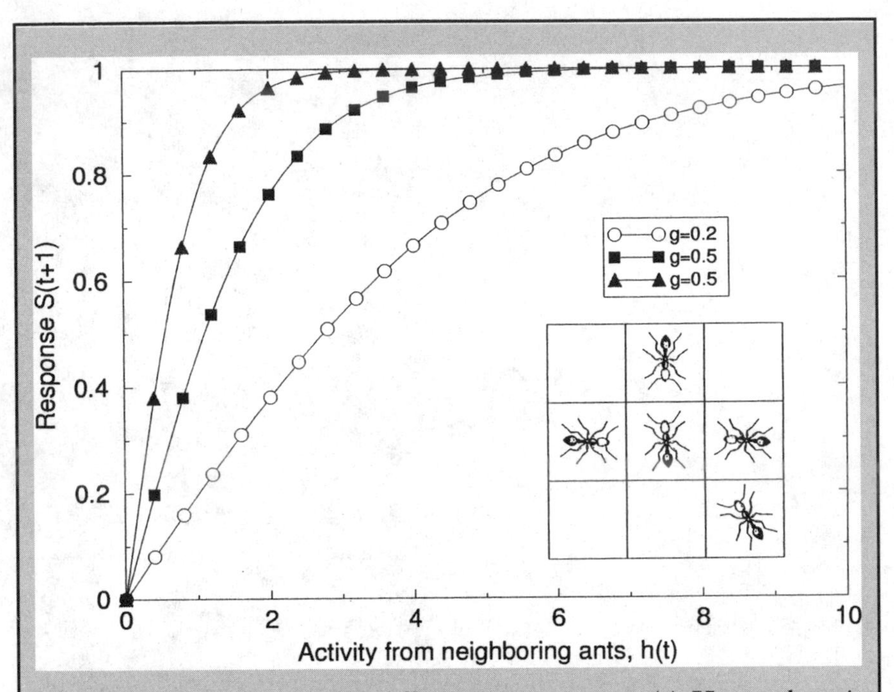

Figure 6.9 Response curves for different gain parameters (g). Here each ant is assumed to be surrounded by a maximum of eight nearest nestmates (inset).

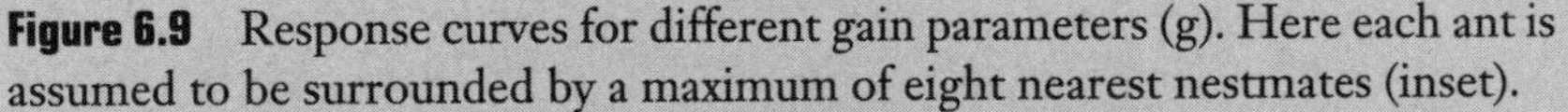

where the sum is performed over the eight nearest sites *and* the central site where the ant is located. This is shown in Figure 6.9, where the shape of this function is shown for different values of g (the gain parameter, as defined in the previous chapter). A given ant (central cell, inset) senses the state of its neighbors (here four nearest sites are occupied) and responds to them. We can see that increasing values of g lead to higher response, but it is always bounded by a maximum value of one.

An additional rule is added to this model: spontaneous activation of inactive ants. Once a given ant becomes inactive, it can become active again in two ways: either by the interaction with another active ant or spontaneously, with some probability P_a. If activation takes place, the state of the ant will be S_a. The state of an isolated ant will change in time according to

$$S_i(t+1) = \tanh(gS_i(t)),$$

which for $g < 1$ gives a decay to zero activity. However, the rule of spontaneous activation allows individual ants to randomly move through space. During these periods of activity, a given ant can encounter an inactive ant, which becomes active through the interaction. At large enough densities, the activation of a single ant will be able to propagate through the entire lattice. A detailed analysis of this model shows that two basic dynamical regimes are present, which we indicate in Figure 6.8b. Here two relevant quantities, the density of ants and the probability of activation, are analyzed. The first enhances the propagation of activity, thus favoring global order. The second introduces random events into the system. At low densities, random events are unable to propagate in a coherent fashion, and we have chaotic dynamics. But once a given threshold is reached, coherent behavior starts to dominate, and we reach the ordered domain. Evidence from experimental data suggests that real ant colonies operate close to the boundary separating order from chaos. Other models involving autocatalysis and spontaneous inactivation show a similar pattern of behavior.[12]

We can think of the difference between chaotic behavior at low densities and periodic behavior at high densities as two different phases. In fact, if we perform different measures, we find a sharply defined transition between these two phases. For a given probability of activation, the density needs to reach a critical value for ordered behavior to occur. However, periodic activity patterns occur over a considerable range of variation in density and sensitivity of ants to stimulation (the gain parameter, g). Hence these emergent patterns are robust.[13]

We can use the information transfer between different units to characterize the boundaries between these two phases.[14] This is an interesting measure, not only because it will detect correlations in the system, seen as a nonlinear phenomenon, but also because information is what ants mainly manipulate. Information transfer in the ant colony is the relevant quantity, and it can be shown to be *maximized* at a critical density (Figure 6.10a). It is easy to understand why. At low densities, activity is mainly maintained by spontaneous activation plus rare ant–ant encounters: we are in the disordered/chaotic regime. In this regime, each pair of ants is essentially decoupled. But at some point the number of elements is large enough to sustain large fluctuations of activity, sometimes involving all the elements in the lattice. As the density is further increased, ants become highly synchronized: they switch between activity and inactivity with a well-defined periodicity. Two given ants are simply doing the same thing all the time (they are both active or inactive): the network is in the ordered regime.

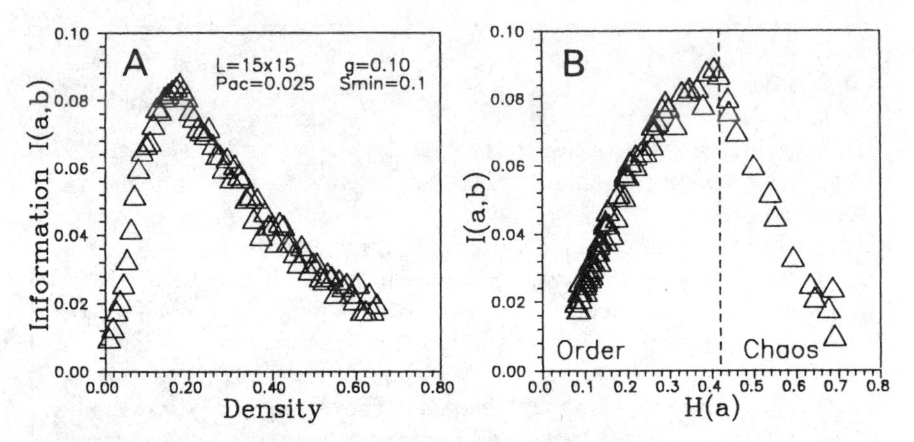

Figure 6.10 (A) Information transfer (I) as a function of the density of ants in the simulation. It has been measured by following two randomly chosen individuals (a and b). A peak is observed at a given critical density. (B) When we plot information transfer versus entropy (H(a)), measured from a single individual, we can see that information is maximal at an intermediate degree of disorder.

At the critical density, random individual activations become able to propagate through the whole colony, but the density is low enough to prevent activity from remaining a long time in the system. In other words, information is properly propagated (information transfer is guaranteed), but strong memory effects are prevented: the system does not remain in a given state for long. This is an interesting result, consistent with Chris Langton's 1990 conjecture[15] about the emergence of computation in natural systems. Langton, then at Los Alamos, suggested that living systems would maximize their computational capabilities at the edge of chaos. Computation, in other words, would require some amount of order, since information must be stored in some stable way. But the manipulation of information also requires some internal degree of disorder. The optimal compromise between both requirements occurs at the transition between order and disorder. This idea is explicitly shown in Figure 6.10b, where the information transfer has been plotted against the degree of disorder, and we can see that intermediate values of entropy yield the maximum information transfer. This is what seems to occur with the observed critical density in real ant colonies.

The Blind Leading the Blind

In far-from-equilibrium conditions, both animate and inanimate systems can form spatial structures. In previous chapters we have seen these structures in various contexts. But few pattern-forming systems in biology are as impressive as army ant raids. The biologist T.C. Schneirla vividly described the raids of the *Eciton burchelli*, one of the best-known species of army ants:

> For an *Eciton burchelli* raid nearing the height of its development in swarming, picture a rectangular body of 15 meters or more in width and 1 to 2 meters in depth, made up of many tens of thousands of scurrying reddish-black individuals, which as a mass manages to move broadside ahead in a fairly direct path. When it starts to develop at dawn, the foray at first has no particular direction, but in the course of time one section acquires a direction through a more rapid advance of its members and soon raids in the other radial expansions. Thereafter this growing mass holds its initial direction in an approximate manner through the pressure of ants arriving in rear columns from the direction of the bivouac. The steady advance in a principal direction, usually not more than 15 degrees of deviation to either side, indicates a considerable degree of internal organization, notwithstanding the chaos and confusion that seem to prevail within the advancing mass. But organization exists, indicated not only by the maintenance of a general direction but also by the occurrence of flanking movements of limited scope, alternately to right and left, at intervals of 5 to 20 minutes depending on the size of the swarm.
>
> The huge sorties of *burchelli* bring disaster to practically all animal life that lies in their path and fails to escape. Their normal bag includes tarantulas, scorpions, beetles, roaches, grasshoppers, and the adults and broods of other ants and many forest insects; few evade the dragnet. I have seen snakes, lizards, and nestling birds killed on various occasions. . . .[16]

Instead of building stable nests, army ant colonies form "bivouacs," temporary clusters of half a million ant bodies packed together. These bivouacs hide in sheltered places, but once the light level exceeds a given threshold the group starts to dissolve and swarm. Out of the initial cluster, a raiding column emerges. The colony starts to walk through the rain forest looking like a giant amoeba magically exploring the forest floor. Yet this swarm is nothing but an emergent pattern resulting from the interactions of blind individuals, whose behavior is determined by chemical and tactile stimuli. Single ants are too simple to be able to direct the group in any predefined direction, but somehow the swarm is able to explore, in a purposeful way, up to 1000 square meters a day. Its foraging trails are sometimes linear, but in some species they often form complex dendritic structures ressembling fractal trees

(Figure 6.11). Swarm raids pose in the clearest way the general problem of collective decision making without centralized control. These raids are in fact the largest organized operations carried out regularly by any animal except humans.

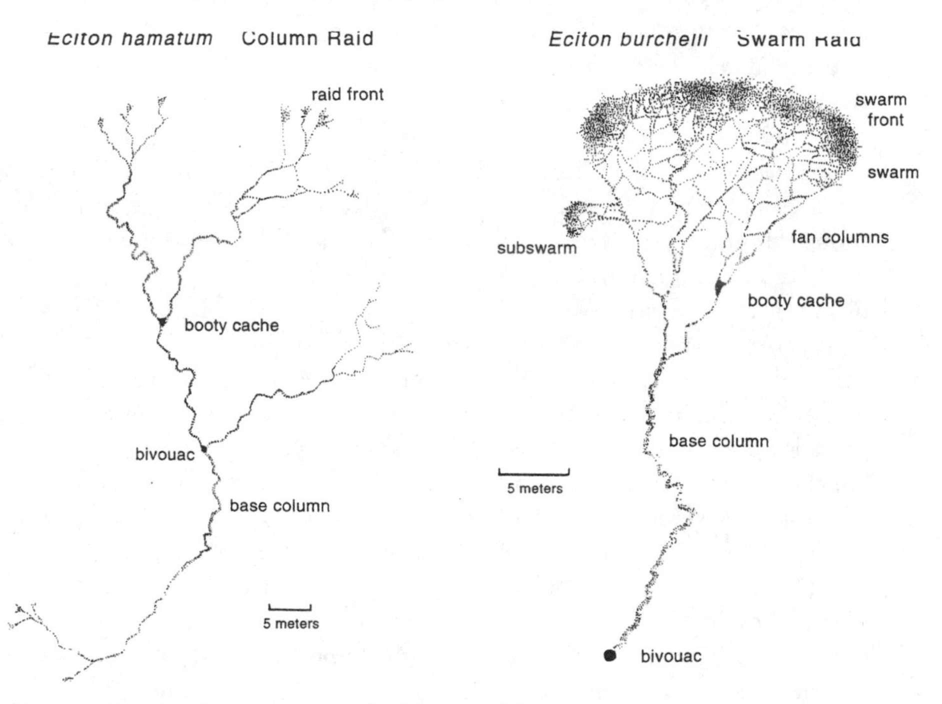

Figure 6.11 Two basic patterns of raiding used by army ants.

Modeling Army Ant Raid Patterns

Jean-Louis Deneubourg, Nigel Franks, and collaborators introduced the following model of army ant behavior.[17] Here army ants behave as discrete units moving on a two-dimensional lattice where a pheromone field is created and maintained by traveling ants (Figure 6.12). To be concrete, let us call $S_i(t)$ the state of a given ant, which will be $S_i = 1$ if the ant is searching and moving away from the nest and $S_i = 2$ if the ant is returning. The trail concentration in a given lattice site is given by $\phi(i, j) \geq 0$.

When ants are searching they leave a unit of pheromone, unless the total amount of pheromone already exceeds a threshold value σ_1. When

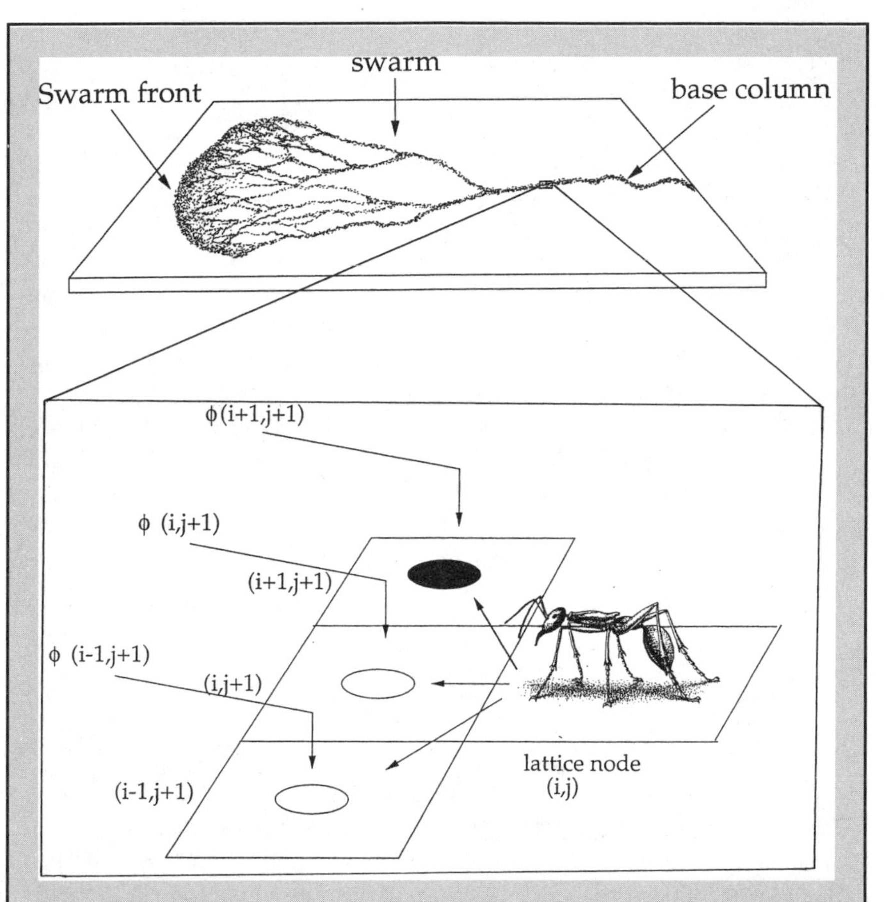

Figure 6.12 Modeling army ant swarms. The model considers a square lattice where individuals move by following the chemical trail in a probabilistic way. A virtual ant located at the lattice point (i,j) will move forward towards one of the three next sites in the lattice. Depending on the concentration of pheromone, the more the concentration the highest the probability of being chosen.

returning from exploration with a food item (returning ants always carry a food item) they leave q units of pheromone; now the threshold is σ_2. The pheromone evaporates at a given decay rate δ. Specifically, at each step we have $\phi(i, j) \rightarrow (1 - \delta)\phi(i, j)$.

The behavior of these artificial army ants involves two basic rules

1. Probability of movement. A given ant located at (i, j) will move with a probability P_m depending on the pheromone field in the three grid points in front of (i, j). If the ant is leaving the nest,

$$P_m = \frac{1}{2}\left[1 + \tanh\left(\frac{\phi(i+1,j+1) + \phi(i+1,j) + \phi(i+1,j-1)}{\phi^*} - 1\right)\right]$$

(and similarly for returning ants but replacing $i+1$ by $i-1$). The parameter ϕ^* represents the concentration of trail pheromone for which the probability of moving per step is 0.5. Here we take $\phi^* = 100$ (changes in this parameter only shift the observed optimal solutions in parameter space but not the obtained patterns).

2. Once movement starts, an ant has to choose one of the new grid nodes. The nodes with higher pheromone levels are more likely to be chosen. There are three probabilities π_L, π_O, and π_R indicating the left, central, and right front nodes, respectively. These probabilities are

$$\pi_O = \frac{1}{C}[\mu + \phi(i+1, j+1)]^2,$$
$$\pi_R = \frac{1}{C}[\mu + \phi(i+1, j-1)]^2,$$

and obviously, $\pi_L = 1 - \pi_R - \pi_O$. Here C is given by

$$C = [\mu + \phi(i+1, j+1)]^2 + [\mu + \phi(i+1, j)]^2 + [\mu + \phi(i+1, j1)]^2.$$

This choice of a sigmoidal-like function is based on experiments. The parameter μ weights the attractiveness of empty nodes. If the total number of ants choosing a particular node exceeds a maximum value A_m, then no movement occurs. Two additional rules are required: (a) lost ants (i.e., those that reach the limits of the lattice) are removed from the system; (b) new ants enter from the bivouac, here located at $(1, L/2)$. This is a fixed number (here we use $N_b = 10$), but if the bivouac site is already at the limit A_m, no new ants are added.

Different species of army ants display different raiding patterns.[2] *Eciton hamatum* organizes itself into a relatively linear column, while *E. burchelli* displays an amoebalike distribution. *E. rapax* is somewhere in between. This diversity was explained by Deneuborug's theoretical model, which predicted that different patterns would result from different food distributions and parameter combinations (according to prey preferences), even though the underlying mechanism to build raid patterns was the same. In Deneuborug's model, artificial ants moved through a lattice depositing a chemical marker. Starting from a virtual bivouac, each individual moves in a particular direction, depositing a small amount of chemical marker. Individuals also differentiate between quantities of pheromone and tend to follow the trails with higher chemical concentrations. Once they find a food source, they return to the nest following the same basic set of rules but deposit larger amounts of pheromone, so that their nestmates will follow the same path. Thus the path grows stronger as long as the food source holds out. The set of rules, though simple, was able to generate the same basic patterns observed in field studies (Figure 6.13) using different parameter combinations of pheromone decay and rates of trail reinforcement.

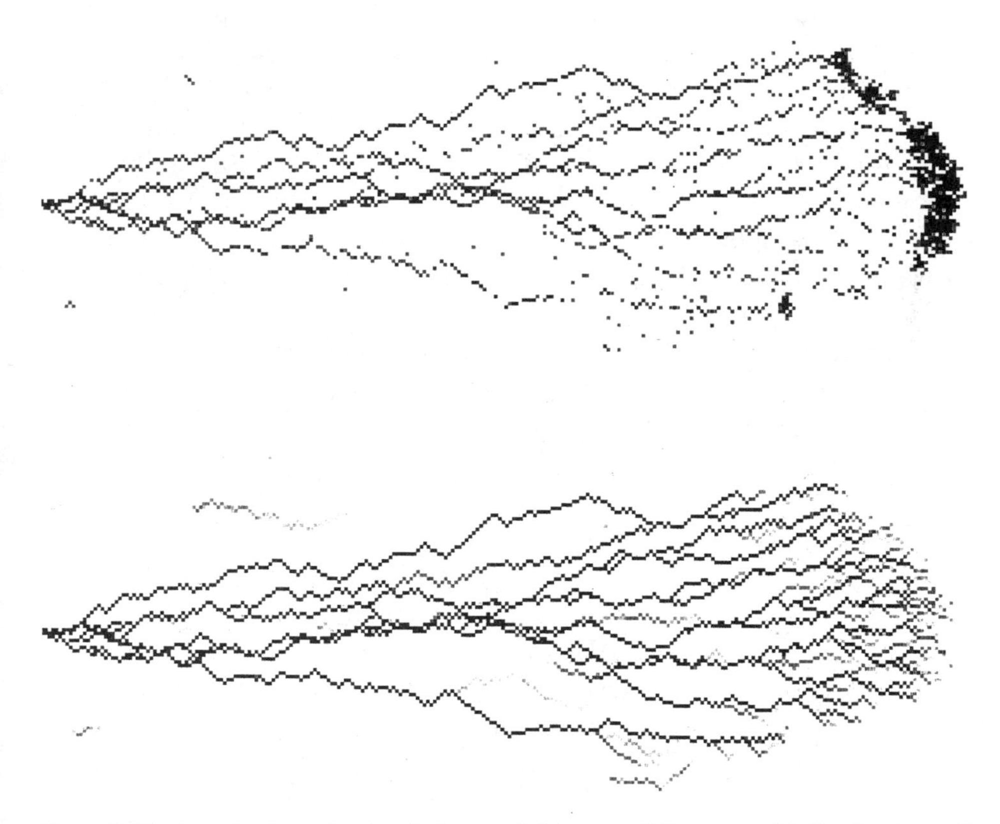

Figure 6.13 Results from the simulation model for two different spatial distributions of food. The results agree with field observations.

Do these patterns have some kind of adaptive meaning? Perhaps the different preferences displayed by different species involve a foraging pattern that is in some sense optimal. We can explore this problem by means of various simulation methods. If some reasonable quantity is optimized by the army ants searching for some given type of food item, then an appropriate search method can find the optimal patterns maximizing some key quantity. Do specific spatial patterns involve optimization in army ant colonies? Are these patterns related to the real ones? Does natural selection tune some specific individual trait that can explain such optimization?

These questions have been carefully addressed through an extensive analysis of the parameter space of Deneubourg's model.[18] Instead of looking at all parameter combinations (which would be computationally expensive) he borrowed techniques from the area of evolutionary computation and treated the problem as a search for optima in a rugged landscape. Here, for each multiparameter combination (a given pheromone decay rate, given detection thresholds, and so on) the total amount of food found and collected over some period of time was measured as a fitness value. The idea is simple: at the beginning of the evolutionary search, we randomly generate a number of possible parameter combinations. These combinations are then used as the parameter sets in independent computer runs of the model. Each run gives a final amount of total food collected. After all simulations are finished, we look at the scores and eliminate those with a fitness lower than the average. We have thus created an artificial selection process. Now we have a set of "surviving" solutions, and we will "reproduce" them. We do so by making copies of the surviving parameter combinations with some small amounts of noise (i.e., we make small changes in the parameters) and also by creating a small number of totally new parameter combinations. The first procedure allows us to explore the landscape close to optima; the second takes us into unexplored domains of the adaptive landscape. This is repeated again and again.

These searches used three different types of spatial food distributions: a lattice with half sites occupied by a single item of food per site (common but small food sources); a very small number of very rich sites; and an intermediate state with many sites with little food and some with a large amount. These cases correspond to the preferences displayed by the army ant species *Eciton burchelli*, *Eciton hamatum*, and *Eciton rapax*, respectively. Each type of spatial food distribution led to a different spatial pattern, and each one was consistent with those observed in field studies for each type of spatial distribution of food. But the most interesting result was that the parameter combinations that gave the optima for each case where rather similar. Different food distributions acted on the same basic parameters to produce the different raiding strategies.

The only important differences arise from the relative significance of trail reinforcement and pheromone decay (which allows the ants to shift from one trail to another and eventually replace some trails with others). When many small sources are present, it is more effective to be able to switch quickly from an already exploited source to a new one. On the contrary, if the colony exploits rich but rare sources, it is important to guarantee that the sources of food are fully exploited at the cost of a reduced exploratory capacity. That the compromise between flexibility and efficiency seems to find a place close to criticality is suggested in particular by the fractal patterns displayed by some species. Here is how E.O. Wilson, in 1971, described the evolution of chemical communication in ants: "the level of accuracy (of chemical communication) has been arrived at a compromise between the utmost effort of the ant's chemosensory apparatus and to follow trails accurately and, simultaneously, the need to reduce the quantity and increase the volatility of the trail substance in order to minimize overcompensation in the mass response."[19]

Assuming that the model indeed reflects what the ants do, this suggests that self-organized raid patterns have been optimized in the course of evolution. Selection acts on the possible sets of emergent phenomena arising from a collection of agents displaying simple local communication. The use of chemical signals allows for exploitation of a limited number of mechanisms of pattern formation, such as symmetry-breaking. It is easy to see that the swarm experiences successive symmetry-breaking bifurcations as it advances, and random events reinforce one emerging path over another. This is especially clear in the fractal-like patterns displayed by *Eciton burchelli* (and the corresponding simulations). The branches tell us that the system breaks symmetry very often, thus showing a high sensitivity to fluctuations. Since raid patterns emerge out of relatively simple rules followed by individuals, it means that colony-level selection has shaped the behavior of individuals in such a way that individuals self-organize to implement optimal foraging strategies. Variation at the level of individual genotypes has been expressed through self-organization in colony-level raiding patterns. The limited set of emergent raiding patterns partitions genotype space into domains, within which genes can vary considerably without affecting the pattern, making these robust to small variations of genes.

Self-Organization in Nest Building

Earlier, we saw how a Turing-like phenomenon of morphogenesis might explain the first stages of termite nest building. But social insects create a wide range of nest structures, from simple to highly complex. Wasp nests are a particularly interesting case (Figure 6.14). There are about sixty different types of wasp nest architectures, most of which are made of plant fibres chewed and cemented together with oral secretions. The internal structure of these nests can be rather complicated, with several stacked combs of cells within an external envelope. Some species build a central or peripheral communication opening that goes through the successive combs, allowing individuals to move from one floor to another.

In general, social insect nests are not just a simple repetition of a given basic motive. As Theraulaz noted, "A beehive is not just an array of hexagonal cells: cells are organized into combs, and each comb is organized into three distinct concentric regions, with a central area where the brood is located, surrounded by a ring of cells that are filled with pollen, and finally a large peripheral region of cells where honey is stored."

Theraulaz and Bonabeau have proposed an elegant approach to the general problem of nest building in social insects.[20,21] Their model differs from previous models in that it is based on individual responses to qualitative stimuli (in the termite nest model, or in the army ant raid system, individuals responded to quantitative stimuli such as pheromone fields). They called their computer model *lattice swarms*. In it, a set of artificial agents (like the wasps in a real nest) randomly move on a lattice and are able to deposit bricks depending on the local configuration of matter. Each individual perceives the state of a local neighborhood formed, for a cubic lattice, by 26 cells (Figure 6.15). The location of the individual at some level z is indicated by the black square. The

Figure 6.14 A wasp nest.

neighboring sites can be empty (no bricks have been deposited) or occupied by bricks of different types. The virtual wasps communicate only though the local environment they perceive: individual actions are directed by the dynamically evolving shape in construction. In such a

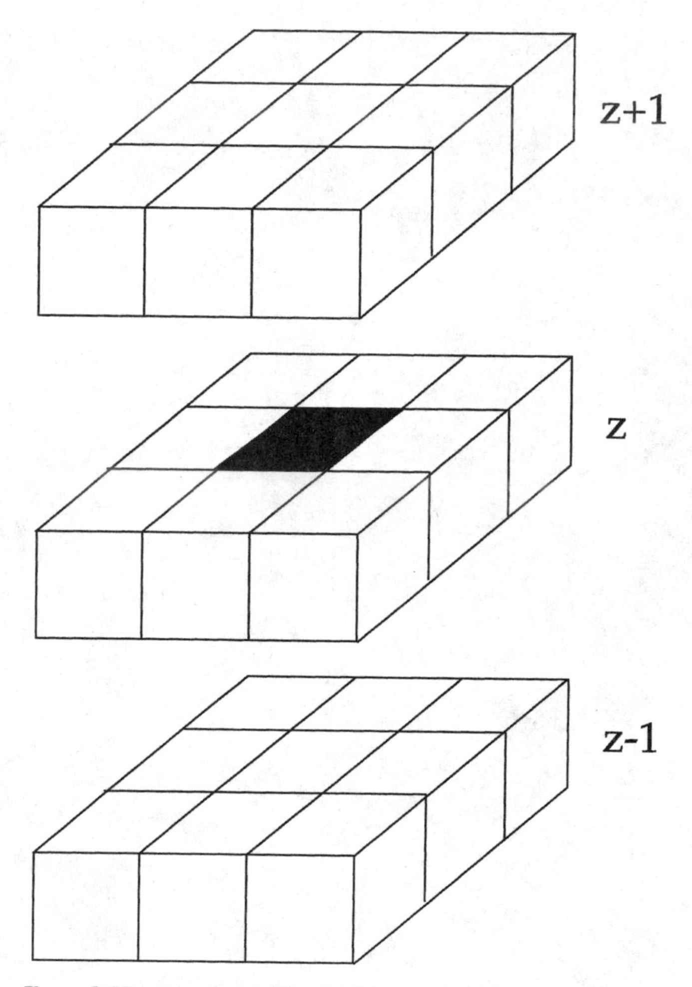

Figure 6.15 Local neighborhood perceived by an individual in the lattice.

swarm formed by many individuals working in parallel the agents do not follow a predefined sequence of behavioral steps. The basic purpose of the lattice swarm model is to help us understand how wasps organize themselves in space and time in order to ensure coherent building.

The agents move at random through the lattice and can drop two types of bricks (indicated here as 1 and 2). The actions performed by the agents are obtained from a predefined look-up table with as many entries as there are stimulating configurations. Some configurations trigger no actions at all. The rule table is a set of matrices such as

$$\begin{pmatrix} 0 & 0 & 0 \\ 0 & 1 & 0 \\ 0 & 0 & 0 \end{pmatrix} \begin{pmatrix} 0 & 0 & 0 \\ 0 & \times & 0 \\ 0 & 0 & 0 \end{pmatrix} \begin{pmatrix} 0 & 0 & 0 \\ 0 & 0 & 0 \\ 0 & 0 & 0 \end{pmatrix},$$

where the elements are ordered from top to bottom (Figure 6.15) and $\times$ indicates the position of the individual within the configuration. Each time an individual encounters a stimulating configuration, it leaves a brick. This particular combination, with only a single brick of type 1 in the top center of the neighborhood, leads to the deposition of a type 1 brick at the center. Another configuration, say

$$\begin{pmatrix} 1 & 0 & 0 \\ 0 & 0 & 0 \\ 0 & 0 & 0 \end{pmatrix} \begin{pmatrix} 1 & 2 & 0 \\ 0 & \times & 0 \\ 0 & 0 & 0 \end{pmatrix} \begin{pmatrix} 1 & 0 & 0 \\ 0 & 0 & 0 \\ 0 & 0 & 0 \end{pmatrix}$$

can generate the deposition of a type-2 brick. As different sets of matrix triplets lead to the deposition of more bricks, various patterns can emerge. Figure 6.16 shows the first steps in the construction of one of these structures. The final pattern corresponds rather well with the nests of the neotropical wasp genus *Epipona* (Figure 6.17). Other look-up tables yield complex structures involving helical ramps connecting different floors, just as in some real termite species.

Theraulaz and Bonabeau found a number of diverse building structures emerging from the coordinated behavior of these swarms, although not every set of algorithms gave rise to such structures. In those that did, called coordinated algorithms, local patterns of bricks from past construction provide randomly moving individuals the necessary cues to coordinate the building process. Although these studies require much further analysis, it is worth mentioning that the close similarities between real and artificial nests strongly suggest that important constraints on nest building have been operating through evolution.

Does Self-Organization Matter?

We have reviewed only a few models of collective behavior in social insects. Some ideas introduced in other chapters, such as pattern formation, symmetry-breaking, and self-organization, have been of great help in understanding the emergence of global patterns out from individual chaos. Although the ideas underlying self-organization were originally

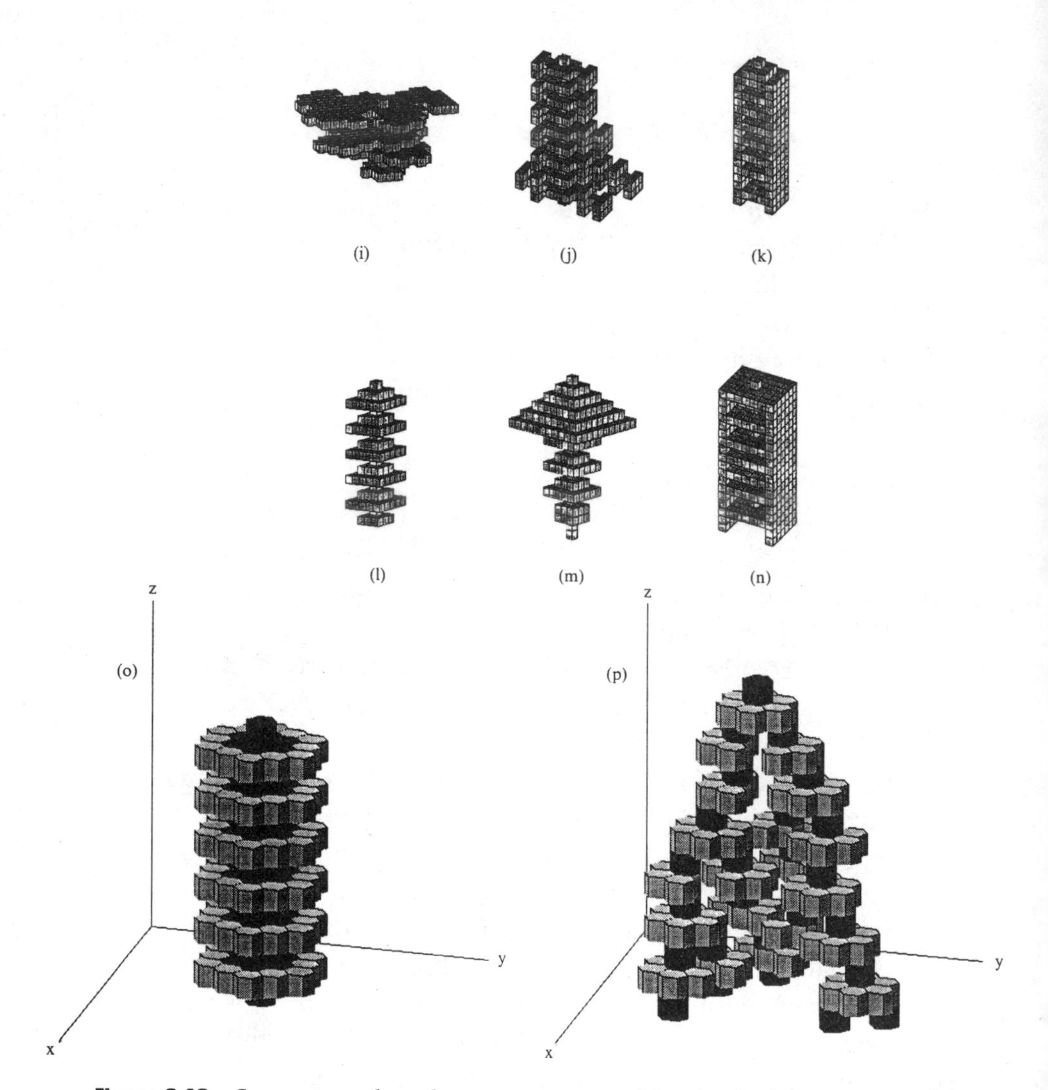

Figure 6.16 Some examples of patterns generated by the lattice swarm model (figure kindly provided by Guy Theraulaz).

introduced in the context of physics and chemistry, these concepts can be as well applied to ethology, suggesting a concise description of a range of collective phenomena in animals, especially in social insects. This description does not rely on individual complexity to account for collective features emerging at the colony level. Instead, it assumes that interactions among simple individuals can produce highly structured collective behaviors. Interactions, not individuals, are the key ingredients of behavioral complexity.

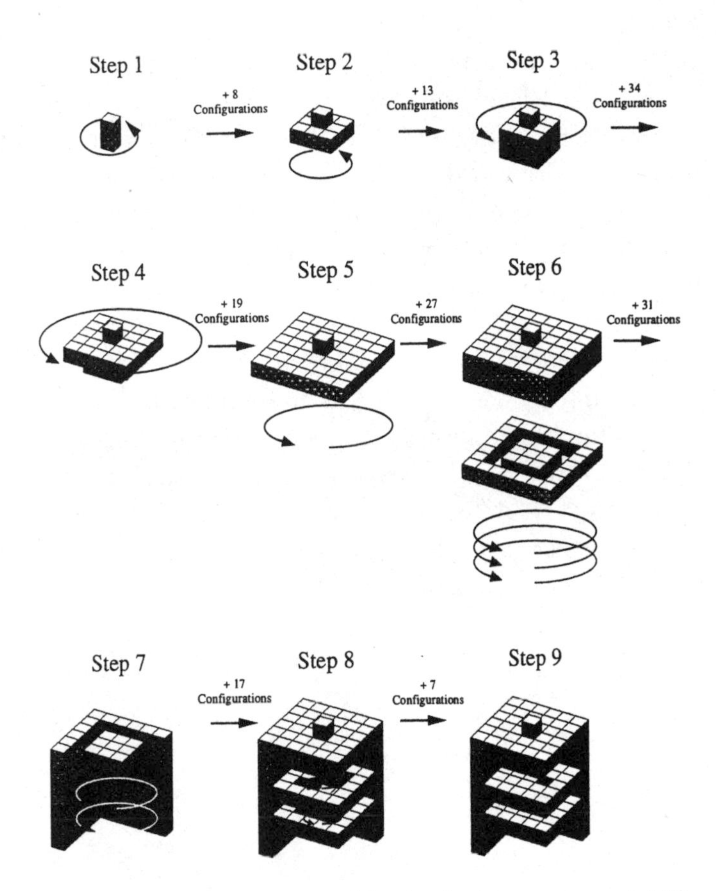

Figure 6.17 Different patterns generated by the lattice swarm model (figure kindly provided by Guy Theraulaz).

How, then, does selection operate on self-organizing phenotypes? As Bonabeau points out

> Selection can operate on parameters or factors that influence colony-level structures, be these self-organized or not. Such factors include response thresholds to stimuli, the behavioral output resulting from these stimuli, or specific properties of chemicals used as alarm, construction or trail pheromones: changing these factors undoubtedly changes global patterns and the conditions under which they can emerge and be maintained. For example, the volatility of a pheromone can affect foraging trails: obviously, this property is essential in defining the efficiency of a colony in a given environment, and may have coevolved with other features, such as colony size, since a volatile trail pheromone requires more individuals to maintain stable trails.

A good reason why self-organized patterns are so widespread is that the same individual-level behaviors may generate different collective responses in different environments. There is thus no need to invoke individual complexity in order to explain the origins of nest complexity. The global structures emerging from the interaction of simple individuals cannot be understood from the analysis of individual behavior: individuals have no idea how to build the nest, and the rules governing its structure cannot be deduced from an analysis of their genotypes. Selective forces operate on a parameter space where there is a limited number of possible dynamical patterns and nonlinear rules. As a consequence, only a limited (but rather diverse) set of higher-level structures can be obtained. The constraints operating on physical pattern-forming systems are also at work here. Only when the interactions generating collective behavior are considered can we start to understand the origins of complexity in insect societies.

The Baroque of Nature

The unsolved mysteries of the rainforest are formless and seductive. They are like unnamed islands hidden in the blank spaces of old maps, like dark shapes glimpsed descending the far wall of a reef into the abyss. They draw us forward and stir strange apprehensions. The unknown and prodigious are drugs to the scientific imagination, stirring insatiable hunger with a single taste. In our hearts we hope we will never discover everything. The rainforest in its richness is one of the last repositories on earth of that timeless dream.

—Edward O. Wilson, *The Diversity of Life*

Space, Time, and Waves

It is difficult not to be puzzled by the diversity of life forms within our biosphere. Some well-known ecosystems, such as coral reefs and tropical rainforests (Figure 7.1), show extraordinarily high diversity. This is what the ecologist Ramón Margalef called "the baroque of nature," meaning that ecosystems contain many more species than would be necessary if biological efficiency were the criterion for their organization.

Rainforests contain a great fraction of our planet's biodiversity,[1] including, according to a Robert May 1990 estimate, over 50,000 tree species worldwide. But the coexistence of such a great number of species in relatively small local areas is not well understood. Rich ecosystems have at least two common peculiarities. First, environmental variables such as temperature and humidity are often relatively uniform in space and time throughout the system, and second, numbers of individuals

Figure 7.1 View of the rainforest interior (drawing by Marshall Hasbrouck).

per species are highly nonuniform. Some species are abundantly represented, but many have very low densities. High species diversity is linked with rarity.

One of the most celebrated theories of classical ecology deals with the outcome of competition between two species sharing a common resource or territory.[2] Competitive interactions can lead to only two basic results: either both species coexist or, if the competition goes beyond some given critical value, one of them is eliminated by the other. The latter case is known as "competitive exclusion" and has been assumed to be the driving force in many ecosystems. The basic idea is simple. A single-species ecosystem will show so-called logistic growth: starting from a given (say small) initial population, the number of individuals will increase to a maximum value, the so-called carrying capacity. No more growth can occur beyond this limit, which is imposed by the available space and resources. Repeating the experiment will yield the same predictable result. But then, if another species is introduced into the ecosystem, how will this two-species system evolve? If space is not taken into account, then the exclusion principle will be at work: for similar growth rates, if exploitation of resources is weak enough, both species will coexist. But if both are strong enough competitors, no coexistence is possible: only one species survives.

But if competitive exclusion operates in real ecologies, why are they so complex? Why doesn't a small number of successful species take over? This is a controversial issue. Even species that exclude each other in the lab often appear to coexist in natural communities. Generally speaking, simple laboratory experiments cannot be extrapolated to the reality of the field. Long-term field experiments, such as those performed by the ecologist James Brown and his colleagues in the New Mexico desert, provide an interesting example.[3] These experiments involved the removal of kangaroo rats from some 20-hectare plots surrounded by fences. The kangaroo rats have important effects both on other rodents and on vegetation. This experiment revealed a complex set of interactions propagating through the food web. These interactions affected competition but also altered the colonization and extinction probabilities of other species. Such cascading effects, as Brown and his team have shown, "have implications for conservation policies, because the extinction of native species, establishment of exotics, and reintroduction of extirpated species will often cause further changes in diversity."

Even in well-controlled field experiments we perceive the complexity of the food web.[4] The specific sequence of community assembly is also very important. Several theoretical ecologists, such as Jim Drake, Stuart Pimm, and Mac Post, have analyzed the effects of changing the sequence of species entering a given growing community. Although the end of the experiment is a mature ecosystem with several robust regularities, the specific species composition can be very different, depending on the specific sequence used in each run. These simulations are usually peformed by using a "pool" of species, which are randomly added one by one. Although some species fail to persist at a given attempt, they can enter the community later. But in some runs some species are successful, and in others the same are unable to persist. Very often, the effects of the removal or extinction of a given species have large consequences for the whole community, and as we will see in this chapter, the effects can be largely unpredictable. But in fact, not everything is so unpredictable. An example is the strong regularities displayed by species-abundant relationships. These are usually plotted as a graph showing the abundance of each species (as measured by the number of individuals per unit area) in order from the most common to the rarest. By following the time evolution of a given ecosystem from, say, a grassland to a mature forest we find that a characteristic change appears common to all rich ecosystems (Figure 7.2). Starting from a species-poor community, with relatively short trophic (food)

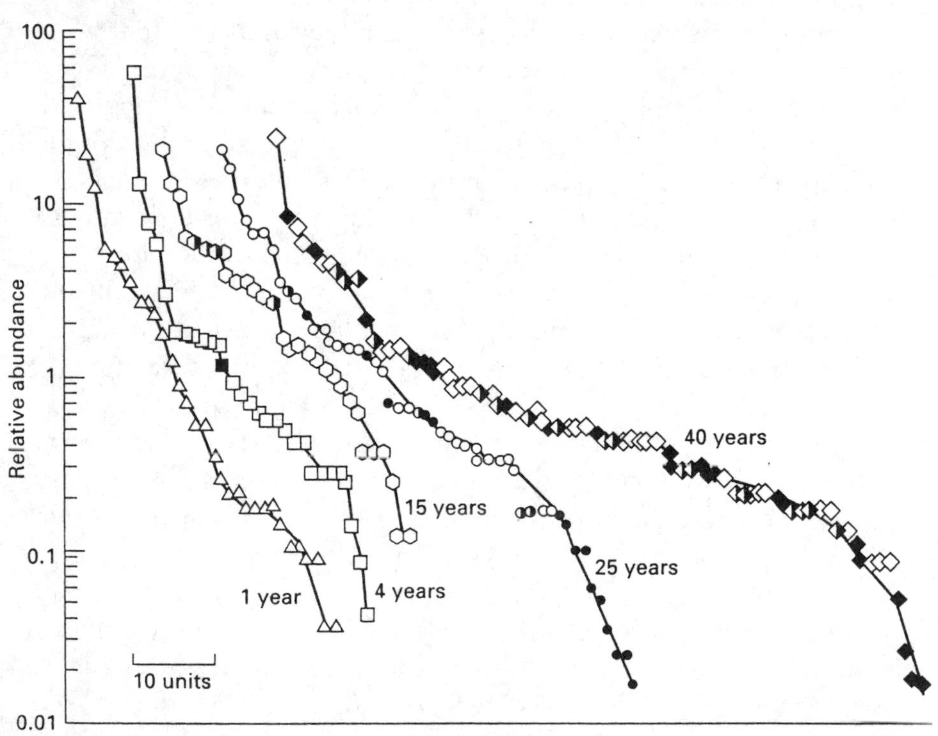

Figure 7.2 Sequence of changes in the rank-abundance vegetation pattern at different times. Here species are ordered from the most abundant to the less frequent, which defines the rank.

chains the system evolves (given a more or less stable environment) toward a species-rich ecology where biomass and productivity have increased at different rhythms. The end of this process features long trophic chains with a high degree of homeostasis. Many rare species are present and a few common ones. But the interesting point is that the final distributions are rather universal and robust.

In a classic study on a rocky intertidal community,[5] Robert Paine showed the effects of the removal of a given "keystone" species (i.e., a species that regulates the abundance of other species). After this removal, the community underwent an extensive reorganization before reaching a new state (Figure 7.3). Yet, this state is statistically quite similar to the original one: although the final species composition was different, the species distribution followed the same law as in the original community. In other words, a new community developed in which the dominance relations between species had changed but the *collective* properties were the same.

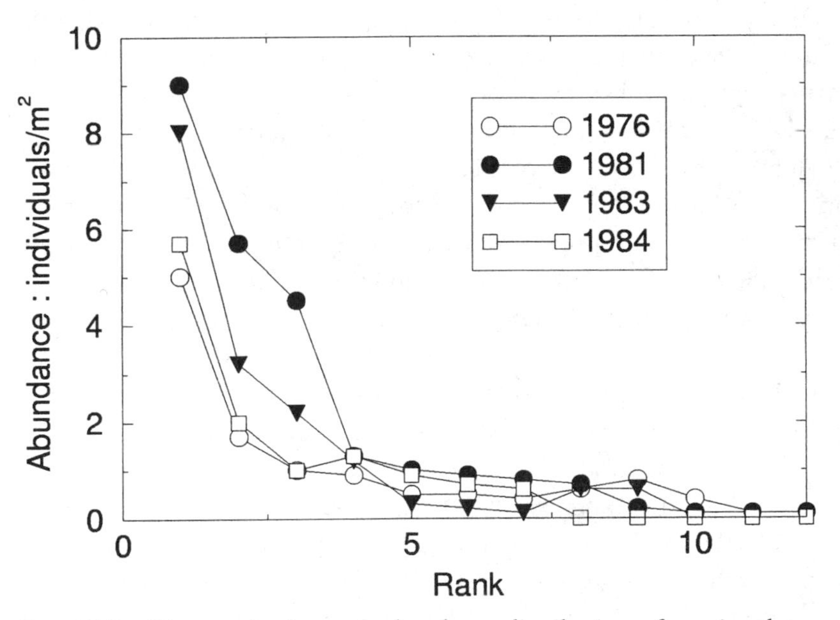

Figure 7.3 Changes in the rank-abundance distribution of species abundances on rocky shore before and after the near extinction of a top predator (see text). The different years are indicated. After a transient change in the distribution, a final state is reached, very close to the original one (data from Brown, 1994).

These and other observations indicate that understanding real ecologies requires a scale of observation far beyond the single-species level, which is simply uninformative about community dynamics. Since real ecosystems are interconnected entities displaying emergent phenomena, uncovering the universal laws underlying ecosystem organization will also uncover the origins and maintenance of biodiversity. Only then can we make rational decisions about the management of natural resources.

Emergence and Spatial Ecology

By returning to the two-species competition problem we can start to illustrate the consequences of spatial constraints on ecological dynamics. Instead of considering our species as confined to a test tube, where no relevant space is available, let us assume that our populations can grow and move on a two-dimensional lattice. We need a simple way of modeling these populations.

One of the most successful approaches to spatial dynamics in nonlinear systems was developed by the Japanese physicist Kunihiko Kaneko, who invented and extensively explored so-called coupled map lattices[7] and whose studies on cell differentiation we discussed in Chapter 3. A coupled map lattice is a discrete spatial grid of "sites," or "patches," which can be occupied by a population, indicated by $X_n(i, j)$ and $Y_n(i, j)$ for two competitors. Here n indicates the time step (generation time); i and j represent location coordinates on a two-dimensional grid. The growth and competition equations are introduced by means of parameters, but now an additional ingredient needs to be added: spatial dispersal between nearest patches (Box 1). It seems that little has changed in relation to the previous description of competitive interactions, but actually things turn out to be very different.[8]

If we run the model in the coexistence regime (where competition rates are low), nothing new happens: both species coexist at all points on the lattice. But something radically different occurs in the exclusion regime (where competition is strong enough). In some places, one of the species wins, but in others, the second species dominates. A spatial structure emerges (Figure 7.4) in which each species looks like the negative of the other. The conclusion is twofold. First, the apparently trivial introduction of space leads to a nontrivial result: the violation of the principle of competitive exclusion, which operates on a local scale

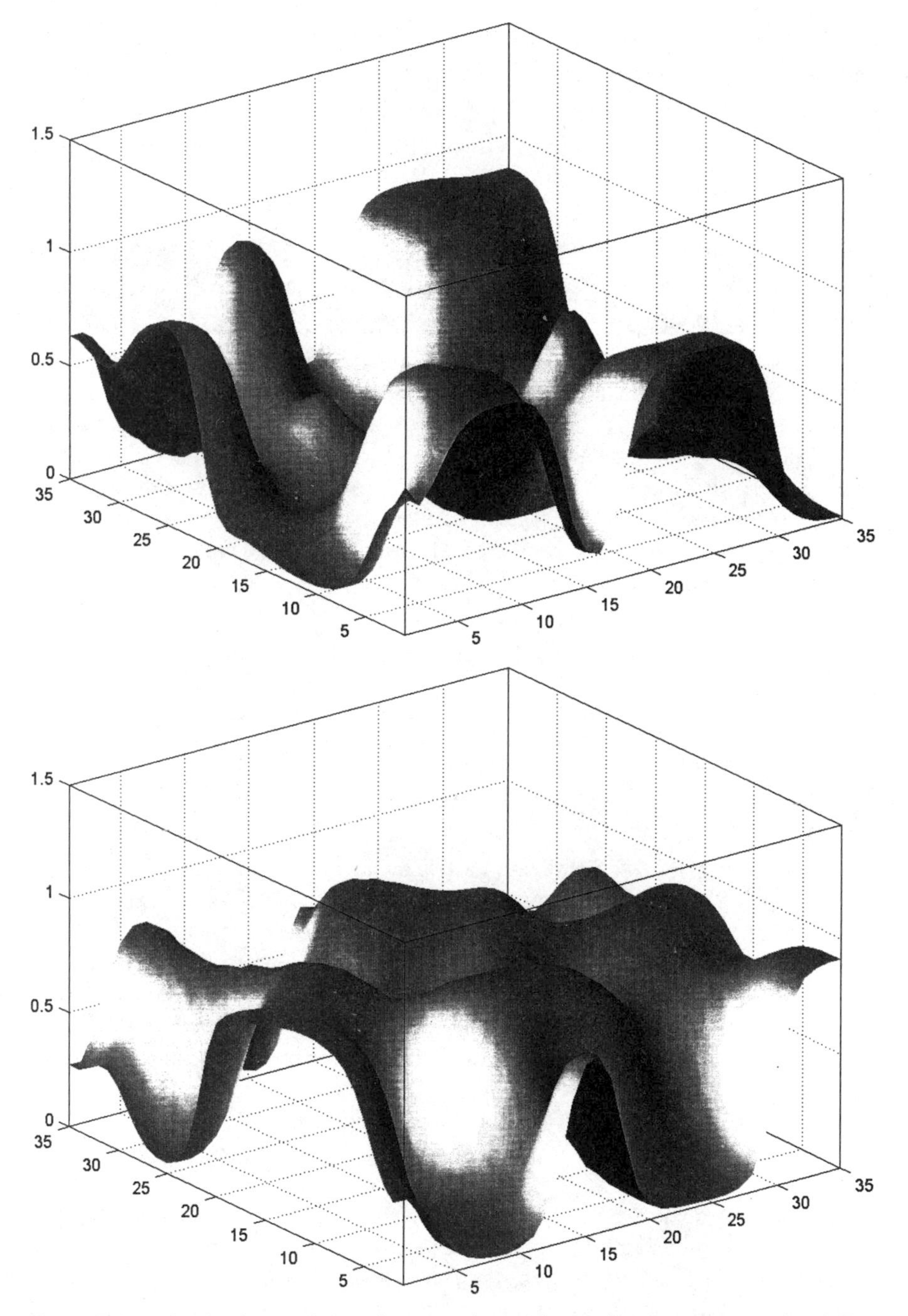

Figure 7.4 The final spatial distribution of two identical competitors under diffusion on a regular lattice. The two species abundances are shown (a and b). We can see that those areas where the first is abundant have very low (or zero) populations of the second, and vice versa.

but not on a global one. And second (and this will be further explored below), there is a critical spatial dimension for the system to be able to sustain a two-species community. Once habitat loss exceeds this critical value, the exclusion principle will operate, and one of the species will become extinct.

Coupled Map Lattices: Two-Species Competition

A coupled map lattice (CML) is a simple description of a dynamical system evolving on a given space with coupling between (here four) nearest sites on a discrete lattice. The discretization of space allows a simulation of this type of system in a very simple way. Two species are considered, whose populations are denoted by $X(i, j)$ and $Y(i, j)$ for each lattice point (i, j). For the two-species competition model,[8,9] assuming that both species have the same growth rates, and compete and move in the same way, the discrete equations describing the CML model are

$$X_{n+1}(i, j) = (1 - D)X'_n(i, j) + \frac{D}{4} \sum_{\text{nearest}} X'_n(a, b),$$

where X' is given by

$$X_n(i, j) = X'_n(i, j) \exp(\mu(1 - X'_n(i, j) - \beta Y'_n(i, j))$$

(similar expressions are obtained for the evolution of $Y_n(i, j)$ by exchanging X and Y in the previous equations). Here D is the diffusion rate (how fast individuals randomly move to nearest positions), μ is the growth rate, and β is the competition rate, which measures the interference between the two species in exploiting local resources. In other words, we first allow interactions to occur, and afterwards, dispersal toward nearest patches takes place. This is the standard formulation of CML dynamics as introduced by Kaneko. Here equation (2) introduces the specific form of competitive interactions. We can see that for $\beta = 0$ and $D = 0$ the model reduces to a set of single-population equations that are closely related to the logistic map (Chapter 1).

For zero dispersal ($D = 0$) it can be shown that the two species can coexist, provided that $\beta < 1$. Otherwise, one of the two species (the one with the largest initial population) will win. If the CML model is run from an initially random distribution of both species, well-defined spatial structures emerge for $\beta > 1$ (Figure 7.4). Here a 40×40 lattice has been used, and we can see that those points where one of the species dominates have a low population value of the second species, and vice versa (Figure 7.4a,b).

By changing the competition strength β (but fixing diffusion D and growth rate μ) it is easy to show that a phase transition occurs at the critical competition rate $\beta_c = 1$ (Figure 7.5). If we define the sum

$$\Omega = \sum_{i,j} |X_n(i,j) - Y_n(i,j)|$$

over all the lattice points and average over many steps, it is not difficult to see that $\Omega = 0$ for the coexistence regime, where $X = Y$ everywhere. But once spatial structure emerges, the local populations will be typically different, and thus the total sum will be nonzero. For a large enough lattice, coexistence will allways occur at the global scale.

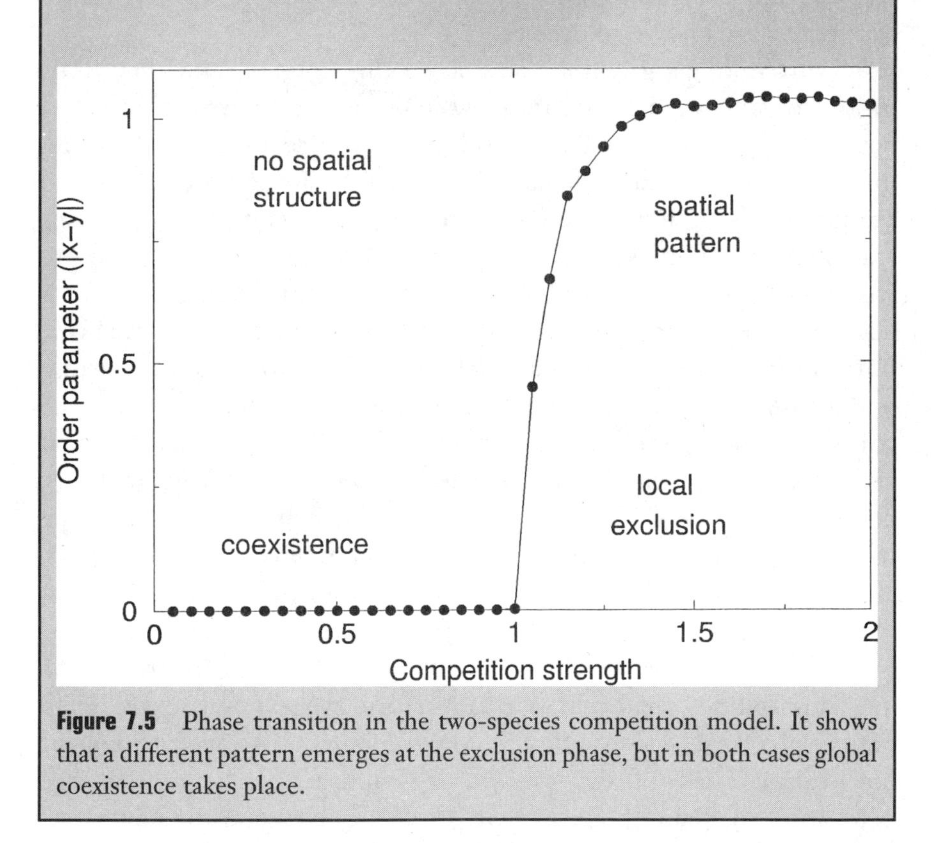

Figure 7.5 Phase transition in the two-species competition model. It shows that a different pattern emerges at the exclusion phase, but in both cases global coexistence takes place.

This result is not an exception but is rather the rule in the spatial dynamics of population models: once space is introduced into the system's description, new properties and phenomena arise. Let us now consider the spatial dynamics of a different system, close to the common predator–prey models of classical ecology. Our model will be inspired by an insect host–parasitoid community. Here the host acts as the prey, and the parasitoid (usually a solitary wasp species) is a particular type of predator that lays its eggs inside larvae of many insects.[10] As these eggs develop into wasp larvae, they devour the living (but usually paralyzed) host. Parasitoids are very common and have been thoroughly studied under laboratory conditions. These test-tube systems are largely unstable: a very common outcome is either the death of parasitoids by starvation (when prey decline so much that they cannot sustain their predators) or the collapse of the whole system. But clearly, parasitoids and their hosts survive in the wild without extinction, although local extinctions are known to occur.

Spatial models give the solution to this enigma, and they do it in a surprising way. Again, these systems can be treated in terms of coupled maps. The first CML models of ecological interactions were introduced in the early 1990s by Ricard Solé, Jordi Bascompte, and Joaquim Valls[11,12] for discrete predator–prey and competition models and by Michael Hassell, Robert May, and Hugh Comins for host–parasitoid systems.[13] Let us start with a two-dimensional domain as described by the previous CML approach. At the beginning, we place hosts and parasitoids at random locations and with random starting populations. We then let the system evolve according to the host–parasitoid interactions together with their diffusion in space (again among the four neighboring patches). After $t = 100$ generations, no clear structure is formed (Figure 7.6a; we show the spatial distribution of the host). But at $t = 250$ some waves start to appear, and by $t = 1000$ (Figure 7.6c) we can see well-defined spiral waves. The characteristic length of these self-organized patterns goes far beyond the local scale on which the rules of interaction operate. The local instabilities that ruin our laboratory experiments translate in nature into large-scale spatiotemporal structures. In spite of frequent local extinctions, globally the system is able to persist through the generation of spatial heterogeneity. And if once again the characteristic length scales of these self-organized structures fall below the critical size, the ecosystem can collapse. Such critical scales play a very important role in the dynamics of species response to habitat destruction.

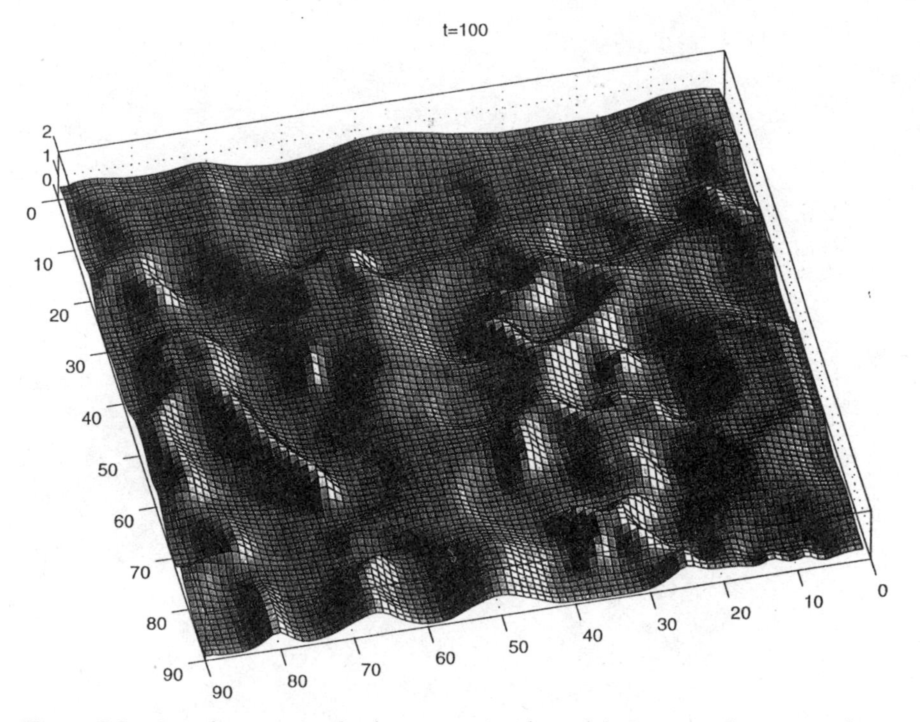

Figure 7.6 Spiral waves in the host-parasitoid model. Starting from a random initial condition, the system evolves (from top to bottom) towards a state characterized by the presence of spiral wave patterns.

Habitat Fragmentation and Extinction Thresholds

In many parts of the world endangered species survive in habitats that have been partially destroyed by human action. This habitat fragmentation is one of the most challenging consequences of human-induced activities. It has been estimated that 43% of terrestrial ecosystems are under human exploitation. Understanding the effects of habitat destruction, a hot area of research, requires the integration of field and theoretical approaches. New theoretical models have been developed in the last decade by several groups, particularly Ilkka Hanski's team in Finland,[14] and their relevance has gone far beyond theory. Some models have been able to predict the effects of habitat loss on the survival of rare species, and these predictions have been used as guidelines for future habitat management.

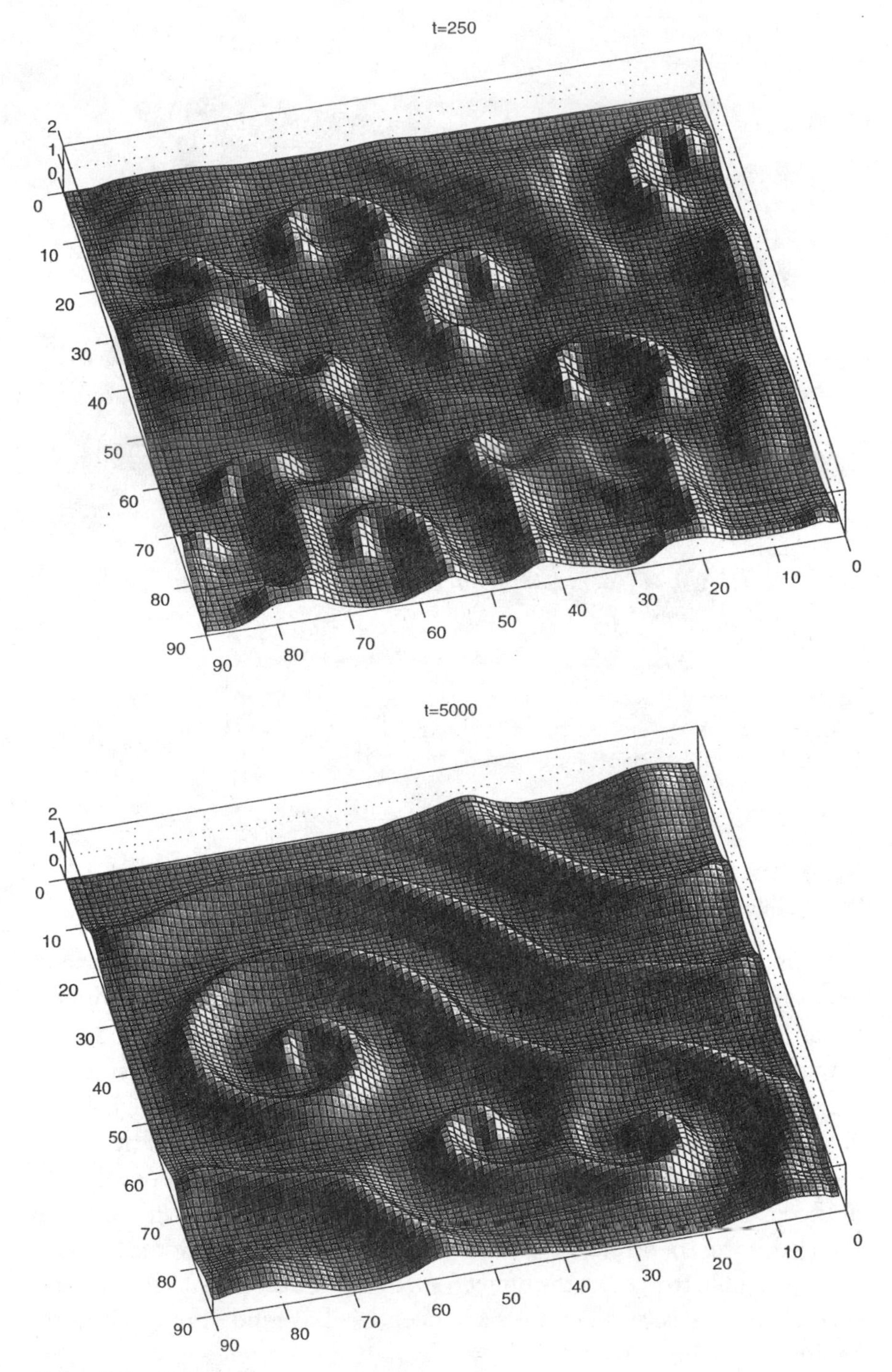

Figure 7.6 *Continued*

There are different ways of approaching the problem of species survival in fragmented habitats. Here we follow a common approximation based on a metapopulation dynamics point of view. A metapopulation is just an ensemble of local populations coupled through dispersal (i.e., individuals can move from one patch to some others). Following our previous examples, our ecology will be a two-dimensional grid of patches with three posible states: destroyed, meaning patches where no population is allowed to survive; empty, meaning those that are not occupied but available, and occupied.

A simple model describes the basic dynamics of metapopulations when patches are locally connected.[15] Here each site is connected to the four nearest patches. Habitat loss will occur at random. As with the problem of percolation on a two-dimensional lattice (see Chapter 2) a number of the lattice sites are randomly selected and destroyed. The metapopulation dynamics follow three basic rules:

—Destroyed sites cannot be occupied.
—If a site is occupied at a given step, there is some probability e of extinction (becoming empty) in the next step.
—A nonoccupied site has some probability c of being colonized from one of the occupied nearest neighbors.

These rules provide an extremely simple but biologically reasonable description of the spatially extended metapopulation. Starting from a given initial occupation and some given fraction of habitat loss, we would like to know how the whole system will evolve. A first thought suggests that as habitat destruction D is increased, the total number of occupied sites (and thus the global population) will continuously decrease, but small numbers will survive. In other words, linear habitat destruction leads to linear population decay.

This is far from true. To understand the outcome of the simulation models, let us first look at the emergent properties arising from the fragmentation mechanism. We saw in Chapter two that there is, in fact, a critical threshold (the percolation threshold) at which some new phenomena arise in a discontinuous fashion. Since percolation has to do with nearest-neighbor connections between sites on a lattice and our species will move only to nearest patches, our metapopulation should react in a nonlinear way close to criticality.

To see how the percolation threshold operates, we can measure the size of the largest *connected* domain in the lattice (L^{*07}) and see how this size varies as habitat loss increases (in the following, black and white squares will indicate destroyed and nondestroyed sites, respectively). This is measured in terms of the largest number of nondestroyed patches that are connected through at least one neighbor. Figure 7.7a shows the result of this calculation from our computer experiment. Above the threshold $D_c \approx 0.41$ (i.e., $p_c = 0.59$, percolation for the nondestroyed sites) there is a linear decay of L^* with respect to D. In this domain (Figure 7.7b) there is a very large patch connecting most lattice points. But once we reach D_c (Figure 7.7c) the largest patch size suddenly decays to very small values: the habitat becomes fragmented into many smaller areas. As destruction is further increased, only very small domains remain (Figure 7.7c).

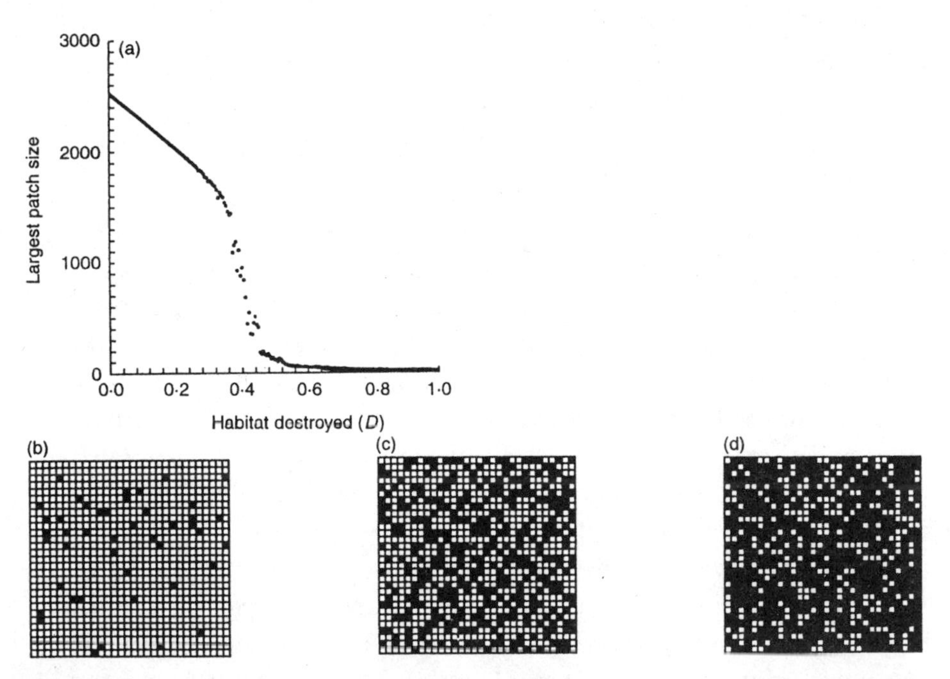

Figure 7.7 Pattern of patch size decay when the habitat is destroyed. Three examples of the lattice fragmentation are shown at the bottom for a subcritical (a), critical (b), and supercritical (c) patch destruction. The largest patch size first decays slowly (and linearly) but sharply decreases at the percolation threshold.

Since the survival of species might be strongly influenced by available habitat and thus be space-dependent, it seems clear that metapopulation dynamics will be strongly influenced by percolation processes. Now we can go on to see the effects of fragmentation on metapopulation dynamics. Starting from a given initial condition with a constant number of occupied sites, we plot the average number of occupied patches as D is increased. An example of this simulation is shown in Figure 7.8. There we can see that the fraction of occupied sites (i.e., the amount of habitat occupied by our species) at first decreases linearly, but it rapidly decays as the threshold D_c is approached. In spite of the theoretically available habitat, the local dispersal rules plus the presence of the threshold determine the outcome. One may argue, however, that if a species is very successful in colonization and perhaps not very prone to extinction, the percolation-like phenomenon might be less important.

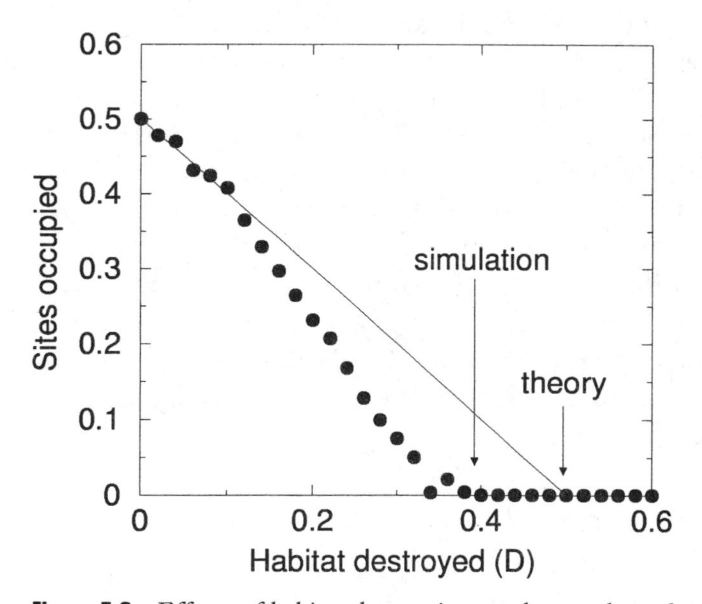

Figure 7.8 Effects of habitat destruction on the number of occupied sites for different levels of destruction. Theoretical arguments (where space is not explicitly included) predict an extinction threshold at a destruction level D=0.5. However, the simulation model, where space is explicitly taken into account (and where percolation takes place) show that extinction takes place much earlier, due to the percolation process.

To extend the previous ideas and explore their consequences in a more general set of situations, let us take into account the fact that real ecologies are not formed by a single species but by many. These species interact, and we can assume (as before) that each species has its own rates of colonization (of empty sites) and local extinction. But now we must include interactions, in the form of a probability that a given species invades a patch already occupied by another species. This probability can be assigned at random just as with colonization and extinction rates.* Now we can repeat the previous simulations using many species. And in order to be more realistic, we can introduce immigration: empty sites can be occupied by species outside the lattice (this is what happens in real ecologies; individuals from outside invade some of the empty areas). An example of how the system behaves when the habitat is destroyed close to the percolation threshold is shown in Figure 7.9. The first half of the plot shows the time fluctuations of (many) species, which sometimes become almost extinct and then recover. These time fluctuations are characteristic of real ecologies. But at $t = 300$ the habitat loss reaches about 40%. The ecosystem collapses: a few species species (usually those that are good empty-site colonizers and are favored by the decay of some good competitors) persist at high levels. We still see a few rare species, but most are immigrants from outside areas that are unable to persist. This scenario becomes more and more unpredictable (in terms of who survives) as the number of starting species increases. As we will discuss in the next section, diversity, stability, and the predictability of species survival are closely related.

Stability and Complexity

The web-like structure of ecosystems has been a matter of analysis and speculation for decades. Ecological networks can be very intricate. An

*Recent studies by David Tilman and coworkers have shown that the results of habitat destruction on complex communities formed by many competitors can result in rather counterintuitive outcomes. One of them is that extinction takes place in order from the best to the poorest competitors as habitat destruction increases. The model also shows that many extinctions can occur generations after fragmentation. They represent a debt, that is, a future ecological cost of current habitat destruction.

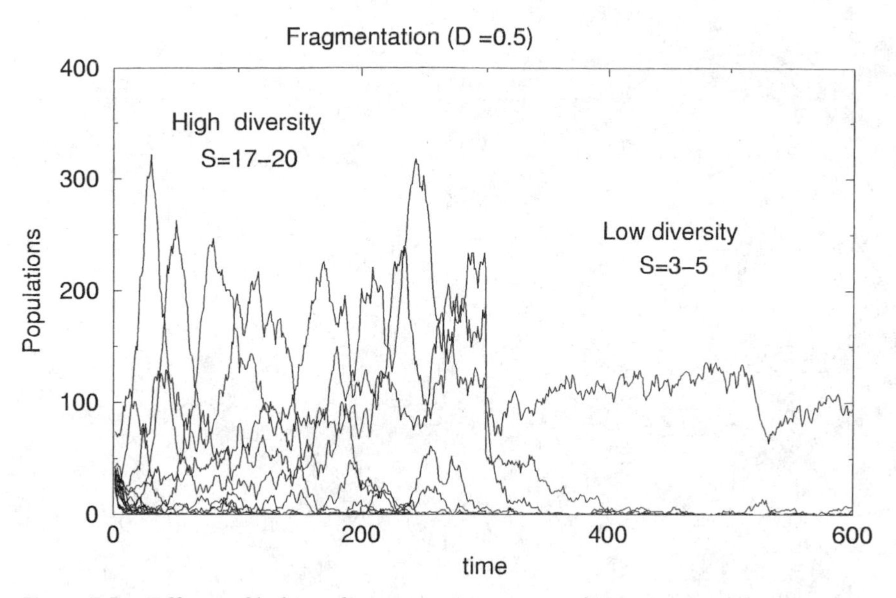

Figure 7.9 Effects of habitat fragmentation on a multispecies model ecosystem: here the percolation of destroyed habitat (which takes place at T = 500) plus the interactions among species lead to a collapse in species diversity.

example is the food web for the northwest Atlantic, shown in Figure 7.10. Observation of these networks reveals that there are some general rules concerning overall organization. For instance, the size of the food web formed by a set of S species present at a given moment and the connectivity of this web (measured as the number of observed links divided by the maximum possible number S^2) each seem to be constrained to some range of values. Here "links" means any direct relationship between two species: they are linked if a predator–prey, competitive, cooperative, or any other type of interaction exists. Other properties, such as the fraction of basal and top species or the food chain length, are also universal.

Why are there universal properties in ecosystems? Why do ecosystems formed by a large number of species appear to be weakly connected, while low-species ecosystems tend to be highly connected?* The answer to these questions requires the consideration of whole ecologies. But it seems difficult to explore these questions without taking into

*Here we mean the abundance of links, not their strength.

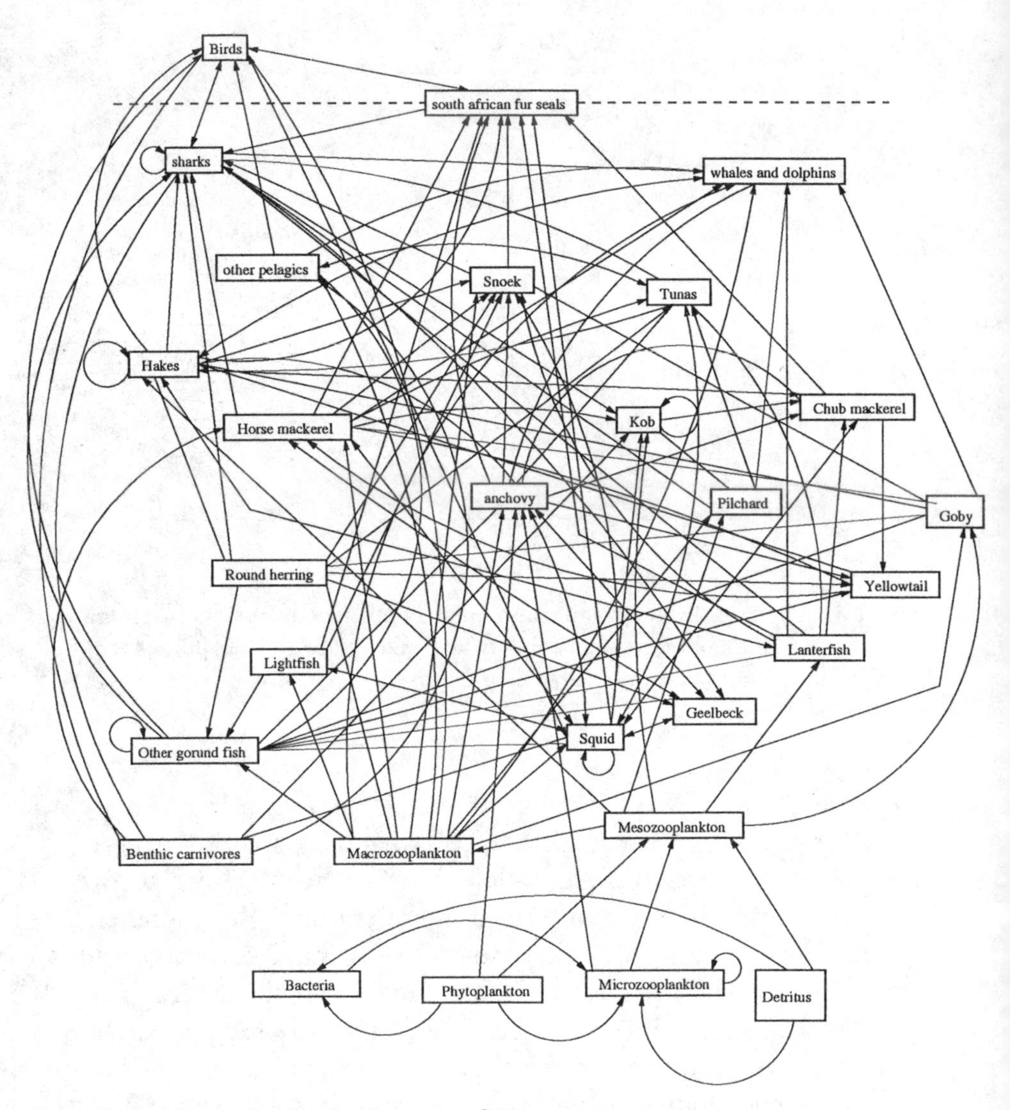

Figure 7.10 A complex food web pattern from a marine ecosystem.

account the vast (and unavailable) set of parameters characterizing each specics and their specific interactions. Or we can try to answer these questions from the perspective of complex systems theory: let us formulate an oversimplified model of a trophic web and see what types of emergent phenomena it produces. If some features of the model correspond to our observations of real ecosystems, perhaps some universal mechanisms are at work.

In 1974, Robert May presented a simple theoretical approach to the complexity–stability problem based on a randomly coupled ecosystem.[16] The model ecosystem was formed by S species, and each pair of species (I, J) had a probability C of being connected. Connections could be positive or negative, depending on their effects on the growth rate of the interacting species. The stability of this random community can be determined through general mathematical arguments. A community is said to be stable if populations of each species reach stationary values and remain there. May's result, based on earlier studies on the stability of cybernetic systems by Mark Gardner and Ross Ashby, shows that for any value of C there is a well-defined upper limit to the number of species that describes a stable community. Here diversity and connectivity are linked through a simple inverse relation: $S \approx k/C$, where k is some constant. This relation separates two domains, one stable ($S < k/C$) and the other unstable ($S > k/C$). The results are shown in Figure 7.11. Here a large number of different coupled networks have been generated for each connectivity, using two different

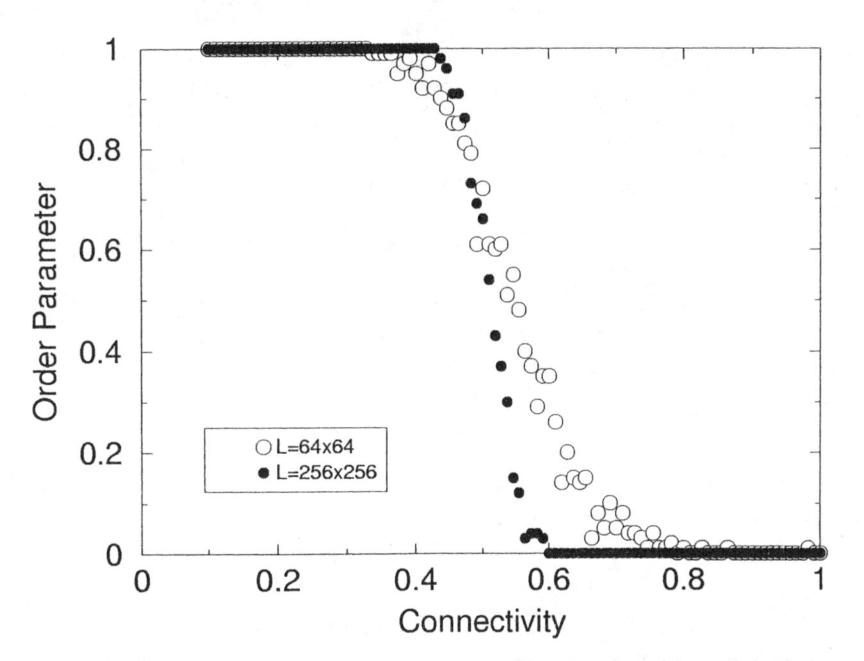

Figure 7.11 Phase transition in the May random ecological model. Here the probability of stability defines the order parameter for the transition. Stability is rapidly lost at a given threshold value of the connectivity, as predicted.

numbers of species (here $S = 64$ and $S = 256$, indicated as white and black circles, respectively). For each ecosystem we check whether it is stable and thus compute the probability of stability for each value of C. We can see that close to a given threshold C^* (as predicted by May) stability decays sharply to zero. No community is stable for larger C values.

Phase Transitions in Ecological Networks

Let us consider a discrete multispecies model, as defined by an S-dimensional dynamical system: $\mathbf{X}_{t+1} = \mathbf{B}\,\mathbf{X}_t$, where $\mathbf{B}$ is an $S \times S$ matrix whose entries are generated at random and $\mathbf{X}_t$ indicates the vector of species abundances, i.e., $\mathbf{X}_t = (X_{1,t}, X_{2,t}, \ldots, X_{S,t})$. Specifically, the matrix $\mathbf{B}$ has connectivity $C \in (0, 1)$, i.e., the total number of nonzero connections is CS^2 (the rest of the links are absent). The weight of the connections follows a distribution with zero mean and variance α^2. It is assumed that all populations are close to their equilibrium point. In this context, the entries B_{ij} of the $S \times S$ matrix symbolize the effect of small departures from the stability value of population j on species i (May, 1974).

Let us denote the probability of stability of the system (in the linear stability analysis sense) as $P(S, \alpha, C)$. Then, the May–Wigner theorem establishes that

$$\text{if } \alpha^2 SC < 1, \text{ then } P(S, \alpha, C) \to 1 \text{ as } S \to \infty,$$

$$\text{if } \alpha^2 SC > 1, \text{ then } P(S, \alpha, C) \to 0 \text{ as } S \to \infty.$$

In terms of statistical physics, this system (for the limit $S \to \infty$) exhibits a phase transition at $\alpha^2 SC = 1$ (Figure 7.11).

The assumption of random wiring is, of course, oversimplified. Yet other, more realistic, models show a similar dependence between the number of species S and their connectivity C. Field studies also generally confirm the relation, although there are some well-defined deviations[17] (see Box 3). There is, however, an additional point of particular relevance for our understanding of natural communities. The previous May's criterion implies that there are two basic domains of stability (Figure 7.12a). Stable communities should be observed at different locations in the lower area of the stability diagram (as shown in Figure 7.12b, where a number of random pairs (S, C) have been generated). This is what we could expect if *global stability* were a natural constraint on real ecologies. But when we look at those ecologies we find that ecosystems are typically distributed *along* the critical boundary separating stable from unstable dynamics (under May's criterion; see Figure 7.12c).

Once again, a critical boundary separating two well-defined regimes (or phases) seems to play a role in determining the properties of complex systems. And in spite of the puzzling appearance of this plot, it has a simple and nontrivial explanation in terms of self-organization far from equilibrium. Let us imagine a real ecology, which can be a small island or a large plot in the middle of the rainforest. Two main forces are at work generating and limiting diversity: interactions between different species

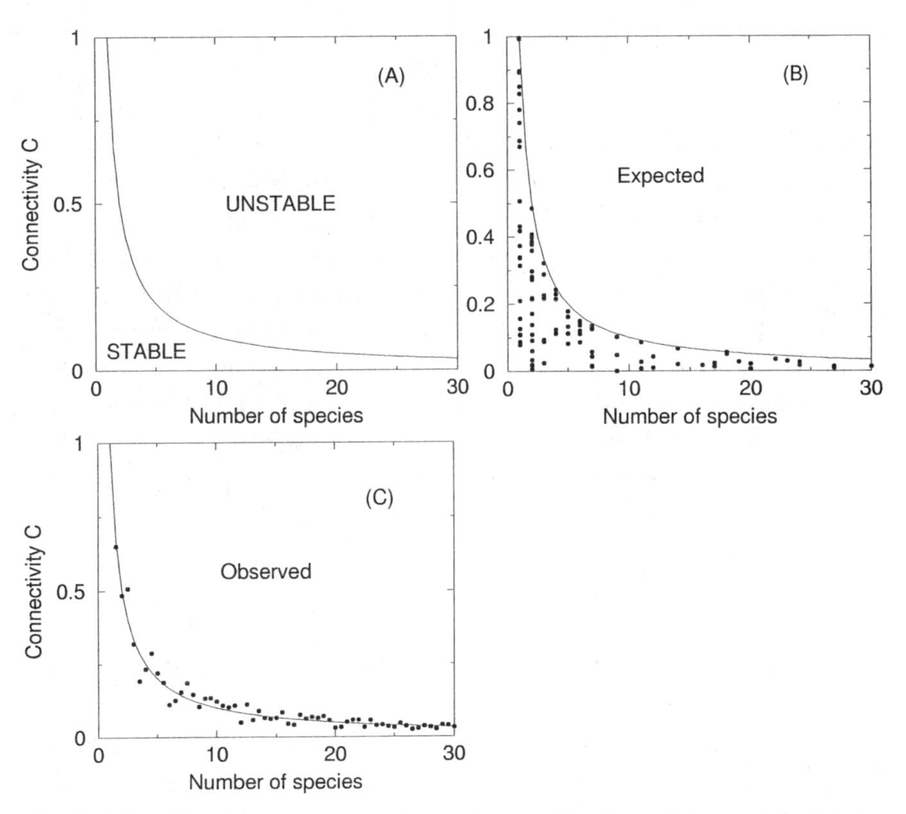

Figure 7.12 (A) the two phases obtained from May's random model; (B) the expected occupation of the stable phase, as predicted if global stability operates as an organizing principle in real ecologies; (C) observed pattern: ecosystems tend to appear close to the critical boundary.

(predation or competition, say) and the constant immigration of foreign species. The first force would tend to reduce diversity (although, as we saw, spatial effects can counteract competitive interactions). But the second strongly resembles a driving force in Bak's self-organized criticality. Immigration pushes the ecosystem toward higher and higher diversity. As suggested by Simon Levin, this requires a minimum immigration rate that is common to terrestrial ecosystems where criticality would be likely to occur.[18] But this driving force cannot push diversity arbitrarily high. Interactions start to operate, and the system reaches some upper limit for the number of species. The limits to diversity would then be imposed *by instability*. Once the critical point (at some location on the critical curve $S = k/C$) has been reached, the ecosystem can change its species composition, but universal regularities still operate.

This suggests that ecosystems operate close to critical points. There are several lines of evidence for this claim. An interesting observation on the presence of thresholds to stability concerning the pattern of extinction dynamics in the Hawaiian Islands was presented by the ecologists Tim Keitt and Pablo Marquet.[19] This pattern involves an extensive record of avian species introductions and extinctions. The Keitt and Marquet study is interesting both because it deals with phase transition phenomena and because of its relevance to conservation. Few native species persist today in the lowland habitats, where humans have introduced new, imported, species. There is a record of the number of introduced species (up to 69) as well as the consequent extinctions. From 1850 to 1920, the system gradually accumulated species with no extinctions. But then after the next successful species introductions, there were numerous extinction events. The pattern in which it happened is made dramatically clear in Figure 7.13a. There we see that extinctions start suddenly around a given number of successful introductions. The distribution of extinction sizes was found to scale as a power law, consistent with a system close to instability (Figure 7.13b).

Other studies also give support to this picture. Evidence for criticality also appears from a remarkable study on the presence of chaotic dynamics from ecological time series made by ecologists Stephen Ellner and Peter Turchin.[20] By analyzing a large number of data sets involving the time fluctuations of populations in both the laboratory and the field, Ellner and Turchin found that their systems appeared concentrated near the transition from stable to chaotic dynamics (see Chapter one). Why? These authors conclude that their results seem to support the

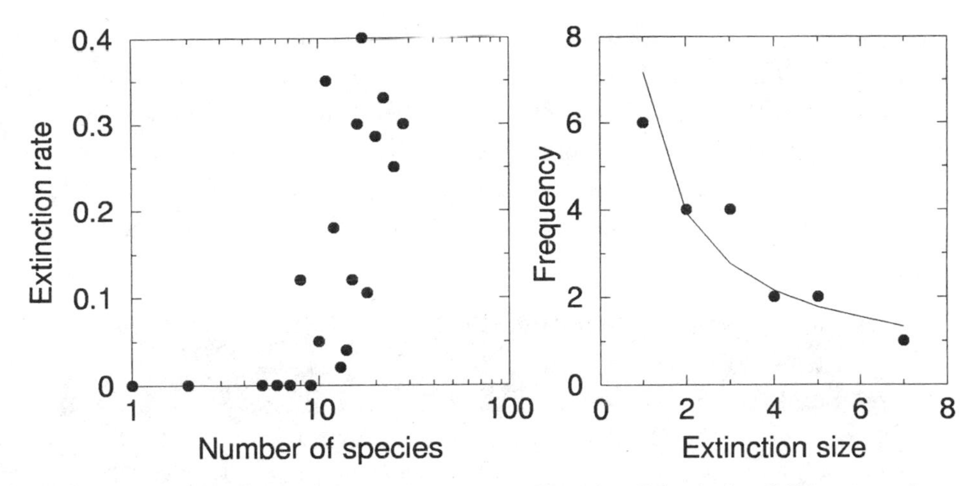

Figure 7.13 Left: observed extinction rates in Hawaiian Islands for different numbers of introduced species. Apparently, as a given threshold is reached, many species become extinct. Right: distribution of extinction events for this data set (data from Keitt and Marquet, 1997).

idea that "interacting populations should coevolve to dynamics at the edge of chaos."

In previous chapters we mentioned that critical points usually involve the presence of power laws. Are there power laws in ecology? In fact, long-tailed distributions are well known from classic studies. The species–rank distributions that we discussed earlier are one example. But instead of this ordering by rank, a more natural way of plotting diversity is by means of species–abundance distributions. Here we plot how many species are represented by one, two, etc., individuals. Available data show that in some cases the number of species $S(I)$ represented by I individuals follows a power law $S(I) \approx I^{-\alpha}$. Here $\alpha \approx 1$. In many other cases, however, the observed distribution is lognormal, characterized by a maximum at small species numbers and by a long tail with a power law decay for large numbers of individuals. But in both cases we see that rarity is the common trait, and species with very large numbers of individuals are rare.* But if a dynamical pattern were involved in these systems such that abundance of species was the result

*In this sense, it is important not to look to power laws as the single trait of complex systems poised close to critical points. Power laws will be restricted to special situations where the driving force is very slow and the system response (the avalanches) rapid.

of some type of "avalanches," most of them small but a few very large, maybe we could formulate a complete picture of how complex ecologies are built and change. In fact, a recent model of multispecies dynamics has shown that most of the previous observations can be described by a general theoretical framework (see Box 3) strongly supporting the idea that complex ecologies might be self-organized close to instability points.

A Stochastic Model for Complex Ecologies

We have seen that complex ecological communities display (a) time fluctuations with highly changing population values; (b) skewed distributions in species abundances (including power laws and log-normal distributions); (c) a scaling relation between species numbers and connectivity, which is found to be (in real communities) $S \propto C^{-1+e}$; (d) edge-of-chaos Lyapunov exponents, i.e., the dynamics appear to be close to the transition from order to chaos; and (e) thresholds to species numbers beyond which the ecosystem becomes unstable and generates extinctions.

Most models of many-species ecologies are deterministic, based on the original formulation by Lotka and Volterra. And they are unable to link statistical data (such as the species–abundance distributions) with other information, like the species–connectivity relation. In a recent study Ricard Solé, David Alonso, and Alan McKane presented a simple model that has been shown to provide a good link between those different aspects.

The model involves a finite number of individuals N and a pool formed by S species. We can imagine this system as a part of a forest, and the pool is the maximum number of available species in the forest. Two main ingredients are introduced: interactions among species and immigration. The model is defined as follows. An $S \times S$ matrix is defined and fixed. This matrix has connectivity C, and each nonzero matrix element Ω_{ij} gives the strength of the interaction of species i on species j. At each time step, with probability $1 - \mu$ we choose two individuals and see to which species they belong, say i and j. Then we look at the matrix values Ω_{ij} and Ω_{ji}. If they are equal, nothing happens (no interaction is allowed). But if $\Omega_{ij} > \Omega_{ji}$ the individual from species j is replaced by an individual from species i. Finally, with probability μ a totally random new individual from the pool is introduced. In this sense, μ represents the immigration rate.

This model displays a very rich behavior. Figure 7.14a shows an example of the fluctuations of population size $n(t)$ of a given species (inset). We can see that the number of individuals of this particular species shows wide fluctuations in time, with some episodes of extinction. The number of species is also shown (main figure), and we can see that a stable value is reached. In spite of continuous turnover of different species, the average number of species present in the ecology is $\langle S \rangle \approx 125$, in spite of the fact that the total possible number of species in the pool is $S = 300$. If we manipulate the system by reducing or increasing the number of species, it returns to the same average value: the number of species tends to increase because of immigration, but internal constraints imposed by the interactions among species impose a limit to the maximum diversity. Once the critical number is reached, no more species will be accepted (except through replacement by an invader). Once this state is reached, we can see that a general relation $S \approx C^{-1+\varepsilon}$ is at work, just as in real ecologies.

Also, the statistics of this model are the correct ones: for different immigration and connectivity values we obtain a range of distributions from log-normal to power laws (Figure 7.14b). When immigration is small and C is large enough, the system displays all the expected features of a self-organized critical state. Since fluctuations occur at the critical

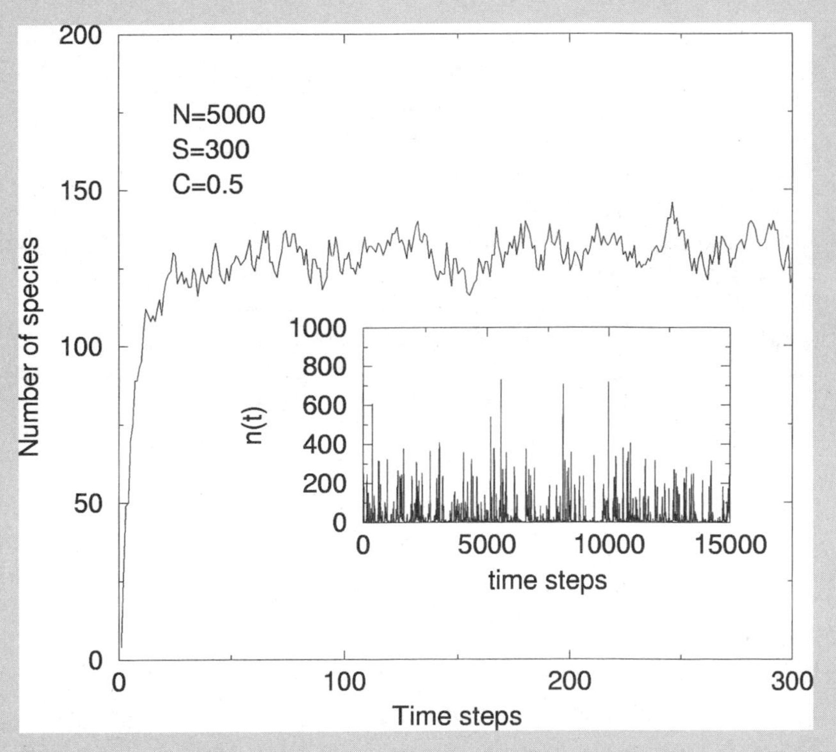

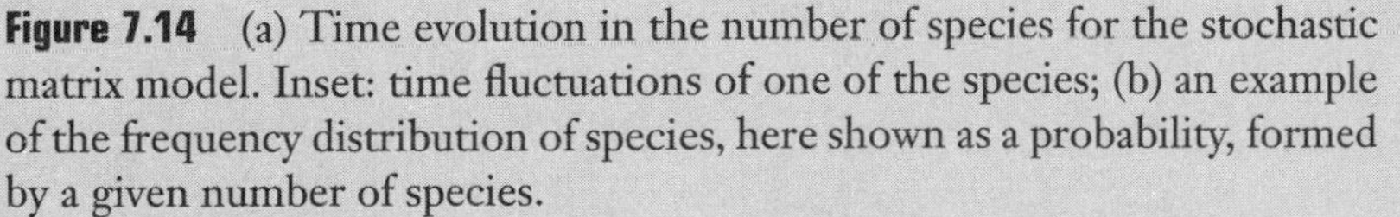

Figure 7.14 (a) Time evolution in the number of species for the stochastic matrix model. Inset: time fluctuations of one of the species; (b) an example of the frequency distribution of species, here shown as a probability, formed by a given number of species.

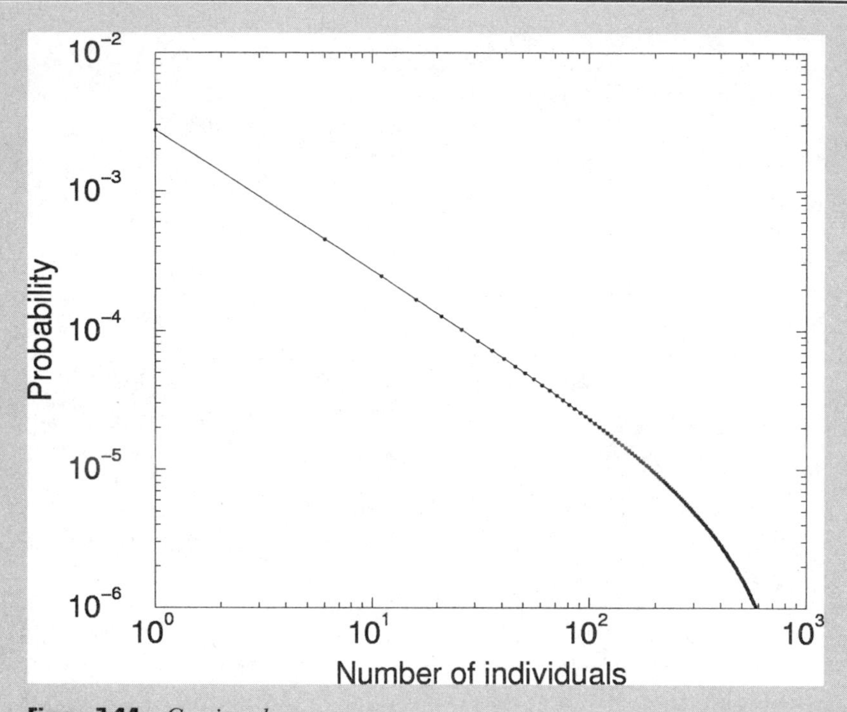

Figure 7.14 *Continued*

boundary of stability, this would explain the results obtained from the analysis of chaos in ecological time series. Also, this model suggests that Bak's idea should be generalized for those systems not displaying a well-defined scale separation between the driving force (immigration) and the avalanches (population changes due to interactions). Since the requirement that the driving must be very slow (see Chapter 2) is a very strong one, one is unlikely to observe power laws in some systems that are nonetheless organized close to instability points. Perhaps a new theory of "self-organized instability" will be required.

We have noted that complexity is often accompanied by the presence of fractal structures. Fractal patterns are common to self-organized critical systems, and if ecologies are linked to instability points, we might ask whether self-similar patterns are detectable in the field. For an answer, there is nothing like visiting the rainforest.

Fractal Rainforests

One of the best-known forest fragments in the world is a 50-hectare plot in the Barro Colorado Island rainforest of Panama. Barro Colorado is a protected rainforest reserve placed in the middle of the Panama canal. Somebody described the people working there as rather like those scientists who led the Manhattan project during the Second World War. They are not trying to build an atomic bomb but are working hard to disentangle some of the most fascinating secrets of our biosphere. The Barro Colorado plot is a well-known part of this project. Tens of thousands of trees in this plot have been identified by species, tagged, and measured over several years. A data set of unprecedented detail has been compiled and is being currently analyzed by many researchers.

One particularly interesting property of the Barro Colorado plot is the presence of fractal features. In Figure 7.15a the reader will detect some familiar structures. This plot shows the distribution of canopy gaps in places where the canopy trees are below 20 meters in height. The 50 hectare plot has been divided into squares of 5 × 5 meters, and black squares indicate low canopy sites. Canopy gaps are the result of tree death and fall. Trees often die standing and do not produce an important change in their neighbors. But sometimes they fall, and then some of their neighbors fall, too, like dominoes. From time to time, large gaps are created by the fall of many trees. When this happens, large black clusters appear in the map. This map was shown to be a fractal object, and power laws are also at work: the frequency distribution of canopy gap sizes gives a very good fit to a power law (Figure 7.16). Other measures, such as the distribution of tree sizes, also show some key features of criticality.

Various researchers have developed models of forest dynamics. By introducing a set of simple rules involving tree birth, death, and growth (one model, introduced by Ricard Solé and Susanna Manrubia, was called "the Forest Game") it is not hard to reproduce most of the quantitative features of the Barro Colorado plot.[22] Although many choices of tree interaction are possible, it was observed that a reasonable choice of tree screening by nearest neighbors and simple rules of gap formation lead to the same basic results in all cases. Our previous work revealed that the fractal behavior displayed by the Barro Colorado data was consistent with a critical state. Recently, a team of Japanese researchers has shown that they could reproduce rainforest gap dynamics using a stochastic Ising model[23] that approximate remarkably well the real gap-size distribution (Box 4). Their results give further support to the conjecture that rainforest dynamics take place close to a critical state. If this is true, then power laws in gap distributions and species abundances would be two faces of the same coin. As species interact

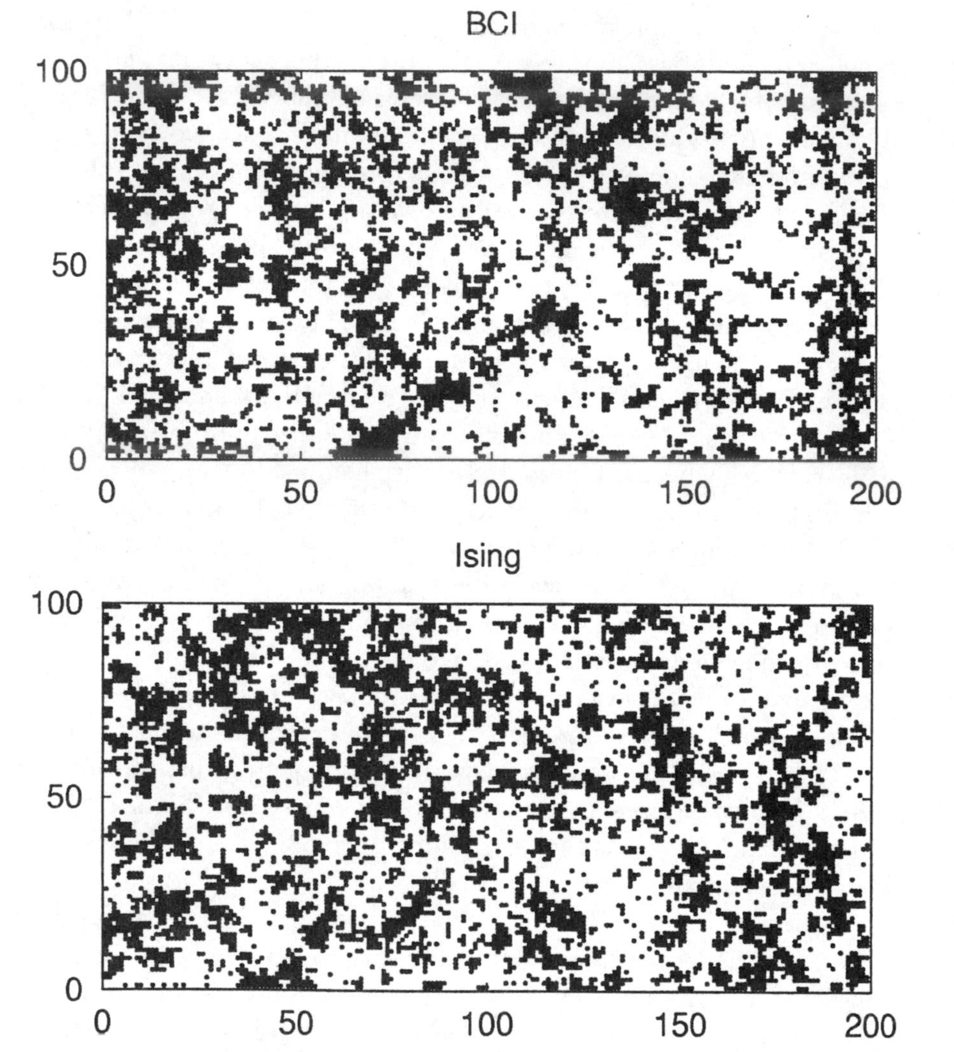

Figure 7.15 (A) spatial distribution of low-canopy points for the Barro Colorado island plot. Black squares indicate locations where the canopy is lower than 20 meters; (B) spatial pattern of spins obtained from an Ising model close to criticality; here the black squares correspond to "down" spins (figure kindly provided by Makoto Katori and Toshiro Takamatsu).

in nonlinear ways, avalanches can occur in the system, leading to the long-tail distributions characteristic of mature ecosystems. And if the system is at the edge of instability, then the species fluctuations would be unpredictable. In this scheme, rarity is a natural consequence of criticality. A small number of species have large populations because of their special species-level characteristics but also because of chance

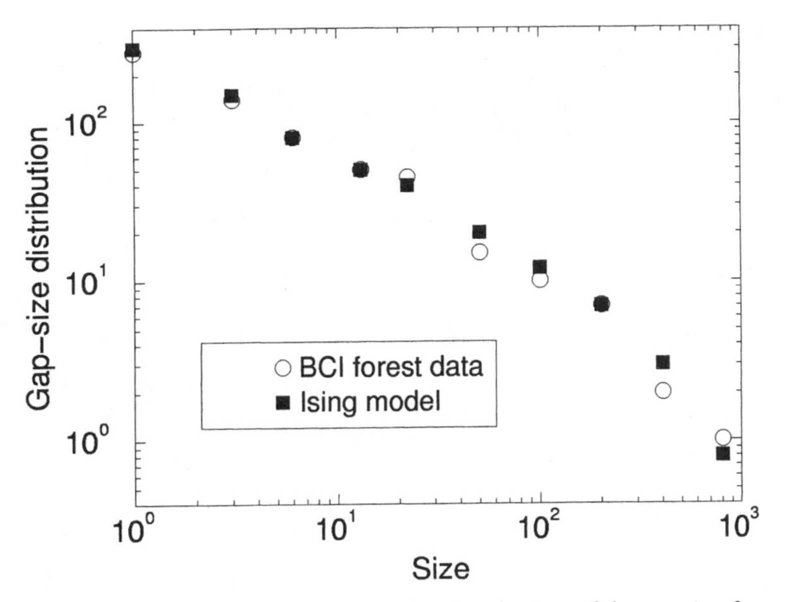

Figure 7.16 Comparison between the distribution of down spins for the two-dimensional Ising model close to criticality and the gap distribution from real data.

events and their interactions with other species. They have successfully "propagated" through the system. But most are rare: internal dynamics lead to a state in which constantly introduced species either propagate weakly (into a system already dominated by interactions and some global diversity–connectivity constraints) or simply fail.

An Ising Model for the Rainforest

In a recent paper, Makoto Katori and coworkers have shown that the forest canopy dynamics of the Barro Colorado plot can, in fact, be regarded as the result of an Ising-like model[23] (see Chapter 2). Using a simple set of rules directly inspired by field data analysis, Katori et al. reproduced some of the most interesting properties of the Barro Colorado 50 hectare plot. In Figure 7.17 the basic rules are described. Here two types of states are allowed (as with the Ising model): nongap points (here white squares) and gap points (black squares, corresponding to canopy below 20 meters high).

Field data provided the estimation of the transition rates (indicated in the figure below the arrows for each of the six possible cases). Here d is the spontaneous creation of a canopy gap (here $d \approx 0.024$), and δ_k

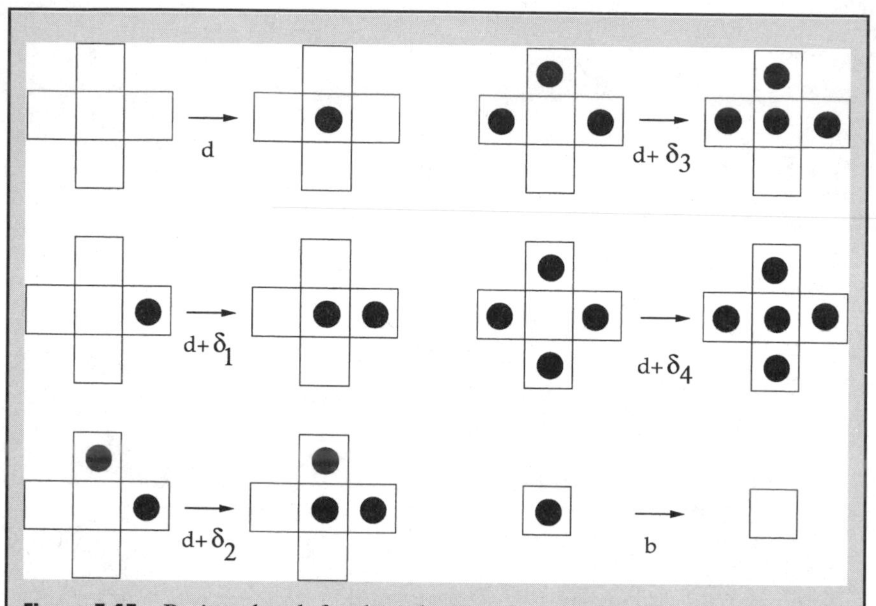

Figure 7.17 Basic rules defined in the Katori's et al. model. Gaps sites are denoted by black circles.

indicates the risk of falling trees due to the presence of neighboring gaps. Katori et al. use the simple (and sensible) approximation $\delta_k = k\delta$, where $k = 1, \ldots, 4$ and $\delta = 0.276$. This choice is based in the observation that the presence of nearest canopy gaps strongly increases the fall of neighboring trees either by direct physical effects or because of the strong modifications of local microclimate. The model is completed by introducing a transition rate from a canopy gap point to a noncanopy point due to tree growth ($b = 0.177$).

The dynamics of this model give quite good results. But Katori et al. go a step further and show that in fact, this model is equivalent to an Ising model close to criticality. They estimated the appropiate temperature for the Ising model configuration consistent with Barro Colorado data. An example of the similarity between the Ising model and its rainforest counterpart is shown in Figure 7.15b, to be compared with Figure 7.15a (from Katori et al., 1998). The quantitative agreement between both plots can be shown by means of fractal measures or by plotting the size distribution of canopy gaps. The later is shown in Figure 7.16, where both field data and simulation are shown. These results give strong support to the early conjecture that rainforest dynamics take place close to critical states.[22]

These ideas are not restricted to rainforests. Many ecosystems around the globe have revealed similar features, although local climate fluctuations and the variability of different physical factors seem to mask their effects. Let us finish this chapter by calling attention to an important aspect of ecosystems related to their intrinsic complexity and emergent properties: the problem of the prevalence of indirect interactions in ecology.

In the chapter on evolution we will raise an important point that is also relevant to ecology: there is a very large number of indirect relations among species.[24] These interactions often run counter to the most obvious expectations. For instance, adding predators of a given species into an ecosystem should surely have a negative effect on that predator's prey populations. But extensive calculations made by Peter Yodzis, among others, on real food webs show that in many cases adding predators can increase the number of prey. Indirect pathways very frequently dominate direct pathways in determining the long-term outcomes of perturbations. Other recent studies also support the view of a patterned, network-dependent response of ecosystems. As an example, by analyzing the links between different species in a laboratory soil ecology in terms of energy transfer, Peter de Ruiter and coworkers have shown that small links can have a large impact on stability, whereas interactions involving an important flow of energy can be almost irrelevant to stability.[25] All these results should give pause to those who still think that ecologies can be managed as linear entities composed of separated modules. The emerging understanding of ecosystems as interacting entities with collective properties arising from their network structure should prevent us from extrapolating single-species properties into many-species realities. Arguments over genetic engineering, for example, consider only the direct effects of the released organisms into some given target species. But ecological complexity does not follow this rule, and the impact of some apparently innocuous perturbation should be explored by means of the tools of complexity. Otherwise, we will be ignoring the salient fact of our biosphere, that it is nonlinear, unexpected, and often unpredictable.

EIGHT

Life on the Edge of Catastrophe

And, indeed, as he listened to the cries of joy rising from the town, Rieux remembered that such joy is always imperiled. He knew what those jubilant crowds did not know but could have learned from books: that the plague bacillus never dies or disappears for good.

—Albert Camus, *The Plague*

Molecular Parasites

Over the last decades we have become more aware of the importance and dangers of emergent diseases. New and old viruses and bacteria come and go, sometimes as phantoms (Figure 8.1). Oliver Sacks, in *Awakenings*,[1] tells the story of one of the most recent but forgotten pandemics: the sleeping-sickness disease. It was apparently first noticed in the winter of 1916–1917 in Vienna and other European cities, and it spread rapidly over the next three years until its distribution was worldwide. The illness itself was rather puzzling: "manifestations of the sleeping-sickness were so varied that no two patients ever presented exactly the same picture." It was as if "a thousand new diseases had suddenly broken loose." Later, systematic work identified the cause of this outbreak as a virus. The pandemic finally retreated, like the plague in Albert Camus's novel. "It took or ravaged the lives of nearly five million people before it disappeared, as mysteriously and suddenly as it arrived, in 1927."

An inevitable outcome of life seems to be the emergence of this type of parasitic entity. Among them, RNA viruses appear to be the most important class of intracellular parasites. Viruses stand right at the border between the living and nonliving, and they seem at first sight to be good candidates for understanding the earlier stages of biological evolution. Parasites contribute to the maintenance and generation of diversity in natural ecosystems and are a constant threat to human populations. Sometimes, they develop strange strategies: the delta hepatitis virus, for example, is unable to complete its replication cycle unless another virus, the B hepatitis virus, is present. The first uses the molecular machinery of the second; it is a parasite of a parasite!

Many RNA viruses are, in fact, late arrivals, and increasing evidence suggests that viruses are often the result of the escape of fragments of functional genetic programs from their hosts. Further mutation and recombination then provide the molecular mechanisms of optimization for both structure and function. This is consistent with what Manfred Eigen calls "the intimate inside knowledge of their hosts that viruses appear to possess." Indeed, these entities are the inevitable result of genome complexity, where the molecular machinery contains everything for making copies and cutting segments of genetic material and providing a safe envelope for escaping fragments. RNA viruses, so called because they use RNA instead of DNA as genetic material, are responsible for a large number of human diseases, ranging from influenza, rabies, and polio to the so-called emerging diseases[2] caused by hantavirus, Ebola, or the HIV-1 virus (Figure 8.2).

These viruses are so successful because they are simple, small, and rapidly replicating, but especially because they possess enormous plasticity and adaptability to changing environments.[3] Their plasticity comes from their extremely heterogeneous population structure (a quasispecies; see below). One consequence of this adaptability is a set of uncommon and apparently bizarre genome replication strategies. This is particularly evident with retroviruses, of which the AIDS virus is the most famous. The retrovirus life cycle involves an enzyme called reverse transcriptase that makes it possible to synthesize DNA chains from RNA templates. Unlike the enzymes responsible for DNA replication in our cells, reverse transcriptase is highly prone to errors. Its error rate has been estimated at about 10^{-4} to 10^{-3} per base, which for an HIV virus means 1 to 10 errors for each replication cycle. Compared with

Figure 8.1 The plague, as displayed by an 1845 wood carving by Alfred Rethel.

the DNA replication, where the error rate is somewhere around 10^{-10}, the AIDS virus shows extraordinary genetic variation.[3]

For eukaryotic cells, variation is dangerous: replication errors can lead to death or cancer. A delicate balance between different genes needs to be preserved. These cells have evolved several ways to repair accidental changes.[4] If repair and conservation mechanisms are so important in cells, why are they absent from RNA viruses? How essential is such a randomness to their well-being? The reason for the enormous variability emerges from the interaction between viruses and their hosts, the cells of the immune system. Before we explore this

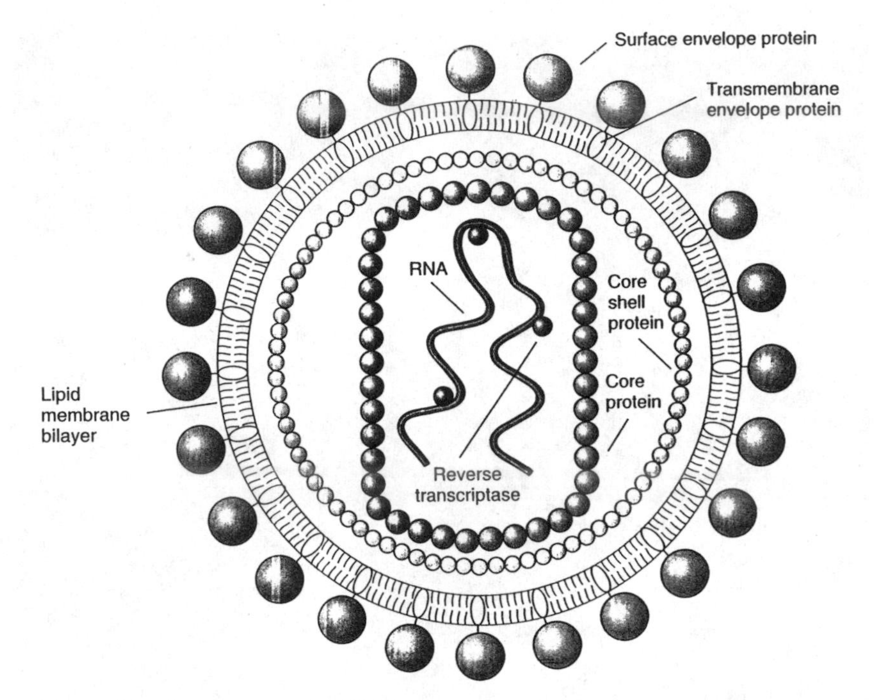

Figure 8.2 Structure of the HIV virus.

relationship, let us consider a very fruitful theoretical approximation to these molecular entities.

Quasispecies and the Error Catastrophe

A long time ago, when our planet was very young, life emerged in some simple form. How life started in our planet (and in the universe) is a matter of intense debate and has recently triggered the emergence of a new research field: the science of astrobiology. Even the specific composition of Earth's early atmosphere is not known. But we have come much closer to an answer since the biochemist Stanley Miller performed his classic experiments in the early 1950s. Miller recreated a primitive atmosphere made up of a mixture of gases heated with water and energized by electrical discharges. The outcome of this experiment was a soup of organic molecules. Most of them were very simple and short, but some were common to known life, including several amino acids, the building blocks of proteins. It was only a small step toward understanding how life arose, but it opened the way to the scientific quest for the origins of life.

Nobody can be sure, after four billion years, what kinds of early life forms appeared, but we can find clues in our knowledge of contemporary organisms and molecular biology. Molecules able to self-replicate from the simple raw materials available in the early earth would have some of the properties characteristic of living mater. The only molecules able to make copies of themselves, and so serve as templates of their own replication, in our current biosphere are polynucleotides, small fragments of nucleotide subunits like those that constitute our genomes. This is what occurs with DNA and RNA in biological systems: the chain of basic molecular units can be used by enzymes as a template. In this way a given sequence can be replicated, although replication, as a physical process, is not free from error.

The most popular theory of the origin of life is the RNA world hypothesis. Before 1982, it was assumed that there was a well-defined division of labor between nucleic acids and proteins. Nucleic acids were the molecules involved with information storage, and proteins (RNA or DNA polymerases) were responsible for their replication. But in 1982 a very important discovery was made: some RNA molecules could also act as catalysts (i.e., molecules that accelerate reactions). These RNA sequences, known as ribozymes, are able to act as catalysts of their own replication under suitable conditions. Although their catalytic power is limited, this finding had tremendous consequences for our understanding of the origins of life: first, the ribozymes found in modern organisms are likely to be the remnants of primitive life forms; and second, they provide evidence for the RNA world scenario, where RNA chains could act both as information carriers and replicators.

Researchers sought other prebiotic structures, such as cell membranes and metabolic cycles, with some encouraging results. Under particular conditions, they could obtain microspheres with a characteristic closed membrane (Figure 8.3), and in some cases metabolic reactions were shown to occur inside these protocells. These simple structures even showed processes similar to cell division (Figure 8.3, upper picture). Electron micrographs also revealed a lipid bilayer characteristic of biological membranes (Figure 8.3, lower picture).

These experiments led to more questions: was life an extremely rare and unlikely event? Are we the result of a highly improbable chain of events, or instead, as Stuart Kauffman claims, are we "at home in the universe"? We are still far away from the answers to these deep questions, but we have accumulated considerable knowledge about

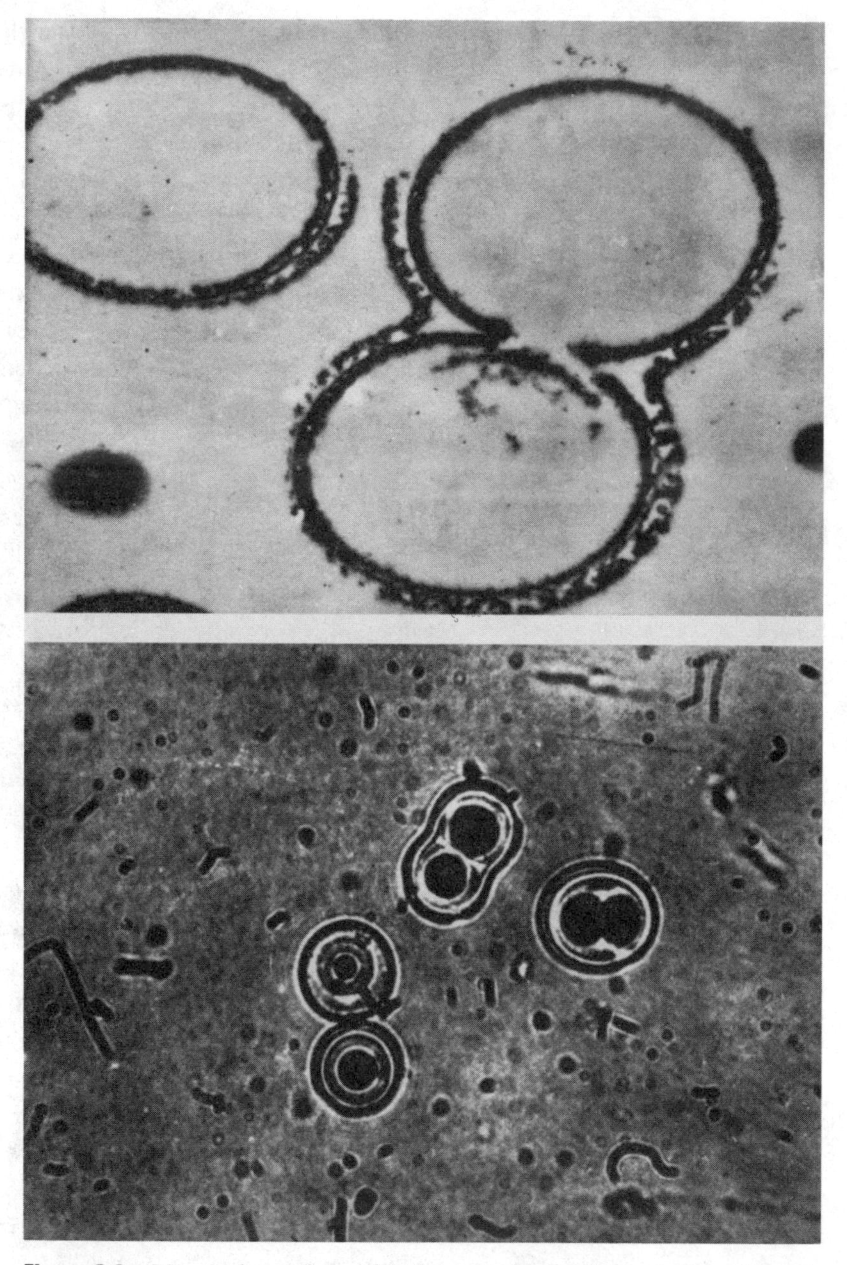

Figure 8.3 Microspheres formed in laboratory conditions (upper picture). These microspheres can display a characteristic closed membrane.

processes of early life. However, their intrinsic complexity and our need to understand the fundamental processes at work have generated a number of theoretical models that have played a key role in the study of the origins of life.

One of these models was introduced in 1977, by the theoretical chemists Manfred Eigen and Peter Schuster, who coined the term *quasispecies* to describe the molecular evolution of primitive sets of replicating molecules.[5] Their work involved approaching the origin of life in terms of selective processes under far-from-equilibrium conditions.

To begin with, let us start with some general definitions. First, we have to consider as our "molecules" sequences of uniform length, ν units each, with κ classes of units. For biological applications, $\kappa = 2$ and $\kappa = 4$ are the most common choices.* There are κ^ν different possible sequences. Even for small chains, κ^ν is enormously large, as shown in the following table:

Organism	ν	$n_s = \kappa^\nu$
$Q\beta$ virus	4200	10^{2529}
Ribosomal 5S mRNA	120	10^{72}
tRNA	76	10^{46}

These sequences define a discrete high-dimensional *sequence space*, which can be imagined in terms of a hypercube in many dimensions. For the one of the simplest cases, $\kappa = 2$ (binary strings) and $\nu = 4$, the binary space is shown in Figure 8.4. A given population can occupy one or several nodes of the hypercube. Here we also show different types of virus particles (here the shape of the virus corresponds to the vesicular stomatitis virus, VSV) with a particular number of traits characteristic of each string sequence.† The total number of possible sequences for any viral genome (4^ν) is a hyperastronomic number. As a consequence, a given quasispecies can fill only a tiny fraction of the available nodes. There is considerable experimental evidence that important constraints limit both the occupation and diffusion of the quasispecies through this space. For instance, the exploration of the sequence space is greately affected by the population size. Rare variants, which are never detected

*$\kappa = 4$ when the four nucleotides are explicitly considered, and $\kappa = 2$ if only purines and pyrimidines are distinguished.

†This is obviously an oversimplified picture, since any virus, even the smallest ones, will evolve on a high-dimensional hypercube.

when population size is limited, become dominant the population is greatly increased.*

Now let us assume that our sequences are able (in some way) to reproduce themselves. Let $q \epsilon (0,1)$ be the probability that a given unit is copied correctly. Assuming that all units are copied independently, the quantity $Q = q^{\nu}$ is the probability of copying the (complete) sequence correctly. In the following, we will assume that $q$is the same for all units.†

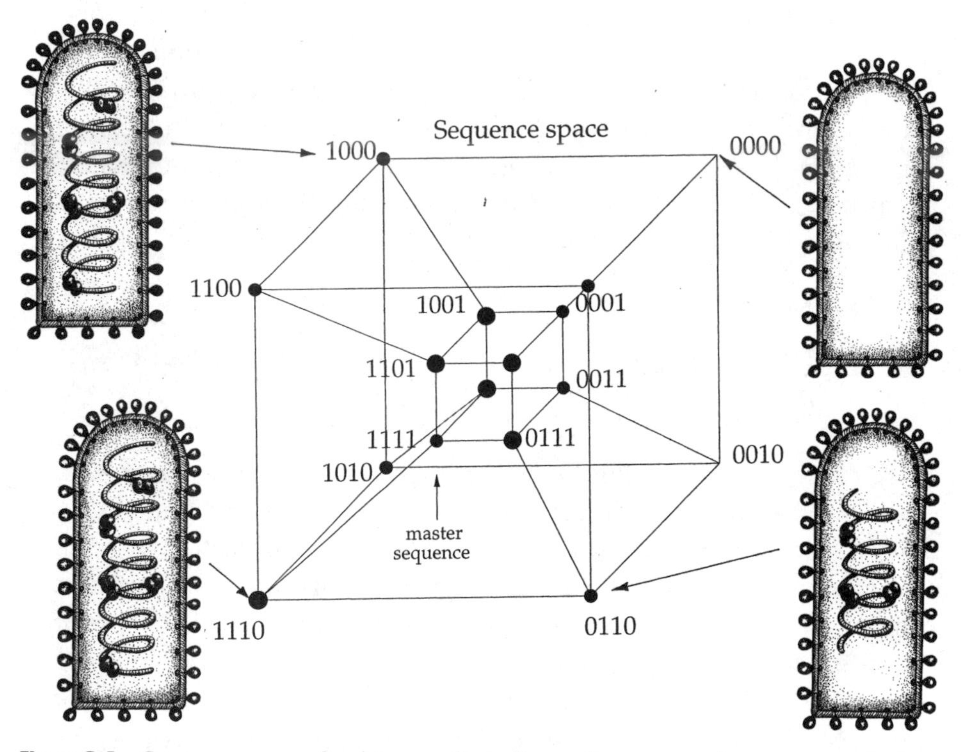

Figure 8.4 Sequence space: for this unrealistically small system, a given sequence is given by a four-bit string. The resulting hypercube is shown here. Each node is surrounded by four nearest mutant sequences that differ in a single mutation. We can imagine that each sequence has an associated phenotype, which would correspond to a virus with different characteristics.

---

*In this sense, although evolutionary changes in virus populations cannot be predicted with any certainty, the probability of new viral disease outbreaks must inevitably increase. AIDS is not the first "new" virus disease in humans, and it will not be the last.

†This approximation ignores the existence (shown by means of molecular studies) of the so called hot and cold spots, i.e., particular units where mutation is respectively higher or lower.

The replication of these molecules implies two basic reactions:

1. Error-free copying occurs when a molecule replicates by using available molecular building blocks and all units are correctly copied.
2. Mutation occurs, with some (small) probability per unit, when a molecule replicates but one or more units in the offspring string are incorrectly copied.

Additionally, we assume that molecules are spontaneously degraded with some small, constant probability. And finally, we assume that the total population size is limited to a fixed value. The last assumption is far from trivial, since a constant population constraint allows selective pressures to act: only those molecules with the fastest replication rates (i.e., fitness) will be able to survive.

The Eigen-Schuster Equations

The standard approach to quasispecies dynamics involves a set of molecules (sequences, strings, RNA virus genomes) that can replicate and mutate.[6] The basic equations are

$$\frac{dx_i}{dt} = (A_i Q_i - D_i - \Phi)x_i + \sum_{i \neq j}^{n} \Psi_{ij} x_j,$$

with $i = 1, 2, \ldots, n$ accounting for the population size of each string. The parameters A_i and Q_i are the replication rate and the quality factor, respectively; $Q_i \epsilon [0, 1]$ is a measure of the correctness of the replication process, and it is maximum ($Q_i = 1$) if no mutations occur; D_i stands for spontaneous degradation of molecules (assumed to be linear); and Ψ_{ij} are the mutation rates, which can lead to transitions $j \rightarrow i$. Finally, Φ is an outflow term that takes into account the removal of molecules from the system. If we introduce the constraint of constant population size, the previous equations become

$$\frac{dx_i}{dt} = (A_i Q_i - D_i - E)x_i + \sum_{i \neq j} \Psi_{ij} x_j,$$

where the mean value of the so-called excess productivity $E_i = A_i - Q_i$ is given by

$$E = \frac{\sum_i (A_i - D_i)x_i}{\sum_i x_i}.$$

Under the constant population constraint, a simple selection process takes place. When mutation rates are very small, the fastest-replicating sequence increases until it reaches the maximum population size. However, if mutation is present, no single-species population structure is found, but a quasispecies instead.

The previous assumptions can be mathematically formalized in terms of the Eigen–Schuster equations (Box 1), which provide an elegant approximation to the dynamics of replication of evolving molecular quasispecies. One of the most interesting results of Eigen and Schuster's study is that there is a critical limit to genome length that gets lower as the mutation rate is increased: this is the so-called error catastrophe.

We can easily model a set of replicating RNA molecules on our computer and see the error catastrophe in action. But instead of a complicated description involving many details, we can follow a much simpler approximation based on a bit string model. In spite of their simplicity, these models have successfully explained some experimental results.[7] Let us consider a fixed number N of molecules, where each "molecule" will be a string of bits (such as 100101010010) of a given length ν. The number of letters of our alphabet (the number of different basic units) can be increased to four to make the model more similar to the real genetic code, but the basic conclusions are not altered.

Let us consider a simple model (first introduced by Jörg Swetina and Peter Schuster) where the fitness of each sequence is given by the replication rate r_i. This rate will be 1.0 for the sequence 111 . . . 1111 (our master sequence here) and $r_i = 0.01$ for all the other strings. Two basic rules will be used, involving replication and mutation. At each time step we randomly choose a string, and with probability r_i we make a copy of it. Since the total number of strings (N) is fixed, each time we remove a string, it is replaced by the new one. The second rule is this: in the replacement step we can introduce mutations in the new string. Any bit can be modified (in two different ways, $1 \rightarrow 0$ or $0 \rightarrow 1$) with a probability μ.

Our intuition tells us that for small μ, the master sequence will dominate the population. Only occasionally will new strings with one or two zeros appear, but their populations will be very small. As the mutation rate increases, our intuition says, the frequency of the master sequence should slowly decrease: a quasispecies is formed as a cloud of mutants around the master, and as mutation grows, the cloud expands continuously.

The simulation experiment, however, shows us something very different (Figure 8.5). To quantify the behavior of the string model we can define a simple order parameter. For each mutation rate, we run our model many times and look at the final step in the simulation. At this step, we check whether or not there is *at least one* string with the master sequence. We calculate how many times a master sequence (in a population of $N = 100$ strings) is observed at the end of these experiments, and this gives us a curve describing the probability of finding this particular string as mutation increases. This is a rather weak

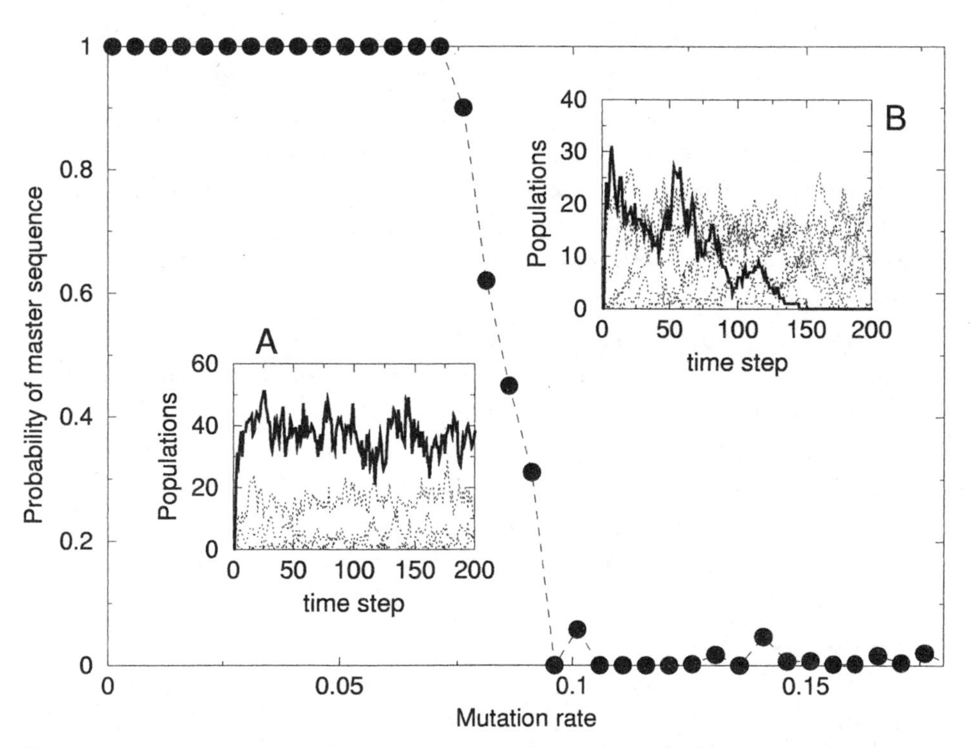

Figure 8.5 Phase transition in molecular evolution: as the mutation rate goes beyond the critical threshold, no master sequences can be found. The two insets show two examples of the population dynamics of strings for (a) a mutation rate below the error threshold and (b) a mutation rate slightly larger than the critical. Thick lines correspond to the master equation abundance and dashed lines indicate other strings.

requirement, and so we might expect a slow decrease in the curve. This is not what we find: there is a rather sharp decay at a critical mutation rate $\mu_c \approx 0.10$, after which almost no master sequences are found. In Figure 8.5 we also show (insets) two examples of the dynamics followed by the bit string model for two different mutation rates. In (a) the mutation rate is $\mu = 0.05 < \mu_c$, and the master sequence population is indicated by a thick line. The other populations are shown as dashed lines. We can see that the master sequence dominates, although it involves only about forty of the strings on average. In (b) we can see the effects of the high mutability: the number of master sequences decays to zero.

This quite puzzling result can be understood once we remind ourselves that the strings "diffuse" in a vastly large sequence space. For small , a small cloud of mutant strings will be observed around the master sequence (Figure 8.6a). The cloud will expand to nearest nodes in the hypercube as the mutation rate increases. But once the error threshold is reached, the master sequence mutates so fast that all offspring strings will be mutants (Figure 8.6b). Beyond the error threshold, these mutant strings will also mutate and diffuse over the whole sequence space (Figure 8.6c). Since the sequence space is so vast, finding the master sequence again is extremely unlikely (the probability is $P_{\text{master}} = N/2^{\nu}$, which for our example means $P_{\text{master}} = 100/2^{15} = 0.00305$). Biologically, this has great consequences for evolving populations of RNA viruses, but also for any kind of biological organization involving replication and mutation. Below the critical mutation rate, selection of the fastest replicating sequences operates. But beyond criticality the system becomes random, and no selective forces can operate (since the newly created sequences will be random). For a virus, to cross the error threshold is to disappear.

The behavior of this system as we tune μ reveals the presence of a phase transition (a simple derivation is given in Box 2). In fact, the behavior of this system is equivalent to a two-dimensional Ising model using an adequate temperature parameter defined in terms of the mutation rate (Figure 8.6, right column). We can even think of the mutation rate as a type of temperature, since it leads to a loss of correlations and to disorder. For low mutation rates, the system maintains ordered behavior by allowing information to be maintained (like the ordered phase in the Ising model, where interactions among nearest units dominate). At the high-mutation regime, correlations are totally lost, and the system drifts through sequence space at random

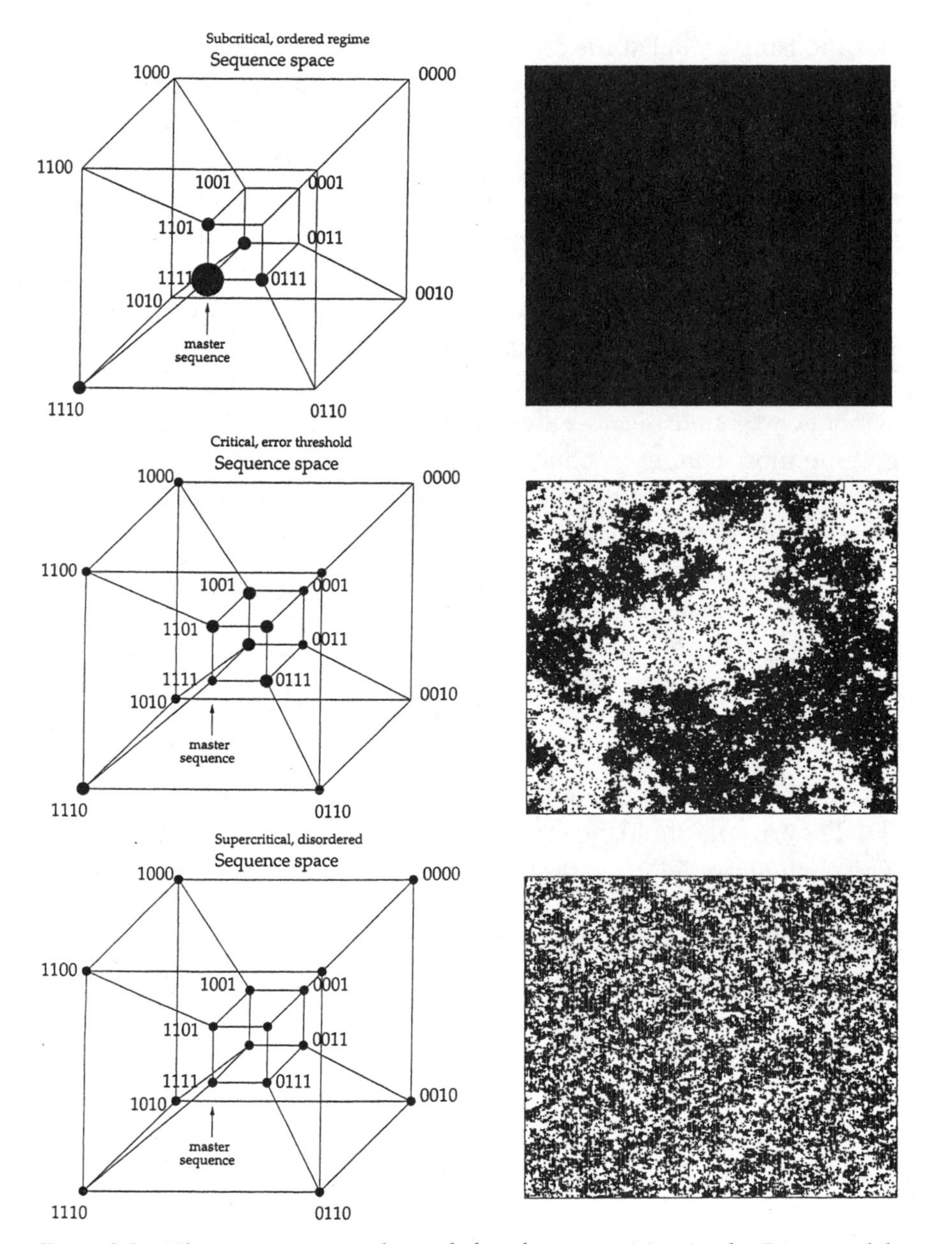

Figure 8.6 The error catastrophe and the phase transition in the Ising model. The right column shows three extreme states for the Ising model (see Chapter 2) corresponding to ordered, critical and disordered states. The Swetina-Schuster model leads to an analogous result (left column). Here at the ordered phase the cloud of strings is rather homogeneous, centered around the master sequence with a small dispersion. As the mutation rate (analogous to the temperature) grows, this cloud starts to expand in complex ways through the sequence space (middle picture). But beyond the error catastrophe, all nodes become equivalent and the cloud just expands trhough all the lattice points.

(for the Ising model at the high-temperature phase, the states of units become independent and simply change at random as if driven by a coin toss). At the critical point, however, the compromise between high mutation and conservation of information leads to a very interesting situation. In the Ising model, we see well-defined correlations in terms of percolation clusters (see Chapter 2). In the quasispecies model, these correlations are not so evident but can be made clear using a simple trick. The master sequence is unlikely to be found, but we can recover it by looking at the whole population of strings. Let us record the most frequent symbol at each position. Using this list of most common symbols, we build the so-called *consensus sequence*, which gives us the average most common sequence representing the whole population. What we find close to the critical point is that the consensus sequence corresponds to the master sequence: the whole population preserves this information collectively. RNA viruses have been shown to replicate at the error threshold, right on the edge of disorder. The quasispecies nature of viral populations was revealed in a classic experiment by the virologist Esteban Domingo and coworkers using the $Q\beta$ virus.[8] Further investigations proved that the mutation rates exhibited by different RNA viruses are always approximately the inverse of their genome sizes, in agreement with the Eigen–Schuster prediction.

The error theshold imposes a sharp limit on genome complexity: for a given mutation rate, chains longer than the one defined by the error catastrophe cannot persist. The previous model allows us to predict the maximum size of simple RNA molecules: about 100 nucleotide units. Since errors are an inevitable feature of RNA molecule-based replication systems, the quasispecies developed above seems unlikely to account for the origin of life in prebiotic conditions. And it can be shown that several competing quasispecies lead typically to a single winning competitor, as in an ecological system with several competitors. How, then, did complex genomes emerge in the early biosphere?

Critical Branching and the Error Threshold

The error threshold condition is easily derived from a simple probabilistic argument.[9] Imagine a string formed by units (nucleotides). Let us indicate by q the probability of accurately copying a given unit, and let W be the probability of no degradation of the chain (here is the probability of decay, i.e., that a molecule is decomposed into units). Three possible outcomes can be considered, as indicated in Figure 8.7:

1. The chain is degraded: this occurs with probability $1 - W$.
2. Correct replication occurs: to this end, the molecule must survive (W), and replication of all units must take place with no errors ($1 - q^{\nu}$).
3. Sometimes mutations will occur such that replication leads to new strings. This implies that no degradation is required (W) and error in at least one unit occurs with probability $(1 - q^{\nu})$.

For a given string A, each of the previous possibilities leads to a different number of copies of A, $N_i = 0$, 1, and 2, respectively, with the probabilities of each event indicated in Figure 8.7. The mean number of correct copies $\langle N_c \rangle$ is then given by

$$\langle N_{\mathrm{c}} \rangle = \sum_{i=0}^{2} iP(i), \tag{1}$$

that is,

$$\langle N_{\mathrm{c}} \rangle = 0 \times (1 - W) + 1 \times Wq^{\nu} + 2 \cdot W(1 - q^{\nu}) = W(1 + q^{\nu}). \tag{2}$$

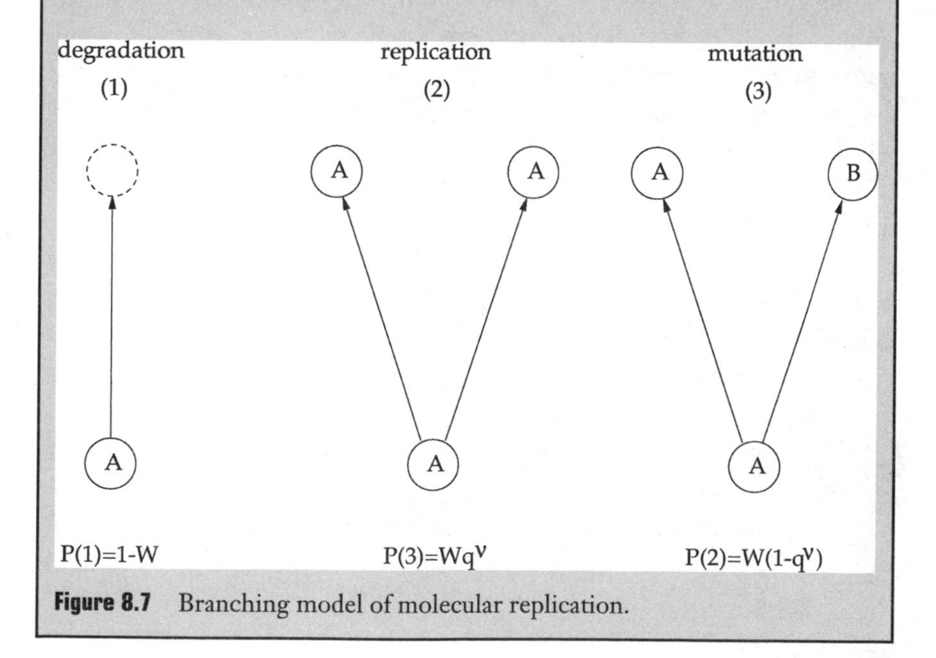

Figure 8.7 Branching model of molecular replication.

The critical condition is given by $\langle N_C \rangle = 1$; i.e., just one string of the same type survives. This critical boundary describes a threshold relation between length and mutation rate:

$$1 + q^{\nu} = \frac{1}{W}. \tag{3}$$

Or taking logarithms after rearranging terms, we obtain

$$\nu = \frac{\log(1 - W)/W)}{\log(q)} \approx \frac{1}{1-q}\log\left(\frac{W}{1-W}\right), \tag{4}$$

which imposes a limit to the maximum length of a sequence compatible with a given mutation rate $1 - q$. In other words, if an RNA soup is considered, and no appropriate mechanisms of repair of mistakes are at work (as happens in higher organisms with DNA repair systems), the maximum length is limited by equation (4).

Beyond the Catastrophe: Origins of Life

A theoretical answer to the information problem was offered by Eigen and Schuster, who suggested that the possible solution lay in *cooperation* among different molecules. They called this solution their *hypercycle* model. A given number of different molecular quasispecies would compete exactly like the ecological competitors in Chapter 7, with an eventual winner. So the potential larger complexity contained in the whole set of species would eventually disappear. The starting point for the hypercycle model is the observation that modern organisms employ a number of catalytic cycles in their metabolic pathways and that cycles of replication involving different nucleotide chains are common in modern RNA viruses. In a cycle, each product of a given reaction is used as a reactant in the next one. The chain of reactions is closed: at some point the product of a reaction is the molecule with which the cycle began.

Figure 8.8a shows a simple hypercycle. Here each type of molecular (quasi) species is denoted by I_i ($i = 1, ..., N$). Each species contributes to the replication of the next species in the cycle through some reaction rate k_i. We see that the hypercycle is not exactly a kind of organism, but a population of molecules with an ecological-like relationship.

Although such a closed structure might look strange at first, thinking in ecological terms allows us to see that hypercycles are common in nature.[10] Food chains often contain closed cooperative cycles. Oak trees and earthworms, for example, are replicators, and the presence of each accelerates the growth of the other.

A mathematical model for the simplest hypercycle is easily formulated (Box 3), and this model allows us to explore the idea in depth. The hypercycle solves the information threshold problem in an elegant way: each molecular species (quasispecies) remains below the error threshold limit, but since replication depends on cooperation among all the species in the hypercycle, the whole amount of information is more than any single species can support. Since each species is forced to cooperate, no competition among different quasispecies is

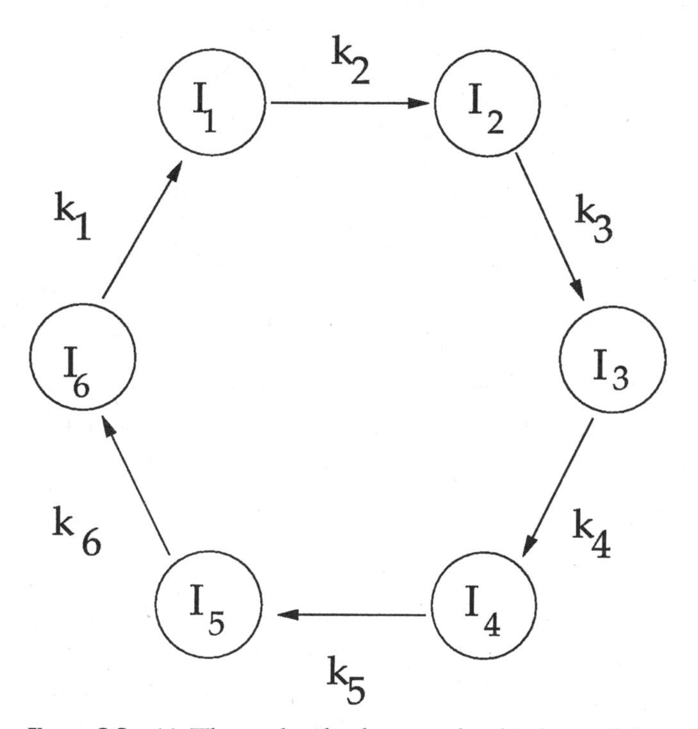

Figure 8.8 (a) The molecular hypercycle; (b) the available states for a three-dimensional hypercycle under the CP constraint are confined to a simplex, here shown as a two-dimensional bounded surface in a three-dimensional concentration space.

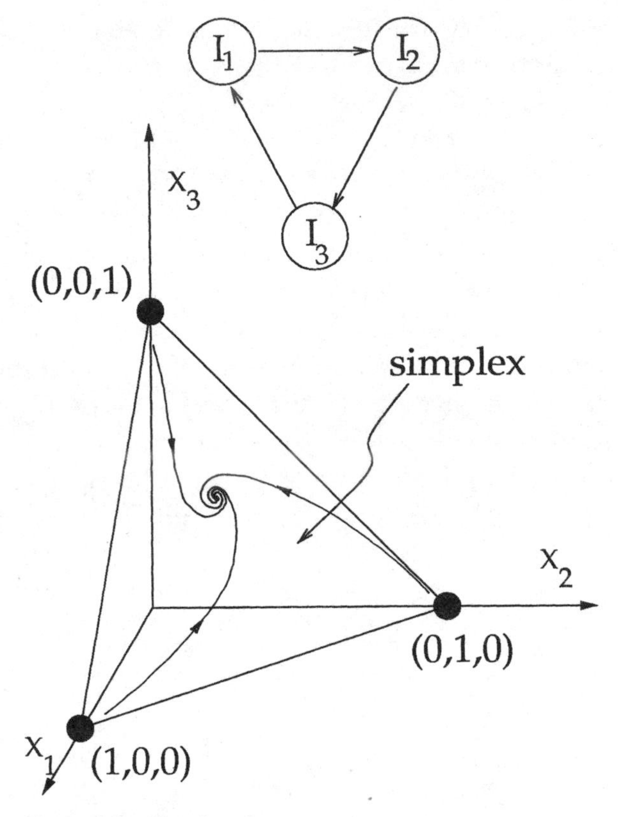

Figure 8.8 *Continued*

allowed (as would happen in a noncooperative set of quasispecies). In short, the coexistence of these different polymers allows the integration of their information and leads therefore to a higher level of complexity.

This theory also allows us to conjecture about the origins of the genetic code. We know that all organisms share the same basic molecular genetic structures. This is an intriguing fact. A more reasonable expectation would be some diversity of mechanisms compatible with an adequate store and propagation of information. Why a single gentic code? Eigen and Schuster showed that although the molecular species constituting a given hypercycle do not compete, hypercycles do compete, leading to the survival of a single hypercycle. A consequence of these strong competitive dynamics is the selection of a single replication system. Such a frozen accident would be responsible for the uniqueness of the modern genetic code.[11]

The Hypercycle Equations: From Order to Chaos

The simplest hypercycle model is a closed loop formed by coupled molecular species (Figure 8.8a). The fraction of each species is denoted by x_i ($i = 1, 2, \ldots, N$), and thus it is normalized, i.e., we have $\sum x_i = 1$! In this sense, the dynamics occur on the surface defined by the so-called *simplex*

$$S_N = \left\{ (x_1, \ldots, x_N) \in R^N : \; x_i \geq 0; \sum_{i=1}^{N} x_i = 1 \right\}.$$

This plane is shown in Figure 8.8b for the particular case N = *3*. Starting from any initial condition, we end at some domain of the simplex (for example, a single point, as indicated in the figure). The basic equations are given by

$$\frac{dx_i}{dt} = x_i(k_i x_{i-1} - \Phi),$$

where Φ indicates the dilution flow, which is regulated in such a way that the total concentration of molecules is constant. Hypercycles can display a wide range of dynamical patterns, from stable points and periodic orbits to chaos.

Yet, hypercycles can be easily destroyed by parasites. By parasite we mean a mutant that uses some resource of the hypercycle without providing any benefit to the system (Figure 8.9). In such a case (which can easily occur) the parasitic molecule will eventually break the closed chain of cooperation. Once one of the elements disappears, the hypercycle cannot persist. This drawback of the hypercycle theory, formulated by Maynard Smith in 1979, remained a problem until a spatial counterpart of this model was developed by two scientists in the Netherlands. As we saw in Chapter 7, space can sometimes lead to counterintuitive results.

Spatial Dynamics of Hypercycles

A reasonable assumption for the hypercycle model is that molecular reactions take place in an explicit spatial domain. The previous presentation involved a set of molecular species with complete mixing: there

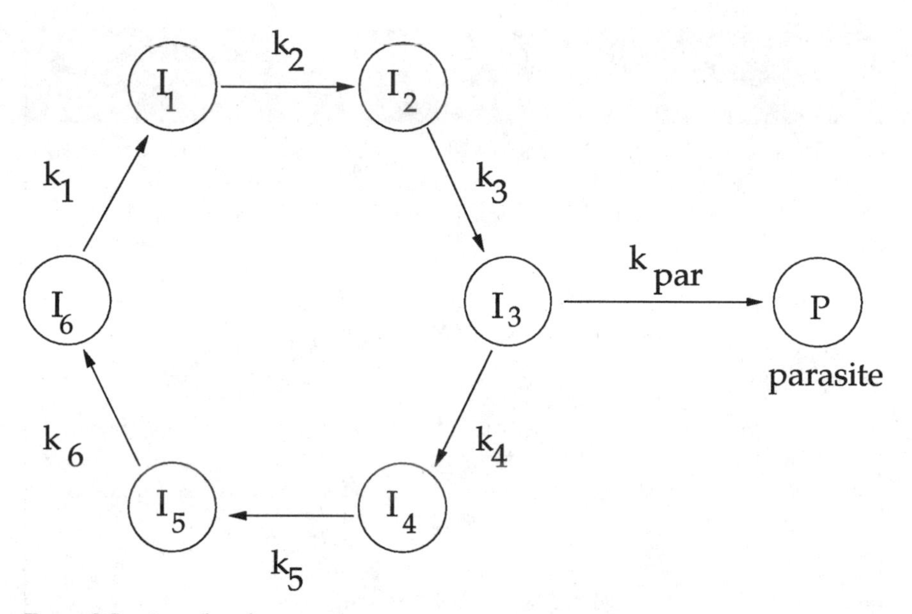

Figure 8.9 A molecular parasite.

was not merely local interaction among nearest molecules but some kind of intrinsic stirring in a reaction tank. The introduction of spatial degrees of freedom, and thus incomplete mixing, seems like a mere complication to the elegance of the theoretical model. But as Paulien Hogeweg and Maarten Boerjlist, at the Bioinformatics Lab in Utrecht, showed, the spatial hypercycle generates large-scale patterns with a very interesting property: the emerging spiral waves makes the hypercycle resistant to parasites.[11]

Two different types of model have been developed: discrete models and continuous models. The basic results are very similar in both cases. In Figure 8.10 we show the basic rules involved in the local dynamics of spatial hypercycles, as defined by Boerjlist and Hogeweg's original formulation. Their model employs three basic processes, representing the basic hypercyclic reactions, which take place on a two-dimensional grid of $L \times L$ sites. Each site is occupied by a single copy of one of the species involved. The rules are as follows.

—Decay: Molecular degradation takes place in occupied sites, which become empty with some probability (in principle, this probability depends on the species under consideration).

—Replication: Empty cells (and only empty cells) can become occupied

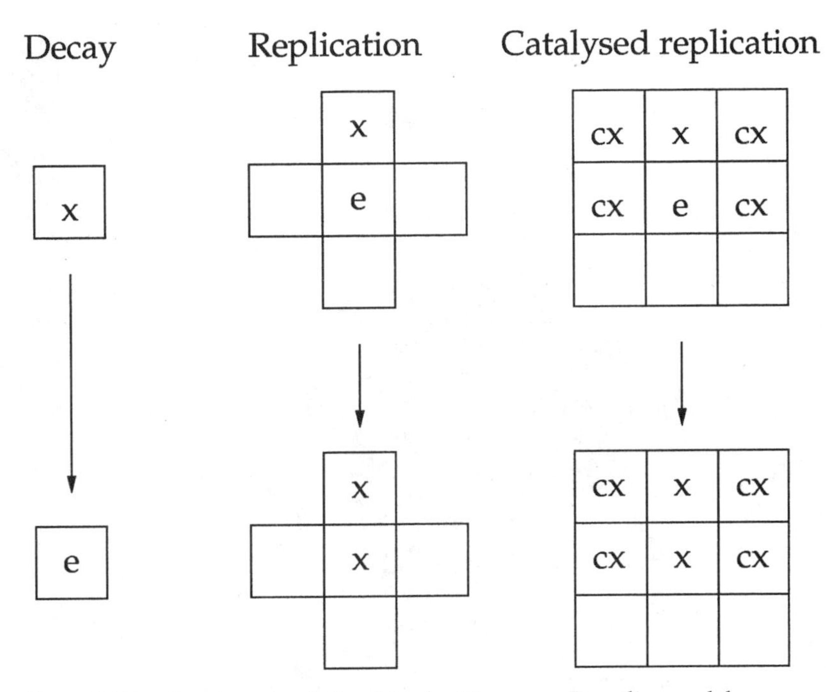

Figure 8.10 Simple rules defined in the Hogeweg-Boerjlist model.

by a molecule of a given species. Specifically, if one of the nearest neighbors of an empty site is occupied by a molecule, it can replicate into the empty site with some probability.

—Catalysis: Replication of a given molecule is enhanced by the presence of catalytic molecules in nearest positions.

The model is completed by adding diffusion of molecules. Similar rules are easily defined in the continuous counterpart (see Box 4). Figure 8.11 shows some examples of spiral waves generated by one of these spatial hypercycle models. Here the gray scale indicates the local concentration of one of the species forming the hypercycle. Not surprisingly, if we study the spatial distribution of population maxima for each species, we find that they are sequentially ordered in space, since the replication of each molecule depends on the presence of the previous one in the hypercycle.

Two main results emerged from these studies. The first is that when a parasite appears in this spatially structured world, its fate is very different from what would have been expected in a homogeneous environment. In that case, the parasite would increase its population

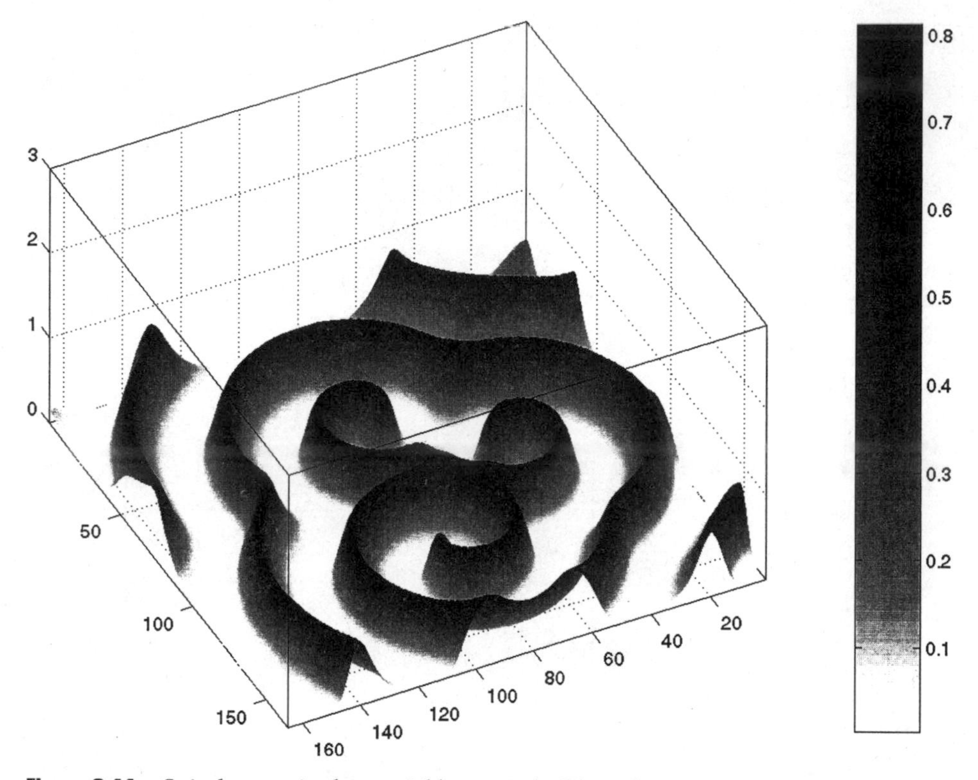

Figure 8.11 Spiral waves in the spatial hypercycle. Here the concentration of one of the hypercycle components is shown.

at the expense of the hypercycle until it destroyed the system. But here, the presence of spiral waves changes our expectations. Boerjlist and Hogeweg found that there is a very unequal distribution of fitness within the spiral wave. At the middle of the spiral we find the molecules eventually generating all the different types of possible replicators of the entire spiral, whereas the molecules at the edge of the spiral disappear. When parasites appear in this system, their populations are forced to grow in the direction of the outward movement of the spirals and so are destroyed once they reach its periphery.

A second consequence of these models is that resistance to parasites implies a positive selection for a strong altruistic property. This contradicts the generally accepted selection theory: it is caused by spatial structuring. Selection appears to take place *at the level of the spirals*: in competition with a parasite, the spirals act as integrated entities. Once again, simple selective pressures acting on particular molecular

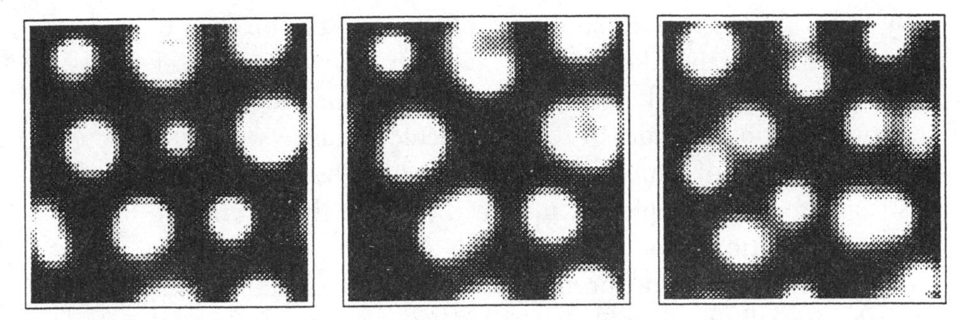

Figure 8.12 Self-replicating spots in the hypercycle model.

species fail to give us an appropriate framework for understanding the emergence of complex dynamics in nonlinear systems.

Let us finally mention that spatial structuring can sometimes lead to a different type of pattern. One of them is remarkable: the formation of clusters able to *replicate* (Figure 8.12). This phenomenon is not exclusive of hypercyclic organizations: it appears to be common in a wide range of far-from-equilibrium chemical reaction systems and has been experimentally observed.[12,13] In Figure 8.12 we show an example of these replication spots in a model of the hypercycle exhibiting chaos.[14,15] These pictures show three snapshots of the system for three different time steps. White zones correspond to higher concentrations of one of the species, and darker zones are domains where it is depleted. We can see that there is growth and division (and decay), and these sequences of growth–replication–decay recur again and again. Although these are not true cells, these results open an immense range of new possibilities for exploring the emergence of complexity in prebiotic evolution.

Random Catalytic Networks

The primordial soup contained many different combinations of molecular species of various types and sizes. This seems unlikely to have provided an adequate scenario for the emergence of life forms. Some molecular species can catalyze the reactions among pairs of other species, leading to new molecular species, eventually with greater complexity. But it is intuitively clear that the likelihood of such catalytic reactions will be very small. Some authors, particularly Stuart Kauffman, Doyne Farmer, and Norman Packard, have explored the probability of the emergence of a coherent, self-sustained network of reactions among a set of molecular species.[16] Such a network is called an *autocatalytic set*.

By that we mean that each member of the set is the product of at least one reaction catalyzed by at least one other member of the set.

The possibility of an origin of life scenario based on an autocatalytic set of polypeptides or RNA molecules leads to some unexpected results. Instead of a mechanism of replication based on a template, it is suggested that simple polymers can catalyze the formation of each other, generating autocatalytic sets that evolve in time to create complex molecular species whose properties are tuned for effective cooperation. Mathematically, this can be formalized through sets of equations where catalysis is represented by a random connectivity matrix that defines which pairs of species react to create new species. These models are usually known as random catalytic reaction networks, since a random connectivity is defined a priori. This model is in fact quite similar to the random matrices defined in Chapter 7.

A very interesting result about random catalytic reaction networks is that as the connectivity of the matrix (i.e., the number of nonzero entries of the matrix) increases, a sharp transition occurs. This transition separates two well-defined domains: a subcritical one, where a small diversity of molecular species can be sustained, and a supercritical one, which sustains a very complex autocatalytic set. With large numbers of molecular species, the transition is sharp, and it takes place as a consequence of the emergence of a reaction graph such that basically all molecular species are connected through a chain of reactions. The rarity of particular molecular reactions is compensated, at the critical point, by a large enough number of possible reactions.

This result is strongly related to the classical Erdös–Renyi theorem on random graphs, where the nodes of the graph are randomly wired. In 1959 Erdös and Renyi proved that in a large graph Γ_N with N nodes and with E randomly assigned arcs, the probability $P(\Gamma_N)$ of getting a single gigantic component (containing most of the nodes) jumps from zero to one as we increase E/N beyond the critical value 0.5.

This surprising result was one of the most celebrated theorems in the theory of random graphs. The Erdös–Renyi theorem is a very important result for complex networks and so has great relevance for complexity theory. Not only chemical networks, but such related problems as ecosystem stability, dynamics of markets, and the organization of the immune system response are all strongly dependent on their connectivity properties. In Figure 8.13 we show three examples of graphs where the nodes (here indicated as black circles) are randomly connected with an increasing number of edges. The ratio of links to nodes is indicated by C. Below the critical value, many nodes remain unconnected. But close to the critical value $C = 0.5$, a phase transition occurs: suddenly, most of the nodes start to belong to the so-called *single gigantic connected component*. This means that percolation occurs in such a way that it is possible to go from almost any node to any other. In other words, the phase transition leads to a sudden shift from a nearly unconnected set to a nearly connected set. If we plot the probability of finding a fully connected set as C is varied, we would see the familiar phase transition plot with a sudden jump at the critical probability.

We can start to see the consequences of this result for the set of polymers. Reactions among polymers resemble links between nodes of a graph. If we introduce these connections (catalysis among different components) at random, we may also expect a phase transition. This is precisely what has been observed. Farmer, Kauffman, and Packard created a model[16] using a set of one-dimensional chains (like the bit strings we discussed before). These strings were assumed to be oriented from left to right (like real biological molecules), and the polymers

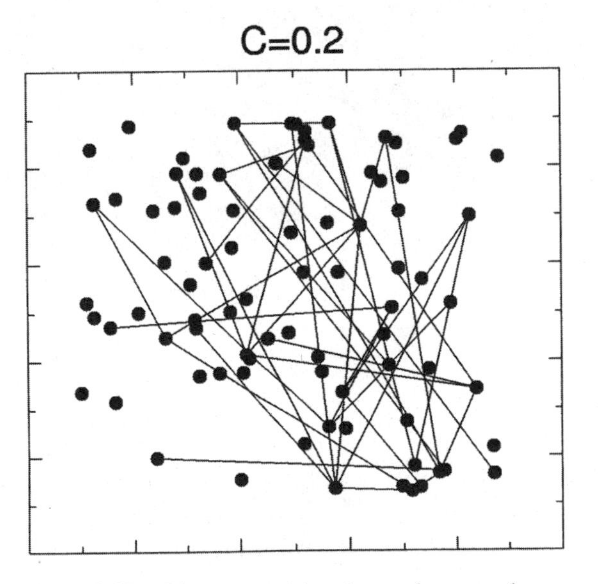

Figure 8.13 Phase transition in random graphs: as the number of edges is increased, we cross a critical point where all nodes are to be connected.

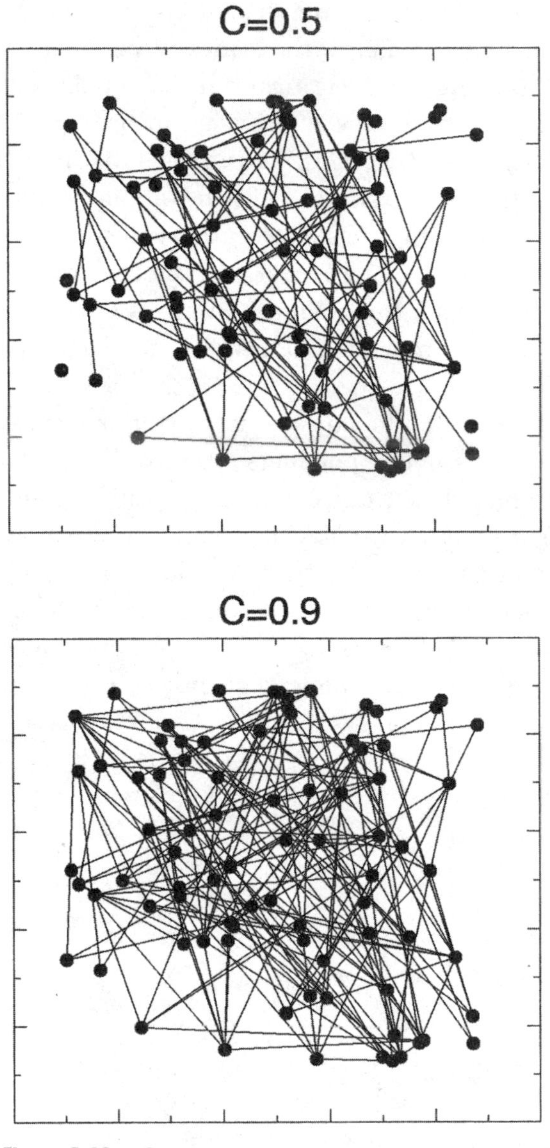

Figure 8.13 *Continued*

could have any length up to a maximum. Two basic reactions were allowed: ligation (joining together) and cleavage (splitting apart).

Reactions were assumed to be catalyzed, so that we have transitions like:

$$a + b \underset{e}{\longrightarrow} c + h,$$

where c is the concatenation of a and b, h represents a reaction product, and e indicates the enzyme catalyzing the reaction. When the three researchers ran their simulations, they got graphs like the one shown in Figure 8.14. We can see, in fact, two superimposed networks, one (continuous arrows) indicating the reactions leading to new molecular species and the second indicating the catalysts to the reaction nodes (dashed lines).

With this simple model of a reaction network, the system was able to increase its intrinsic diversity as more and more new polymers appeared and played some role as catalysts. As the diversity increases, at some point every reaction needed to make any component of the reaction network will be catalyzed by some member of the network itself. The explanation is simple: as the size of the polymers grows, the

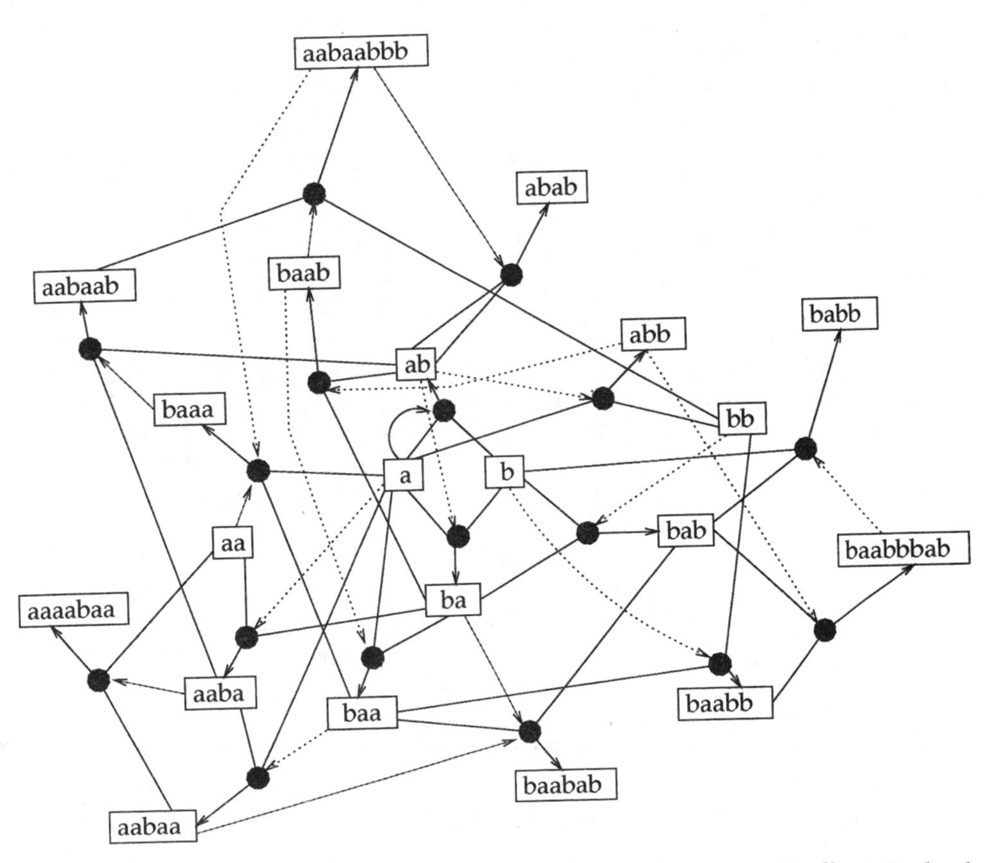

Figure 8.14 An example of autocatalytic set obtained from the Farmer-Kauffman-Packard model.

number of catalyzed reactions grows faster than the number of reactions required to make the polymers. An autocatalytic set spontaneously emerges. This network is able to maintain its complex structure even though some components come and go with time. Not surprisingly, these catalytic sets display several universal properties (including some scaling properties close to criticality).

A Simple Model of a Random Catalytic Network

A very simple toy model of a random catalytic reaction network exhibiting a phase transition has been proposed by Bartolomew Luque.[17] Let us assume that we have a system involving N molecules and a maximum number S of potentially available molecular species, with $N \ll S$, and let $\Omega = \Omega_{ij}$ be the reaction matrix, defined for each (possible) pair of species as the result of the reaction, with $\Omega_{ij} \in (0, 1, 2, ..., S)$. Here a zero means no reaction. The matrix elements are randomly generated. As an example (for a very small pool of $S = 5$ molecular species) the reaction matrix would be

$$\Omega = \begin{pmatrix} 0 & 3 & 0 & 1 & 0 \\ 5 & 1 & 0 & 0 & 0 \\ 0 & 5 & 1 & 1 & 0 \\ 0 & 4 & 0 & 0 & 2 \\ 0 & 0 & 4 & 3 & 0 \end{pmatrix}$$

This matrix has connectivity $C = 11/25 = 0.44$, where 11 nonzero matrix elements have been divided by the total 25 possible elements). Using this matrix, we can see that the interaction of species two and species one (in this order) leads to species three. But the interaction of one and two (in this order) leads to species five. We can understand these differences by thinking that the first molecule acts as an enzyme and the second as a receptor. We can also see that there is no need of symmetry. Species four interacting with species one leads to another molecule of species one, but the symmetric reaction is not possible.

The matrix has some given connectivity, measured as the fraction of nonzero elements divided by S^2 (as defined in the previous chapter). The rules are very simple: at each step we choose two molecules belonging to species i and j, respectively. If the matrix element Ω_{ij} is zero, then nothing happens. If $\Omega_{ij} = \kappa \neq 0$, then a new element belonging to species κ must be added. In order to do so, we randomly choose one of the N elements of the system and replace it by a new unit belonging to species κ.

To see the presence of a phase transition, we can run the model many times and explore the effect of different connectivities in the internal dynamics of the species. Starting from an initial condition involving a small number of species, we use the previous rules and let the system evolve over a large number of steps until it reaches a stationary state. Afterwards, we follow the dynamics over $T = 10^3$ steps and look at the different species that have appeared over this interval. For low connectivities, the system is unable to generate diversity and becomes frozen, so that the number of species present is low. The populations will remain frozen at their steady values. For high connectivities, it is clear that the matrix will generate very high diversities and disordered behavior, since S is large and two given elements are likely to react and give a *different* species. As a consequence, we will see an ever changing population of molecular species. The interesting point is that there is a critical threshold C^* where the system jumps from order to disorder.

Autocatalytic sets of this type could have played a major role in supplying the complex chemical prerequisites for the origin of life. Perhaps autocatalytic sets of proteins and small RNA molecules were both formed in the primitive biosphere and eventually began to interact with each other. These rich sets of molecules would have eventually brought forth long nucleotidic chains with template activity. Since these systems have some universal features (such as those arising from the phase transitions in random graphs) we might ask whether every prebiotic system will have some universal features. This problem has been analyzed by Walter Fontana, of the University of Vienna and the Santa Fe Institute, by means of a highly abstract model based on the so-called λ calculus, a formal system that allows one to explore how strings (such as RNA molecules or mathematical functions) interact with other strings.[18]

In Fontana's system, called AlChemy (for algorithmic chemistry) networks of strings spontaneously emerge and proliferate due to their capacity to replicate as causally closed sets. The runs of these simulations lead to various outcomes, including networks with unbounded growth. But they also show some other interesting adaptive patterns: some systems are able to repair themselves when some components are removed, and others can adapt to the introduction of new components. All of them, in sum, are close to real life.

Real Life, Artificial Life

The origin of life is one of the most fascinating topics in science. Ancient civilizations believed that life was generated from inanimate matter, and for a long time their observations were consistent with this idea. Mice and flies seemed to arise spontaneously from rotting meat. Modern science has deeply altered this view, and different lines of evidence support the idea that life emerged in the early history of Earth and quickly expanded through all environments, from friendly to extreme. Darwin himself recognized the necessity of a starting point in the evolution of life.

Despite the failure of many experiments recreating the primordial soup to generate simple living systems, most biologists think that it must be possible to create life in the lab. This is an old dream that will someday come true. In the meantime, we will have to explore the universe by means of powerful telescopes, molecular methods, and mathematical models. Such models can help us design new experimental setups; more importantly, they can open our minds to alternatives that have not succeeded on our planet but are perhaps present in other worlds. Renewed efforts toward the search and understanding of life in the universe have led to the emergence of astrobiology, a new interdisciplinary area of research. Astrobiologists search for evidence of life from planetary atmospheres, geologic processes, and the analysis of recreated primitive organic soups. But models are very important: the generic properties of hypercycles and the presence of phase transitions leading to sustained autocatalytic networks suggest that replicating, evolving systems are everywhere subject to well-defined laws and constraints.

When we land on another planet, say Jupiter's moon Europa, we may find strange and bizarre life forms. We can only speculate, but we can also explore the forms life might have taken, using theoretical models and computer simulations. Artificial life,[19] a novel area of research created by Chris Langton, has been very successful in providing insight into some deep problems of real biology. Artificial worlds do not need to be constrained by biologically sensible ingredients. Artificial life systems often involve entities exchanging information in some abstract way, with no direct reference to real biology. Yet these systems very often display properties common to real life. Some lead to evolving ecologies where parasites, hyperparasites, cooperation, and sexual differentiation spontaneously emerge. Such an apparently universal behavior is remarkable: why should artificial systems behave like natural living systems? Perhaps because under a very general set of conditions life is a common phenomenon that follows universal laws. These laws are part of what complexity theorists and astrobiologists are looking for. Perhaps they will show us that we are at home in the universe.

NINE

Evolution and Extinction

> But we should not despair. There is real order in all this apparent chaos. Life has had a long and complex, but ultimately comprehensible, history. There are patterns repeated over and over again as new species come and go, and as ecosystems form and fall apart. These organizing principles of life's history are the processes of evolution. We apply them to fossils to render that order.
>
> —Niles Eldredge

Life on Earth has been changing in many ways since the first living organisms started to invade marine and terrestrial habitats, and the fossil record, like a great script, tells us some part of what happened. From those few recovered pages we have to reconstruct the whole masterpiece.

For a vast time, our planet's biosphere was dominated by bacteria and other simple single-celled organisms. These 3900 millions years are known as the Pre-Cambrian Period, and the remains of life forms from that period are fragmented and scarce. Between 530 and 520 million years ago most of the basic life forms emerged, defining the beginning of the Cambrian Period. Whether this explosion was really so brief or the result of a slower process is a matter of debate, but there was a time when multicellular life forms had little diversity of form, and a time afterwards when animals displayed an amazing range of morphological patterns.

There is an extraordinary gold mine of organisms that first emerged during the Cambrian explosion:[1] the Burgess Shale in the Canadian Rockies (Figure 9.1). These old remains of life's history in the Rocky

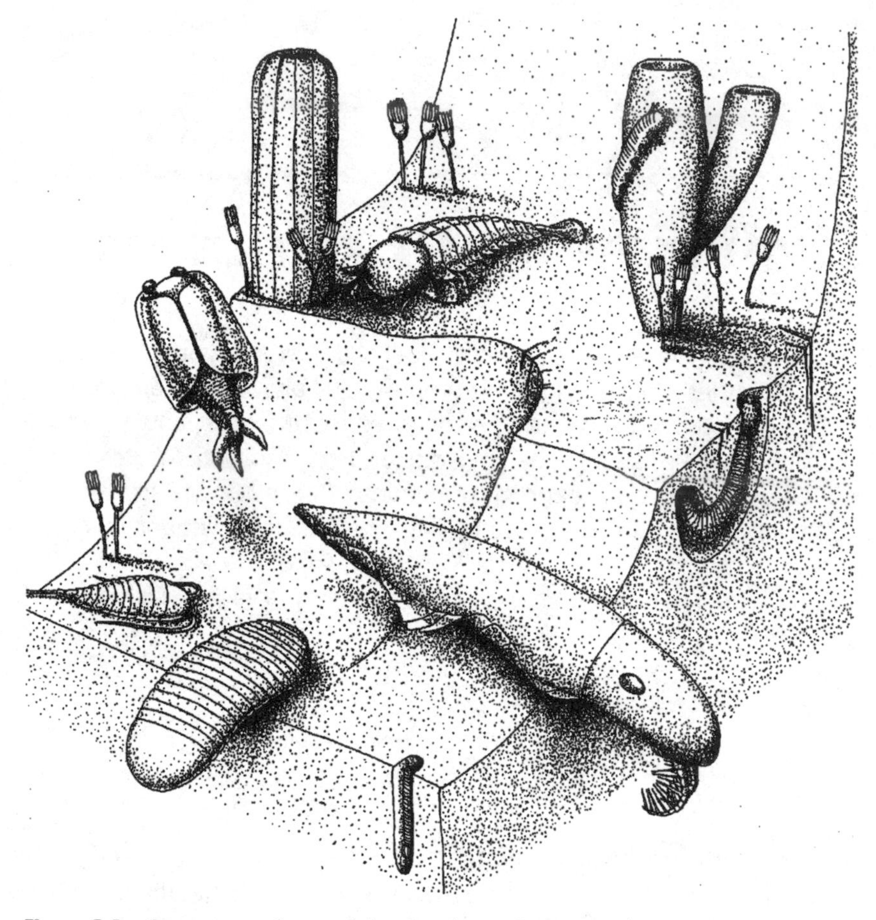

Figure 9.1 Some organisms of the Cambrian fauna of Burguess Shale.

Mountains are so rich and unique that they completely changed our view of the evolutionary history of animals. The fossils of the Burgess Shale were discovered by Charles Walcott, a paleontologist from the Smithsonian Institution, in 1909, and their interpretation and consequences for evolutionary theory have been explored in some recent books, including Stephen Jay Gould's book *Wonderful Life*.[2] The Burguess Shale faunas are well-organized communities, with predator and prey species having a wide range of sizes and morphologies.* The ecosystem was remarkably modern, with well-defined groups of

*We use the term "fauna" for convenience, but it is not universally agreed that the Ediacarans were animals. They may have belonged to an entirely different, now extinct, kingdom.

suspension-feeders, deposit-feeders, and carnivores. All these species seem to be linked by a complex food web, and it seems likely that different ecological niches were subdivided to a considerable degree. The basic rules that organize a complex community today were already present in Earth's first animal communities.

In our modern biosphere we identify several major animal groups, which we call *phyla* (Figure 9.2). The 32 recognized phyla define an amazing diversity of living animals. We easily recognize that the different animals do not make a continuous spectrum but cluster into limited categories of form, known as *body plans*. All organisms within a phylum share a common body plan, i.e., the early phases of their morphogenesis are basically the same. It is also possible to detect similarities between some body plans in different phyla, although this often requires careful comparison of larval forms. Surprisingly, most (if not all) of these body plans were generated during a period of anatomical creativity that is absolutely unique in Earth's history.[3]

To point out how extraordinary this episode was, the science journalist Bob Holmes used the following analogy: "glass skyscrapers, gothic cathedrals, Georgian terraces . . . imagine that all the architectural styles that human ingenuity could ever devise appeared during one 35-year period, sometime in the middle of the 15th century." The Cambrian explosion asks for an explanation. Did it occur because certain special conditions were present? Was it inevitable? How relevant were historical effects? Was there another explosion afterwards, some time in the Phanerozoic? Is the outcome of the Cambrian event something totally contingent, as Stephen Jay Gould has advocated, or are there universal laws operating on the possible set of morphological types?

The Summer of Ediacara

The Cambrian "big bang" is one of several major transitions that include the origin of life, the origin of cells, and the origin of social behavior. It also involves the problem of extinction: some of the life forms that emerged at the Cambrian explosion are not present today. And it did not mark the first appearance of multicellular forms. The so-called Ediacaran fossils, first found in 1940 at the Ediacara Hills in South Australia, dominated Earth for a few tens of millions of

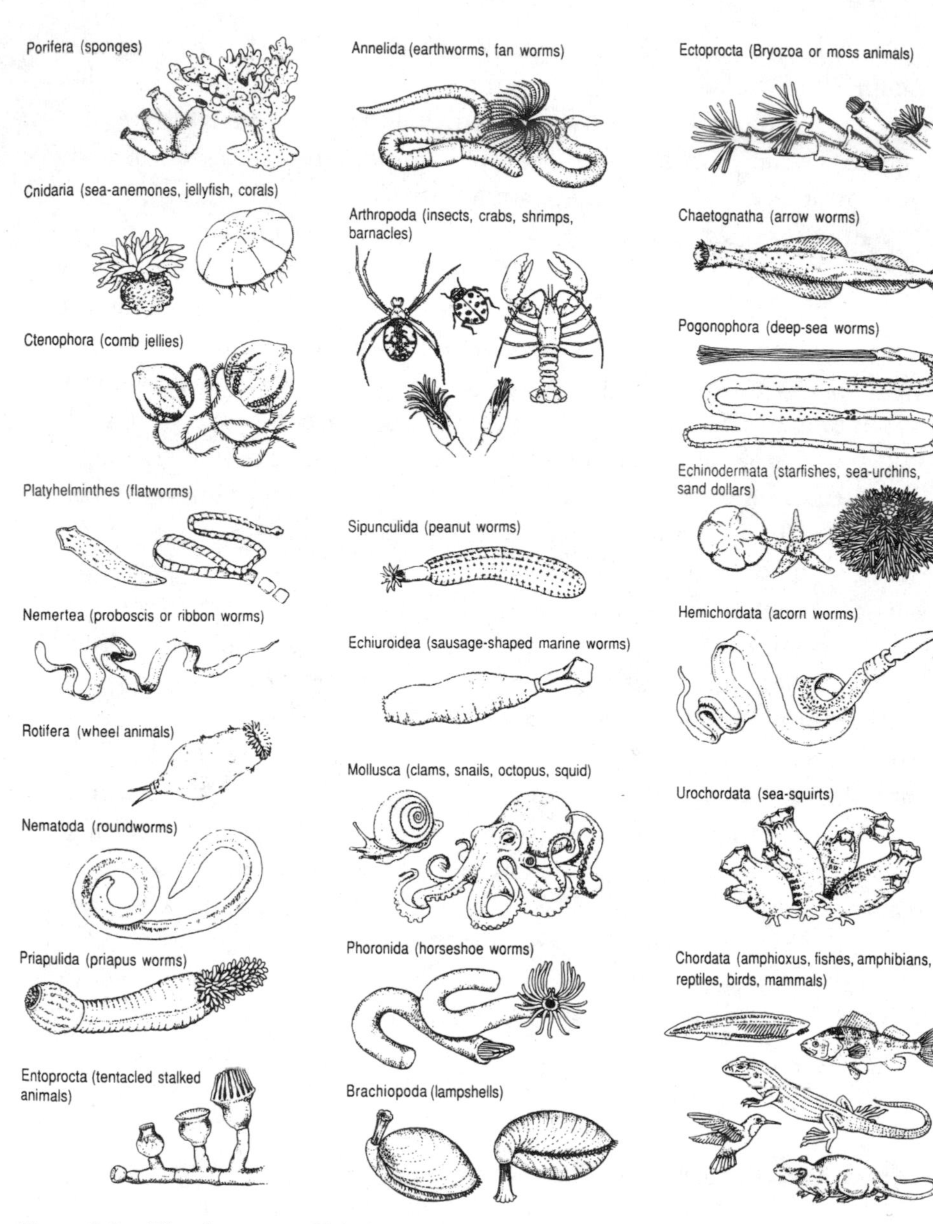

Figure 9.2 The diversity of life forms.

years before the Cambrian. Ediacaran faunas (Figure 9.3) were quite bizarre, and some of the collected fossils are still controversial. But they failed to survive. The summer of Ediacara ended in the late Precambrian. It has been suggested that these faunas were doomed when a new class of creatures appeared: the predators. The Precambrian

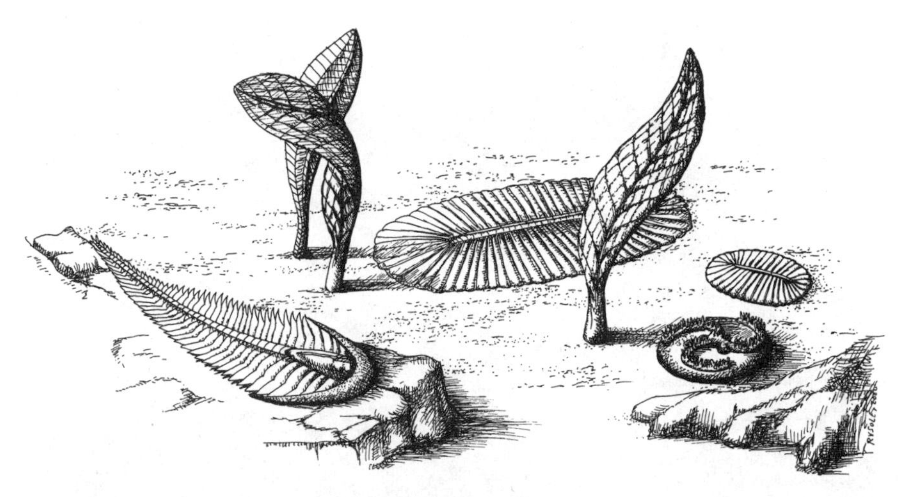

Figure 9.3 Some organisms of the Ediacaran fauna.

faunas were probably dominated by soft-bodied creatures nourished by photosynthetic algae or bacteria living within their tissues, much like corals today. The story of Ediacara is the tale of a failure that took place when modern ecologies entered the scene.

The explosive appearance of disparate animal clades remains a puzzle for paleontologists and evolutionary biologists. The battle to understand how it took place is being fought in a complex arena: it involves not only fossils but our knowledge about how embryos develop and how genes interact. And together with origins, the extinction of once successful organisms and ecologies is also a deep issue.

Natural Selection

Charles Darwin's theory of evolution is of course one of the landmarks of modern science.[4] Darwin tried to formulate an account of biological change that was simple, elegant, and nontrivial. The resulting theory has been misused by many people through its history, and Darwin himself suffered from the wrong interpretations of opponents as well as would-be supporters. In Figure 9.4 a cartoon from the satirical magazine *Punch* parodies the evolution of life forms, from the the simple worm (and chaos) to Darwin himself.

Many people have trivialized or misunderstood the theory of natural selection. It has two important components. First, Darwin understood, from the long tradition of artificial selection applied by humans to both animal and plant species, that there is an intrinsic source of

Figure 9.4 Punch's cartoon on Darwin.

variability in living organisms. The underlying source of this variability, the chemical basis of inheritance, was then unknown. We know today that random mutations and other changes at the DNA level are transmitted to the offspring, resulting in heritable variation. Second, population size is limited by the finite nature of available resources. Inspired by the writings of Malthus, Darwin realized that resource limitations operate as a mechanism of selection on organisms' intrinsic variability. The result of these two components is that variation is always at work, but only some of the variants survive. As individuals tend to produce offspring similar to themselves, these varieties will become more abundant in subsequent generations. The changing patterns of gene frequencies in a population would be slow and essentially continuous. In this sense, evolution by natural selection would be a historic process of successive accumulations of changes. Darwin's book *On the Origin of Species* extrapolated these mechanisms to the emergence of new species. Natural selection would favor the increasing divergence of new varieties able to exploit new aspects of the environment. The increasing divergence of new intermediate forms would eventually create distinct species.

This powerful theory received a tremendous impetus during the twentieth century. Novel molecular techniques and new theoretical models improved and expanded Darwin's ideas in unexpected directions. Thousands of molecular geneticists today are engaged in disentangling cellular processes taking place at microscopic scales. But although such knowledge has strongly modified our view of life and has crucial biomedical applications, extrapolations have often led to an oversimplified picture of evolution as a molecular process. Scientists have often claimed to have found *the gene* for a given process, as if whole processes could be represented by elementary genetic interactions. And if they don't make such claims directly, the media can be relied on to put these words in their mouths. Although some theoretical approaches (such as Kimura's neutral theory of evolution[5]) based on random events at the molecular level are remarkably good at explaining some important long-term molecular evolutionary patterns, it is unlikely that the complexity of whole organisms (let alone communities) can be reduced to such low-level properties.

Other evolutionary biologists, such as Stephen Jay Gould, have strongly advocated contingency in the evolutionary process. Gould interprets the Cambrian explosion in terms of a sudden, revolutionary diversification event followed by decimation (Figure 9.5). Instead of a growing tree of increasing diversity (Figure 9.5a), Gould suggests a bush with many branches at the beginning and fewer after the explosion. Which species survived this initial diversification, he argues, is determined entirely by accident. The Burgess Shale faunas thus tell us that historical contingency has profound implications even for our own place in the world. But as Conway Morris suggests,[1] Gould's claims are rather easy to accept, while others are probably exaggerated. Contingency is an obvious consequence of the fact that real life is highly

Figure 9.5 Two possible trees of evolution.

nonlinear and involves a huge number of variables.* Each of us is alive because of a long, random chain of unlikely events.

The problem is to what extent the overall properties of living systems are the result of contingencies instead of emergent, universal phenomena. This question arises at many levels. Here we will discuss two of them: morphological variation and the network-like structure of real ecologies. Concerning the latter, it is important to remember, as we discussed in Chapter 7, that species are basic units of more complex systems with emergent properties. Modern population genetics lacks any trace of such ecosystem-based dynamics: species are considered almost as isolated entities coupled with an external environment. Still worse, species are considered as boxes of genes, with the individual genes treated almost as isolated entities whose frequencies change as a consequence of natural selection.

Nemesis: Is Bad Luck the Main Force?

The fossil record of life reveals that about 99.9% of all species that have appeared on our planet have become extinct. In fact, before Darwin's time researchers had known that most of the fossil groups that had been discovered belong to extinct groups of organisms. Extinction is the fate

*Some authors have in fact made the mistake of identyfing this intrinsic unpredictability of complex systems made of many parts with chaos. However, the interest of chaotic dynamics resides in the deterministic and low-dimensional character of most well-identified chaotic systems. That a multidimensional system involving many variables and with both stochastic and deterministic ingredients is unpredictable in the long run does not contradict elementary common sense.

of most groups, and this tells us that no biota are infinitely resilient. As paleontologist Peter Ward said, "Extinction is the fate of all species. And it is not an abstraction. At some moment in time there was but a single dinosaur left on earth, or a single ammonite, swimming on the sea . . . and when that last individual dies, the unique genetic information that makes up its kind will disappear."[6]

If we look at the time series of extinction rates through the last 550 million years (the so-called Phanerozoic Period), the picture looks highly uneven (Figure 9.6). Sometimes there are large-scale events: mass extinctions. These were defined by the paleontologist Jack Sepkoski as substantial increases in the amount of extinction suffered by more than one geographically widespread higher group of organisms during a geologically short period of time. The history of life is punctuated by mass extinctions, five of them so important that they have received considerable attention from paleobiologists. They involved such great shifts in biotas that they serve as benchmarks of the geologic time scale.[7]

There is good evidence from all continents for large asteroid impacts. Some impact craters are small, but some are very large. How im-

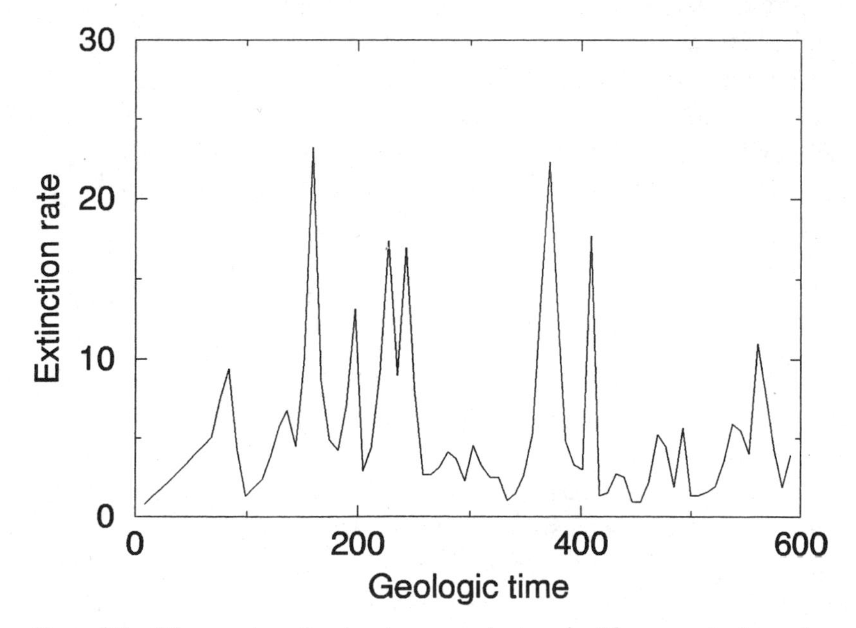

Figure 9.6 Time series of extinction rates during the Phanerozoic; here the time scale is in Millions of years.

portant are these impacts in causing mass extinctions? Over the last two decades, an impressive body of evidence has been amassed in support of a large asteroid impact close to the end of the Late Mesozoic, about 65 million years ago, defining the so-called K-T boundary (separating the Cretaceous from the Tertiary). This mass extinction involved the disappearence of the dinosaurs as well as other important groups such as the ammonites.

A group led by the physicist were doing field research in Gubbio, Italy, looking for evidence of reversals of earth's magnetic field, when they found that a clay layer close to the K-T boundary contained remarkably high levels of iridium. This is a rare metal in the earth's crust but very common in meteorites and asteroids. The team's analysis of clays in other locations revealed that the anomaly was widespread and clearly associated with the end of the Mesozoic era. A further (though not immediate) implication was the idea that the mass extinction at the K-T boundary was caused by a devastating asteroid impact.[8] In a remarkable set of papers, the Alvarez's and other researchers, such as Michael Rampino of New York University, revealed evidence for a causal connection between such an impact and the mass extinction. They even suggested that if the remains of an impact crater were preserved, it would display a diameter consistent with an object about 10 kilometers in diameter. Such a prediction received, ten years later, the support of an amazing finding: there is a well-preserved 65-million-year-old impact crater on the northern coast of Mexico's Yucatán Peninsula, north of the town of Mérida. The Chicxulub crater, buried under half a mile of sediment, is 170 kilometers wide, consistent with impactor from an object 10 kilometers in diameter.

The success of these results was followed by an especially interesting suggestion, first advanced by David Raup and Jack Sepkoski, that the fossil record seems to display a periodic pattern of extinction events.[9] By analyzing the available compilations, Raup and Sepkoski found evidence for a 26-million-year cycle of extinction. Such a long period could be related only to an astronomical cycle, and Rampino and others suggested the existence of some unknown stellar companion, which was named Nemesis ("the star of destruction"). This and other ideas promoted a rich discussion that still continues. In particular, Raup advocated external events as the basic causal mechanism,[10] although the evidence was controversial for both the periodic pattern and the connection between asteroid impacts and species loss. Some authors, such as Mark Newman, of the Santa Fe Institute, have further developed the idea that extinctions are basically caused by external stress by using explicit, simple mathematical models involving external perturbations.[11] Newman's model shows that some of the statistical features of the fossil record can be recovered from purely external causes. The direct link between asteroid impacts and extinction events is far from obvious, but for certain events, such as the K-T extinction, the evidence is too strong to be denied. Other important changes due to volcanic activity or worldwide climate modifications have been identified in recent years.

The study of paleoclimates clearly indicates that our planet has not exactly been a stable, warm pond. A complete theory of macroevolution, therefore, should take into account external fluctuations, of either astronomical or geological origin. But again, ecosystems are not linear entities smoothly reacting to the external environment. And several observations show that (at least at some scales) biological events, acting alone or together with physical changes, may have been of equal importance. Maynard Smith[12] cites some evidence for the prevalence of biotic factors, including the surprisingly constant diversity (as measured by numbers of genera and families) within particular groups over many millions of years. Such constancy must be related to a feedback between extinction–speciation rates and standing diversity. In fact, some authors, such as Sepkoski, have used simple models of species competition to reproduce the main traits of global diversity in the Phanerozoic. On the other hand, Gaia theory asserts that the global regulation of the planet's climate involves a strong biological component. If this is true, then even if external events play a role, the (presumably nonlinear) response of our biosphere could be very important.

The Race of the Red Queen

In 1973, Leigh Van Valen proposed a theoretical approach to evolution in multispecies known as the Red Queen hypothesis.[13] This model offered a theoretical explanation for the observation that the probability of a species becoming extinct is approximately independent of its length of existence. In other words, the fossil record suggests that a species may disappear at any time, irrespective of how long it has already existed. Intuitively, we would have expected species within any group to become longer lived. Take mammals, for example: careful analysis of their extinction rates shows that modern mammal species are just as likely to become extinct as were their ancestors living 200 million years ago. But if evolution leads to improvement through adaptation, aren't modern mammal species more durable than their ancestors?

Van Valen's interpretation is that species do not evolve to become any better at avoiding extinction. If adaptation improves species progressively through time, we should expect a decreasing probability of extinction: older species would last longer. Van Valen suggested that constant extinction probability would result from a constantly changing biotic community, in which species continually adapt to each other's changes. The name for this conjecture refers to the Red Queen's remark in Lewis Carroll's *Alice Through the Looking Glass* (Figure 9.7), "Here, you see, it takes all the running *you* can do, to keep in the same place." Van Valen's view of evolution is that species change just to remain in the evolutionary game. Extinctions occur when no further changes are possible: if genetic variability is insufficient, the player is removed from the ecological game.

On the other hand, some aspects of the fossil record suggest that some amplification processes are at work, with well-defined scaling laws (Figure 9.8). The most remarkable of these are:*

Figure 9.7 Alice and the Red Queen.

*We should be conscious of the many limitations of the fossil record. Some of these data are poor and have several sources of sampling error and, in this sense, must be taken with some caution.

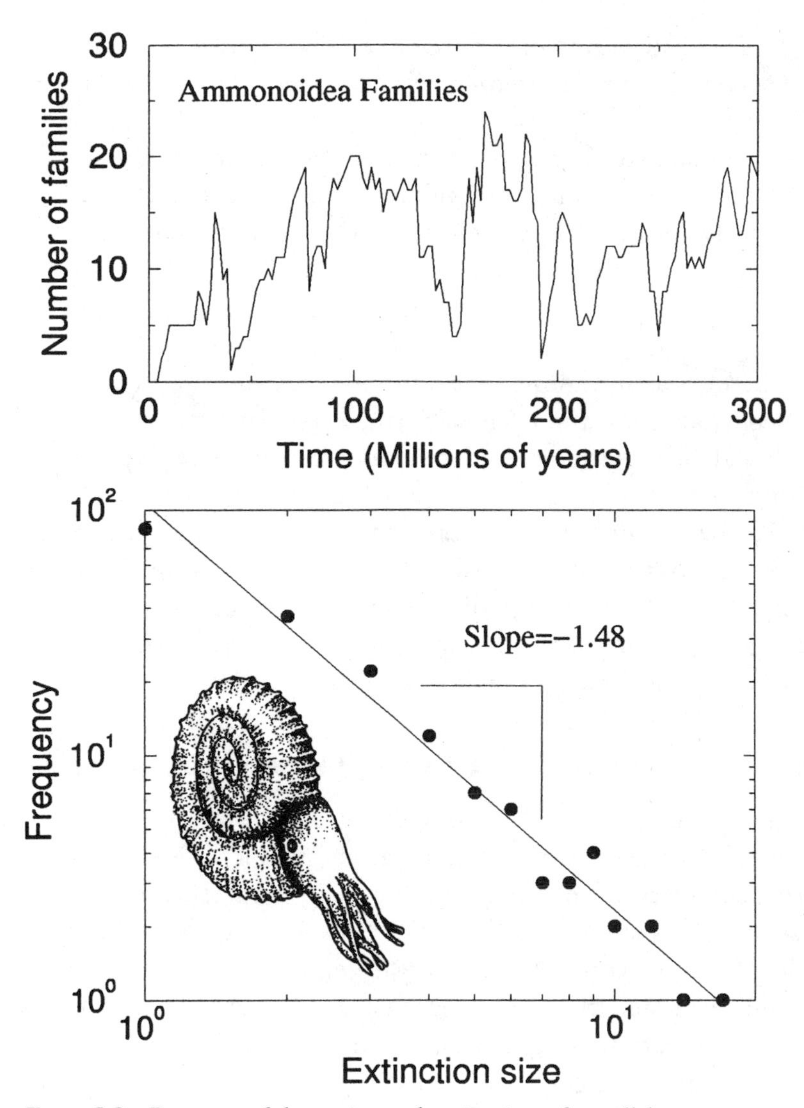

Figure 9.8 Patterns of dynamics and extinction of a well-known group of marine organisms: the Ammonoidea. These molluscs flourished over a long period of time, although they approached complete extinction a number of times (upper plot). The frequency distribution of extincctions is shown in th elower plot (in log-log scale). A power law is observed.

1. The extinction pattern of species (or families or other taxonomic units) is clearly punctuated. Long periods of time show low extinction rates, but from time to time there is a sharp rise in extinction levels. The time series seem to exhibit fractal features,[14] although this evidence is controversial.[15]

2. The distribution of extinctions follows a power-law decay.[16,17]
3. The lifetime distribution of family durations follows a power-law decay.
4. The statistical structure of taxonomic systems also shows fractal properties. Bruno Burlando, at the University of Geneva, in Italy, showed, for example, that the number of genera containing *S* species follows a power-law distribution. In other words, the tree of life exhibits fractal branching.[18,19]

These are all intriguing observations. They suggest, first, that there is no obvious separation between small and large extinction events (as should be expected if different mechanisms were operating for extinctions of different sizes), and second, that scaling laws are at work, both in the dynamical pattern and in some structural features, such as the organization of organisms into groups. Not surprisingly, these observations led to a group of scientists, including Stuart Kauffman, Per Bak, and Kim Sneppen, to suggest that the evolution of life takes place at the edge of chaos.[20,21]

Coevolution in Rugged Fitness Landscapes

The first attempt to understand large-scale evolution in terms of a complex adaptive system with interactions among different species was introduced by Kauffman and Johnsen, who used previous theoretical ideas on fitness landscapes.[22] Kauffman extensively developed the appealing intuitive concept of the great geneticist Sewall Wright (who used the term "adaptive landscape" instead of "fitness landscape"). The basic idea is that single species can be characterized in terms of a string of genes defining its genotype. Strings have an associated real number. This number is the *fitness* of the string in terms of the phenotype* it produces, and the distribution of fitness values over the space of genotypes defines the *fitness landscape*. Adaptation is then thought of as a process of "hill climbing" toward higher, nearest peaks. We can imagine a simple situation (Figure 9.9) where fitness requires the specification of only two traits, whose characteristics can be measured (the x and y-axes of the plot). Imagine that these two characteristics describe the

*Biologists define genotype as the set of genes in the genome and phenotype as the set of traits characteristic of the organism under consideration.

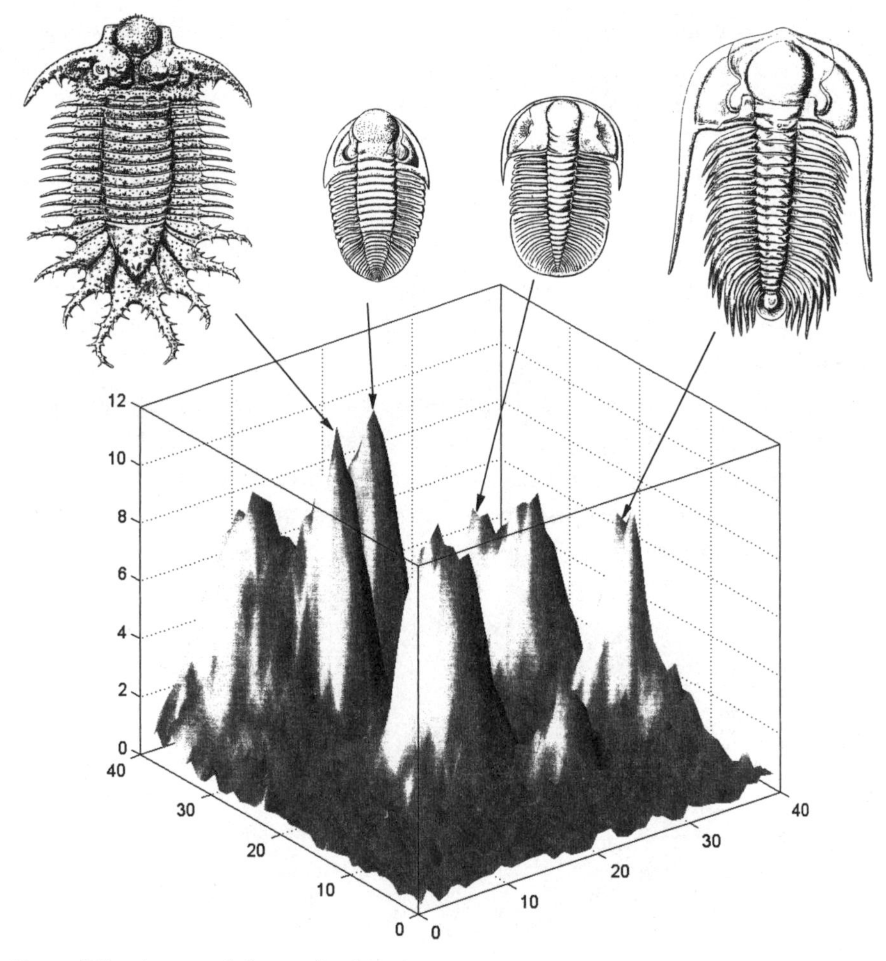

Figure 9.9 A rugged fitness landscape.

shape of the organism and that different combinations are allowed. The fitness landscape gives us an idea of how optimal these combinations are, and for any fixed environment there is a set of peaks corresponding to best fit combinations. In Figure 9.9 we have represented one of these simplified landscapes whose peaks are artificially linked with some possible morphologies of known fossil trilobites.*

Depending upon the distribution of the fitness values, the fitness landscape can be more or less rugged. The ruggedness of the landscape is a crucial property, strongly constraining the dynamics and leading

*This is only a graphical representation and has nothing to do with real trilobites.

to universal phenomena. But if we are modeling macroevolution, we must take into account many different. Each is characterized by the number of genes N and by another parameter K, which is related to the degree of ruggedness. The parameter K indicates how many other genes influence any given gene. In other words, changes in one gene depend on K other genes.* Imagine, for simplicity, that we use a binary definition for each trait. Each possible genotype will be a string of bits for which we can define some fitness value. The fitness landscape is now restricted to a hypercube in an N-dimensional space. For the very simple case of $N = 3$, a simple three-dimensional cube is enough (Figure 9.10, right). The fitness of a given string is obtained by means of a table of values like the one shown in Figure 9.10. For each particular combination of zeros and ones, each value makes a specified contribution to fitness. The fitness of the organism is the average fitness of its genes (i.e., $w = (w_1 + w_2 + w_3)/3$, see Figure 9.10, last column in the table).

This NK model and its statistical properties have been widely explored. Species evolve by means of so-called *adaptive walks*. Here,

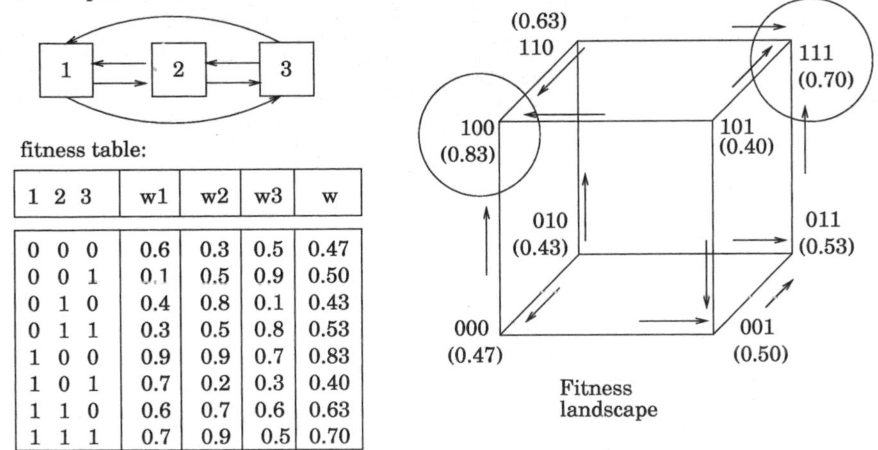

1 2 3	w1	w2	w3	w
0 0 0	0.6	0.3	0.5	0.47
0 0 1	0.1	0.5	0.9	0.50
0 1 0	0.4	0.8	0.1	0.43
0 1 1	0.3	0.5	0.8	0.53
1 0 0	0.9	0.9	0.7	0.83
1 0 1	0.7	0.2	0.3	0.40
1 1 0	0.6	0.7	0.6	0.63
1 1 1	0.7	0.9	0.5	0.70

Figure 9.10 An example of the NK landscape model. Here $N = 3$ and $K = 2$. The possible final equilibrium states are indicated by circles. The possible adaptive walks are indicated by means of arrows.

*In Chapter 3 a related type of NK model was used, also due to Kauffman (1993), describing the dynamics of genetic networks. There no fitness was introduced, since no phenotypes were considered.

for a single species, we choose a given trait and flip a coin (i.e., mutate the bit). Then we look at the fitness table, and if the average fitness of the new configuration is higher than the last one, an adaptive walk occurs: the species moves in the fitness landscape. If not, no walk is allowed. This simple procedure causes species to climb "hills" in the landscape until they reach a local peak. Afterwards, nothing happens. For $K = 0$ (known as the Fujiyama landscape) no interactions among different traits are present, and the landscape is very smooth, with a single global maximum and an expected number of walks $L_w = N/2$ to reach the optimum. This is a highly correlated, simple landscape. At the other extreme, when $K = N - 1$, the landscape is fully random. A statistical study of this landscape reveals several interesting properties.[22] For instance, (a) the number of local fitness optima is maximum; (b) the expected number of fitter one-mutant variants drops by one-half at each improvement step; (c) adaptive walks to optima are short, with a characteristic value $L_w \approx \log(N)$.

Now the problem is how to obtain a more complete picture of an evolving system formed by many species in interaction. For this we can use the so-called NKC model. The new parameter C introduces the number of couplings between different species. Again each species is represented by just a string (instead of a population of individuals), which stands for the average characteristics (the phenotypc) for that particular species. But now the previous table must be extended in order to take into account that each trait receives inputs from C other traits belonging to different species. These traits are chosen at random among the S species involved (Figure 9.11 a,b, Box 1).

In computer simulations the Kauffman–Johnsen model shows a wide variety of dynamical behaviors as the parameters are tuned. One particular property is the appearance of two well-defined regimes. The first is the high-K, chaotic, phase, where changes in the ecosystem are constantly taking place and the system does not settle down to a number of local optima; one might talk about a chaotic Red Queen phase. The other is the low-K, ordered (or frozen), phase, where all species reach local optima (the so-called Nash equilibria in economic theory). At the boundary between these regimes, species in a finite system reach local peaks, but any small perturbation generates a *coevolutionary avalanche* of changes through the system. The distribution of these avalanches follows a power law, as expected for a critical state. Kauffman and Johnsen mapped these avalanches into extinction events, suggesting

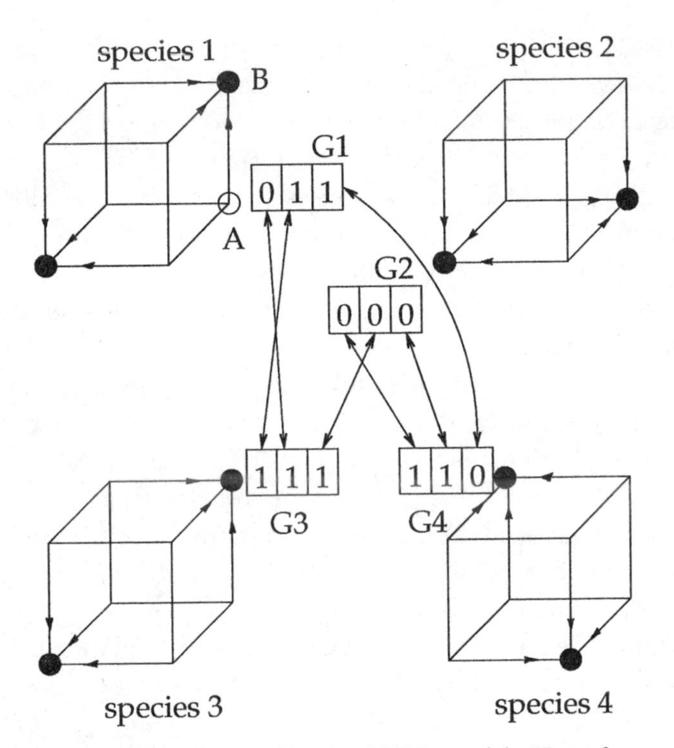

Figure 9.11 Changes in the NKC model. Here four species are considered, defined by $N = 3$ bits. Nash equilibria are indicated as black circles and the current states by means of white circles. (A,B) Interactions among genes of different species are shown as arrows. The jump of species one towards its local peak leads to a change in the other landscapes in such a way that species 3 and 4 have now new local maxima to be reached (lower picture). (C) The two phases for the NKC model with $K = N - 1$.

that the number of changes in species is proportional to the extinction of less-fit variants. If this analogy is used, then the obtained scaling relation for the number of avalanches $N(s)$ involving s changes is $N(s) \propto s^{-1}$, which does not agree with the value reported from the fossil record. Kauffman has recently developed a variation of this model[23] that allows changes in the parameters (i.e., allowing connections among species to coevolve). This version has been shown to self-organize itself to the critical state, with avalanches following the correct $\alpha = 2$ exponent (Figure 9.12). The final picture that emerges from this model is that as species tune their own landscapes (by readjusting their parameters and thus landscape ruggedness) they poise the entire ecosystem close to the critical boundary.

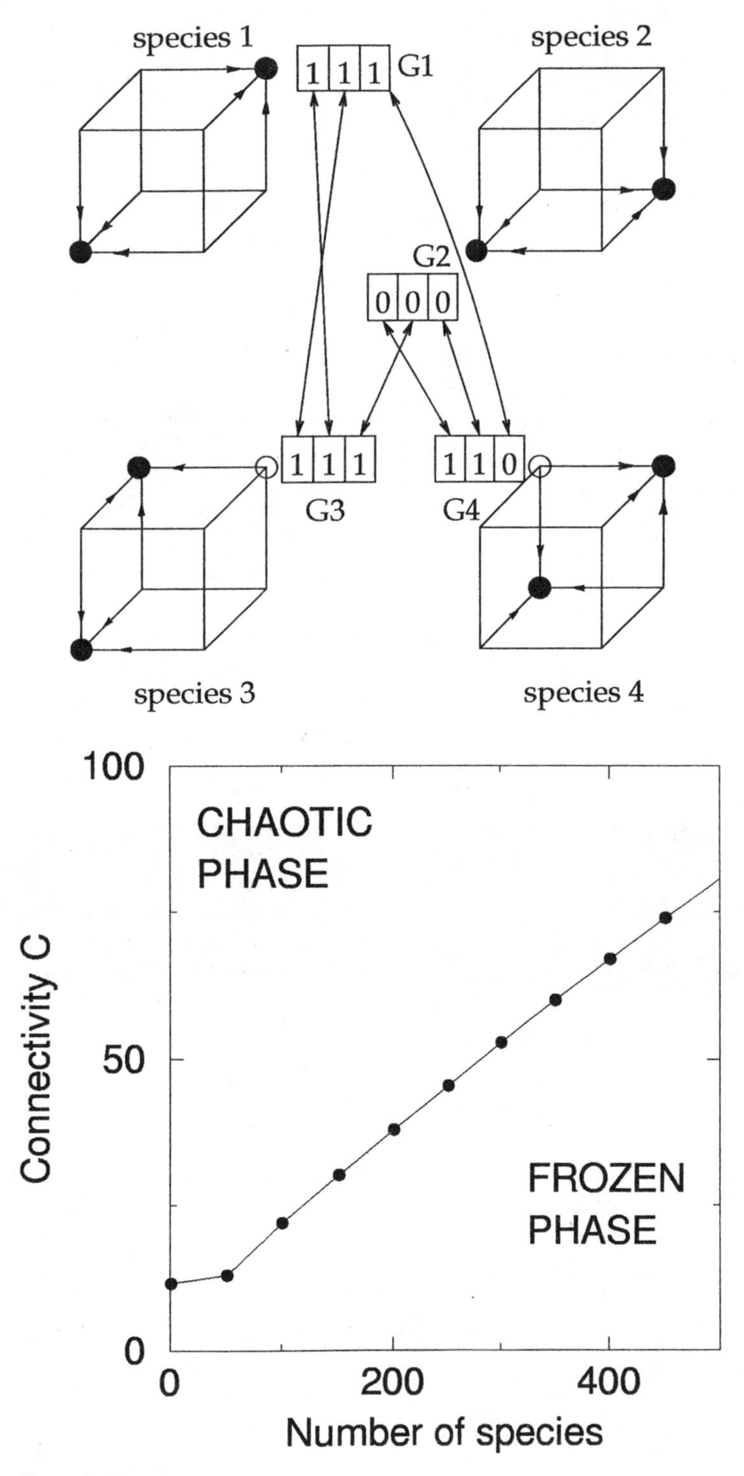

Figure 9.11 *Continued*

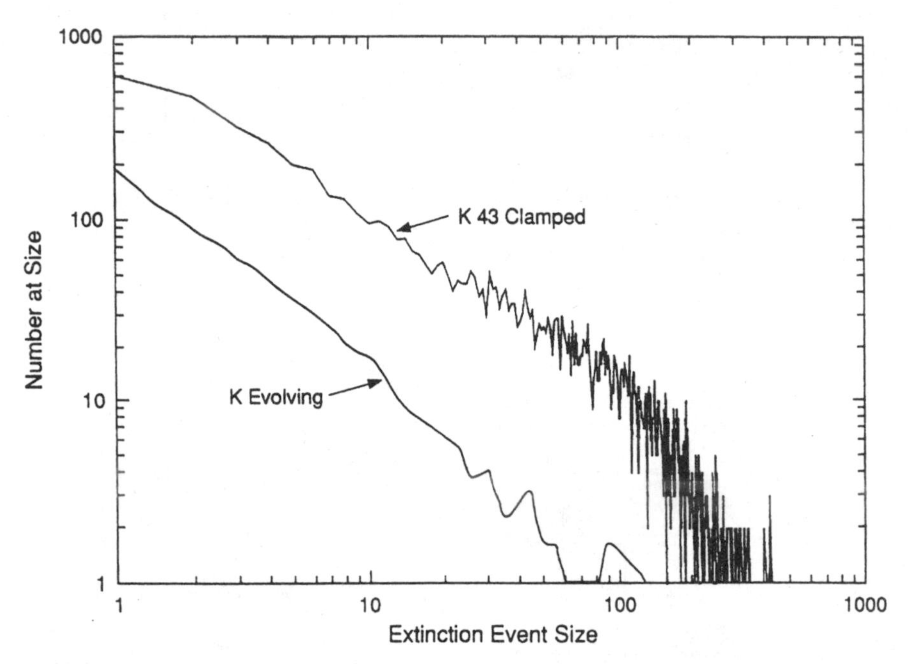

Figure 9.12 Scaling in the NKC model. Here two cases are shown: the fixed-parameter model (so parameters have been previously tuned) and the evolving-K model, where connectivity changes through the evolution.

Phase Space of Rugged Fitness Landscapes

The NKC model involves three basic parameters describing (a) the number of traits required to characterize a given species (N), (b) the number of so-called *epistatic* interactions among genes in the same species (K), and (c) the number of interactions among traits of different species (C). In this model it is assumed that all individuals can be considered equivalent and thus only one genotype (instead of a population) represents the whole population. In Figure (9.12a) we show a simplified view of the landscapes of three coupled species. Here $N = 3$, and the local peaks are indicated by black circles. Some value of K is assumed to be defined. The specific set of traits displayed by each species is indicated as a string. The connections between traits of different species are indicated by means of arrows. We can see that species two, three, and four are already at their local peaks, but species one is not (A, white circle). It is close to a local peak (B), and thus in the next step it will perform an adaptive walk toward the peak at B. In Figure 9.12b, a possible result of the adaptive walk is shown: species one is now at the local peak as well as species two. But species three and four are not located in fitness peaks: their landscapes have been modified by the adaptive move of species one. This is what occurs at the chaotic regime: changes in a given species *propagate* through the system. There is, however, another regime (characteristic of small connectivities): the frozen, ordered, regime where species reach their local peaks and remain there.

Let us consider the NKC model and see how the NC-plane for $K = N - 1$ (i.e., random landscape) shows two well-defined phases. A simple derivation has been obtained by Per Bak and coworkers.[24] The basic idea is the following: we know that for this landscape the number of walks L_W required for a given species to reach a local fitness peak is on average $L_W \approx \log(N)$. Let us assume that all species are located at local peaks and that a perturbation is performed: one of the species (say species one) is moved to a random position in its fitness landscape. This species will start to climb toward some local peak. If the interactions among species are weak, the other species will remain unaffected by the adaptive walks of species one. But if C is large enough, then the other species can see their landscapes modified and start to change, too.

Each adaptive walk of species one involves a change in a given trait. But in fact, the fitness of other species depends, through C, on the values taken by the genes/traits of species one. If any of these genes/traits are among the ones that changed through the walk, the affected species will be set back in evolution. The question is, What is the critical condition that defines the combination of N and C that can trigger a "chain reaction" able to percolate through the system? The critical condition is easily obtained. The probability that a given trait in a random species depends on species one is C/N. The critical condition is that at least one change in a species occur. This means that

$$L_W = C_{crit} N = 1,$$

i.e., when on average one out of C randomly chosen genes is among the L_W changed genes. This gives the critical line in NC-space

$$C_{crit} = \frac{N}{\log(N)},$$

which is shown in Figure 9.12c.

Networks, Unpredictability, and Ecosystems

Kauffman's results support to the conjecture that evolution could take place close to critical states. This idea was further explored by Per Bak and Kim Sneppen, who developed a simple model of evolution displaying self-organized criticality. This and other models have confirmed that under a wide range of conditions, critical dynamics are likely to occur in model ecosystems and thus perhaps also in real ecologies. Mark Newman has also developed similar models, some involving both internal and external dynamics, others purely externally driven evolutionary dynamics. The latter reproduces some of the patterns found in the fossil record, a result that has generated a considerable debate. In this context, the extent to which nonlinear phenomena are important in driving extinction events is a hot issue. As some authors suggest, such as Princeton's ecologist Simon Levin, critical states are likely to be relevant in terrestrial ecosystems where immigration/speciation events are especially important, but clearly, external events have played an important role in shaping our biosphere, and it is far from clear that critical phenomena might be operating at all scales. Although we take a partisan view in favor of ecologically based dynamics and emergent phenomena, it is important not to forget that alternative models can provide a complementary (instead of antagonistic) understanding of macroevolution and extinction.

The previous chapter presented several aspects of the nonlinear behavior of complex ecologies. Some field data seemed to support the idea that ecologies might display some features characteristic of criticality. The presence of similar phenomena at such different scales (ecological and evolutionary) prompts us to ask whether we might construct an ecologically based network theory of species changes. The question is relevant in many ways. One very important implication is the possibility that micro- and macroevolutionary changes would be essentially decoupled. Another is that the presence of a network-like structure in real ecologies immediately suggests that new, emergent properties and unexpected outcomes may arise from the interactions among species (see Box 1).

The network-like structure of ecosystems can sometimes lead to unlikely chains of events. Our example is about the large blue butterfly that became extinct in Britain in 1979. It was one of five European species of *Maculinea*, a genus that was well known in Europe for its attractiveness and rarity. The large blue lived in warm, dry grasslands in England, where the female laid a single egg on a flower bud of wild thyme. After hatching, the larva would eat the flowers and seeds for two or three weeks, then fall to the ground and wait to be discovered by a red ant. The caterpillar produced a secretion similar to that of ant larvae. The ants carried the caterpillar into their nest and fed it. About a year later, it matured and crawled to the surface as an adult.

The decline of this species was not understood at first; only later did field analyses reveal the reasons for its extinction.[25] The basic chain of events is summarized in Figure 9.13, where the arrows indicate causal connections among the components (either interactions or transitions). It all started with the introduction of the myxoma virus. This virus was intended to control the growing populations of rabbits in different parts of Europe. The strategy was spectacularly successful: Rabbits (H) declined very quickly to very low levels. Browsing by rabbits had promoted the growth of certain grasses (*G*), which otherwise would have been overgrown by woody shrubs (*G*). The ants (A) built their nests in the roots of the grasses, and under these conditions they took care of the caterpillars (C), which eventually metamorphosed into adult butterflies (B).

As the rabbits declined (Figure 9.13b), browsing fell too, and the grasses were overgrown by the woody shrubs, reducing the population of ants. This was followed by the decline and eventual extinction of the blue butterfly. Although direct interactions among members of an ecosystem can be understood in terms of classical coevolution and adaptation, this chain shows that indirect interactions can play a very important role. The response of some species to the perturbation of another species far away in the food web can be highly complicated and sometimes unpredictable. In these cases it is the network structure, more than direct species–species interactions, that really matters.

It is difficult to see these cascades of ecological interactions at work in today's ecologies, and the problems only increase when one is dealing with the fossil record. But several authors have suggested that cascade effects have been important in the past. This would be the case, for example, as suggested by Jablonski, of the probable collapse of marine food chains at the end of the Cretaceous.[26] Another example was the extermination in North America at the end of the Pleistocene of the large herbivores such as mastodon and mammoth, which began about 18,000 years ago. These were the most attractive human prey, and their extinction may have brought extensive vegetational changes that in turn would explain the concomitant disappearance of many other vertebrates.[27]

It is easy to construct a very simple network model of large-scale evolution involving a set of N species.[28] In this model, interactions

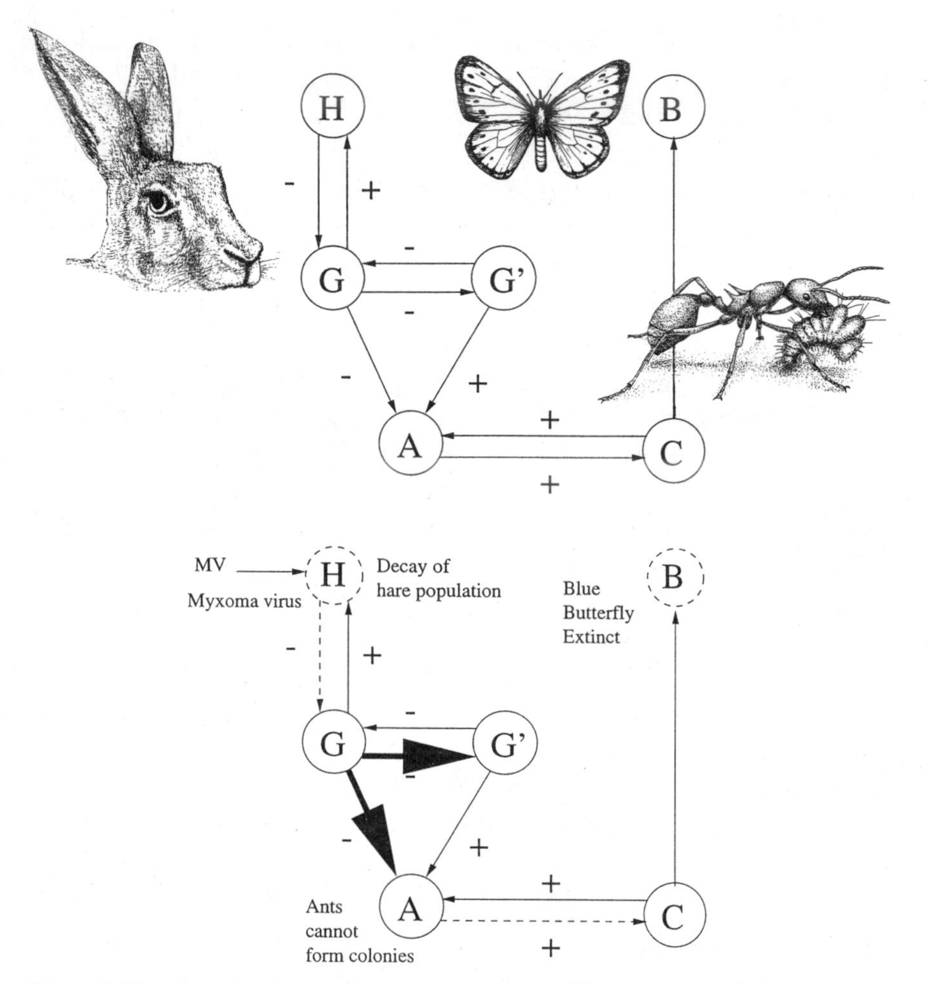

Figure 9.13 Complex interactions in ecosystems. The example of the blue butterfly (see text) provides a clear illustration of the relevance of higher-order interactions. Here the players are rabbits (H), two types of grasslands (G and G'), a species of ants (A), the caterpillars of the blue butterfly (C) and the adult butterfly (B). The effects of each component are indicated prior to the introduction of the myxoma virus (upper picture) and after it (lower picture). As the rabbit population declined, the effect propagated through the food web and triggered the extinction of the butterfly.

among species are introduced by means of a connectivity matrix, and evolution is represented through changes in its elements. The "state" of each species is described by a binary variable $S_i \in \{0, 1\}(i = 1, 2, \ldots, N)$, for the ith species. Here $S_i = 1$ if the species is alive, and $S_i = 0$ if it is extinct. So the whole ecosystem can be described in terms of a simple directed graph where the connections are initially set to random values. Each species receives input from some others. Input can be either positive or negative, depending on whether the receiving species is favored (e.g., by prey or cooperator) or harmed (e.g., by predator or parasite). In this model species are described by their sets of connections.

We then introduce a few rules:

(i) Random changes in the connectivity matrix. At each time step we pick up one input connection for each species and assign it a new, random value. This rule introduces changes into the system. These changes can be evolutionary responses (such as the adaptive walks in the Kauffman–Johnsen model) or environmental changes of some sort. In this sense, we incorporate the two sources of change (intrinsic and extrinsic).

(ii) Extinction. At each time step we compute the sum of the inputs for each species, and this sum defines the condition for extinction: if it is negative, the species is extinct ($S_i = 0$) and all its connections are removed. Otherwise, it remains alive ($S_i = 1$).

(iii) Diversification. As species disappear as a consequence of the previous rule, their empty sites are refilled by diversification. Each extinct species is replaced by a randomly chosen survivor. The connections of the survivor are simply copied into the empty site.

In analyzing this model (mainly in collaboration with Susanna Manrubia, now at the Fritz-Haber Institut in Berlin) we have found good agreement with the fossil record. It shows strongly nonlinear behavior, with avalanches of extinction as well as the correct power law distribution of extinction sizes. Figure 9.14 shows the temporal dynamics of extinctions. We can see that small and very large events are generated by the same dynamical rules. Most often, the extinction of one species has no consequences for the others. But from time to time, a species with many positive inputs to others becomes extinct, and its removal can have a destabilizing effect, which can propagate further, leading to a mass extinction. Since most large extinctions result from the removal of such cornerstone species (see last chapter), one might conjecture that large extinctions are triggered

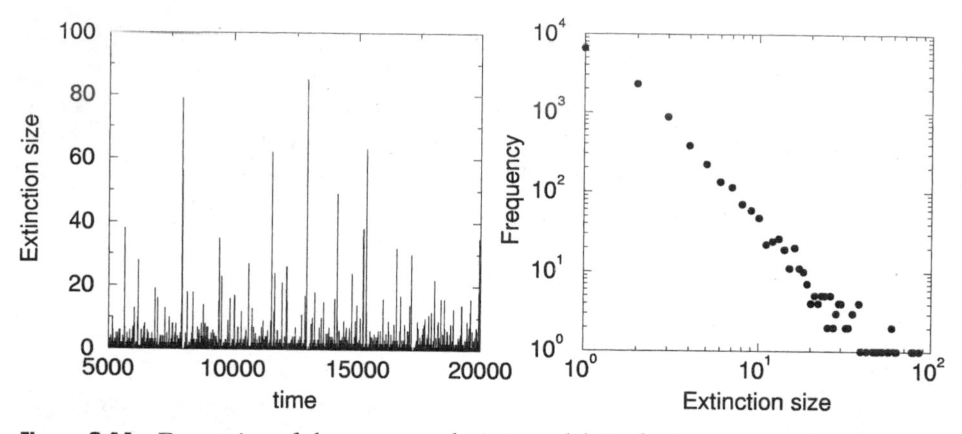

Figure 9.14 Dynamics of the macroevolution model. Left: time series of extinctions. We can see small and very large events (mass extinctions), which are generated through the same rules. The right plot shows the corresponding distribution of extinction sizes.

by external events that affect the stability of some key species, one that has a large-scale influence on the global dynamics of the biosphere.

The model also recovers other relevant properties. Species diversify, so we can follow the trees generated by successive speciation events. These trees describe genera in the model and have a power law distribution. In other words, we have a fractal taxonomy, just as in Burlando's studies of real data. We can also follow the dynamics of these genera and see how they change in time. The statistics for the duration of genera reproduce the observations of the fossil record. Another relevant result involves the well-known observation that the diversity of life shows an increasing trend (see Figure 9.15). In terms of families or genera, the diversity of organisms has shown an increase since the beginning of the Phaneozoic. The metazoan fossil record shows a first phase involving a logistic-like pattern of diversity (with a plateau punctuated by some declines associated with mass extinction events) followed by a further increase after the Permian extinction, 250 million years ago. Usually, scientists are not very worried by their model's transient behavior, but in this case it seems reasonable to ask how the model can reflect the apparent two-step evolution of the historical record. The result is rather interesting: the transient behavior of the model[29] looks very similar (Figure 9.16). Typically, a first increase in the number of genera (b) is observed (with some drops associated to large extinctions, as shown in (a)), and often, after a very large extinction, we see a further increase. Later on (inset, Figure 9.16b), the model reaches still higher plateaus.

This model and other related models[30] imply something important about evolution. There has been a long debate over the last decades concerning the basic mechanisms operating at small and large time scales. Some authors (particularly among population geneticists) believe that the rules operating at the small scale can be directly translated into the process of macroevolution. Others, such as Stephen Jay Gould, see different processes at work at different scales. No well-defined mechanism for such decoupling has been proposed. But the network organization of ecologies immediately provides one possible source of decoupling. Since species are not isolated entities, changes in the food web structure can propagate through an ecosystem, eventually leading to large-scale extinction events. The dynamics of these changes are not species-dependent but *network dependent*. The global features of the food web are what matters, not the specific properties of the species constituting the web. In the sandpile and other models with critical dynamics most changes are unable to propagate. They are like small

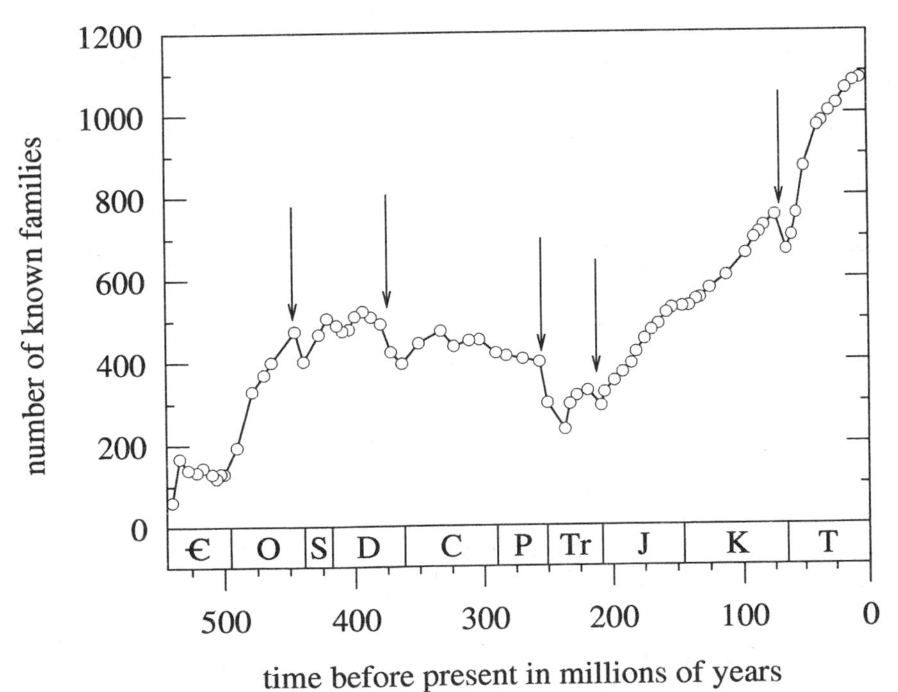

Figure 9.15 Pattern of diversification in the fossil record.

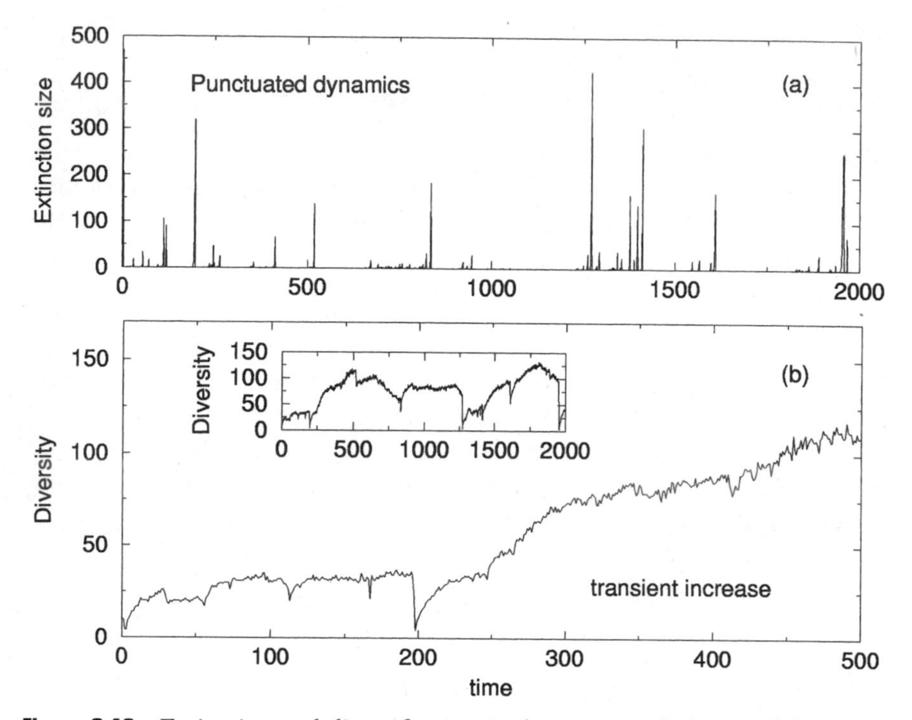

Figure 9.16 Extinction and diversification in the macroevolution model. (a) extinction events through time; (b) pattern of diversification, as measured by the number of genera present at each time step (compare with Figure 9.15). Inset: a longer simulation showing that very large drops in diversity can occur.

avalanches involving a single grain of sand. In this sense, individual adaptations can evolve, as can interactions between pairs of species (such as predator prey relationships) without affecting other parts of the network. But in the long run it is the network and the emergent properties arising from network dynamics that determine the collective behavior of ecologies. As with Bak's sandpile (Chapter 2), we cannot understand the behavior of the whole by looking at gravitation and friction acting on individual grains. Instead, the sandpile at the critical state must be analyzed as a collective phenomenon. Natural selection and adaptation are likewise operating on species in complex ecologies, but the complete picture requires us to consider the interactions among species. In the long run, the effects of selection pressures on single species are like gravity and friction operating on grains of sand: we need to take them into account, but they cannot explain the avalanches.

Complexity and Evolution in Silico

The fossil record of life, while incomplete, is rich enough to provide us with more than a glimpse of past ecosystems. Still, we cannot travel in time to see those past ecologies at work: as in astrophysics, we cannot test our theories by repeating the experiment. Evolution happened only once. If time travel were possible, we could "see the life of the past in all its strangeness, complexity and richness."[1] The only empirical approach able to cope with large-scale evolution has been provided by the remarkable long-term experiments led by Richard Lenski at the Michigan State University.[31,32] Lenski and his team followed the evolutionary change of several bacterial populations over thousands of generations. They found that different traits, such as cell size and fitness (here measured in terms of replication rate) evolved in sudden jumps. In fact, periods of stasis followed by rapid changes were quite common. More recently, Paul Rainey and Michael Travisano, at the University of Oxford, explored the same problem by studying the evolution of diversity in a spatially structured environment, which they get by simply shaking the containers, which then stratify into regions with different chemical and physical properties.[31] They showed that different mutants appeared and adapted to different environmental conditions. The mutants competed for resources and created a simple evolving ecosystem: diversity was shown to increase and then persist through time. Since these experiments allow the researchers to store ancestral forms in laboratory freezers, they have what Lenski calls a "frozen fossil record" from which evolution can be rerun again and again from different starting conditions.

Time travel is possible only in our imagination, but perhaps we might do something similar: generate a whole world of interacting and evolving creatures, able to recreate macroevolutionary patterns and perhaps behave like real life. This "life as it could be" scenario, named *artificial life*,[34] is currently a hot area of research. Artificial life originated with the work of the great Hungarian mathematician John von Neumann, who formally explored the problem of constructing a self-replicating automaton. But the relevance of artificial life to real biology and evolution became clear only with the work of the ecologist Thomas Ray.[35]

Ray's ideas started to develop when he was studying complex ecologies in Costa Rica. His training was not in computer science or physics but tropical ecology. Inspired by his field observations and the Japanese game of go, he started to think of the possibility of constructing an artificial ecology of digital organisms able to self-replicate and mutate. The model, called *Tierra*, is a set of computer programs that compete not for food but for processing time from the computer. The virtual organisms start from a short ancestral program and quickly evolve to diverse forms of increasing length and behavioral complexity. When Ray first ran the program, the earliest mutations were shorter programs that outcompeted and eliminated the original code. At some point, a creature of a very short length emerged, and Ray realized that it was nothing but a parasite: it couldn't reproduce by itself but was dependent on others to do so. Further evolution generated hyperparasites and some large extinction events. Afterwards, social behavior flourished: each organism required the help of at least one other to reproduce. And it is worth mentioning that Tierra's artificial world (and other similar systems like Chris Adami's *Avida*[32]) display jumps in their evolutionary activity (Figure 9.17).

Richard Lenski and Chris Adami have recently taken advantage of the opportunities offered by digital organisms to test generalizations about living systems that may extend beyond the organic life that biologists usually study.[36] Specifically, Using the Avida program, they explored the effects of single and multiple mutations on the fitness of two different types of previously evolved individuals. The simple types were

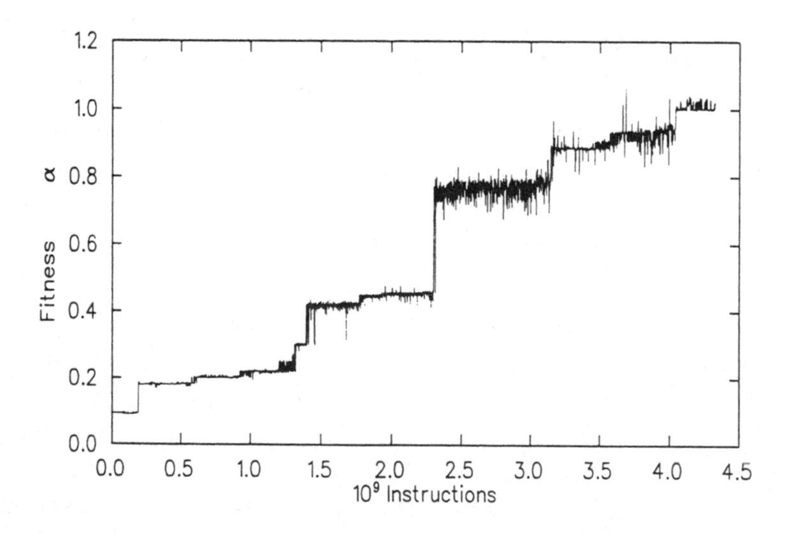

Figure 9.17 Dynamics of fitness in an artificial life simulation of evolution.

generated by allowing them to replicate in an environment favoring only faster replication; the complex ones grew in an environment where they were rewarded for performing certain mathematical functions.

The artificial life model allowed manipulation of these digital organisms and exploration of the effects of mutations on their fitness. Lenski and Adami found that the complex organisms were more robust than the simple ones with respect to the average effects of single mutations. In fact, the simple organisms, which basically only replicate, were prone to exhibit lethal responses to mutations, since these often disrupted self-replication. In complex organisms, however, where genomes were larger and a variety of different functions was present, mutations had a smaller effect. Moreover, it was shown that interactions among mutations (the so-called epistatic interactions) were present and were especially pronounced in complex organisms, as in real life. These experiments revealed that interactions are likely to be a general feature of any genome-like structures that store and manipulate information. These include the life forms on our own planet as well as those we may find elsewhere.

Back to the Cambrian

If life is a common phenomenon in the universe, its evolution may be inevitable. Errors in replication and information processing are unavoidable, and thus variation is likely to be a universal feature of primitive ecologies. If larger genomes and epistatic interactions are a universal feature of any type of evolving system, then the emergence of complexity may be a universal phenomenon. What about the explosion of life forms? As we saw at the beginning of this chapter, the Cambrian "big bang" was an unprecedented event that led to the emergence of most animal phyla. All the basic animal body plans were created in just 10 million years, and nothing like that took place again on Earth. Even after the most devastating mass extinction events, when large numbers of species were wiped out, the recovery of diversity did not include the creation of new body plans. This includes the end-Permian extinction, some 250 million years ago, which paleontologist Douglas Erwin calls "the mother of all extinctions." About 90 percent of all ocean species disappeared during the last million years of the Permian Period. On land, more than two-thirds of the reptile and amphibian families vanished. Afterwards, a rapid radiation took place at the family level, but no new classes or phyla emerged. Why?

Again, complexity theory offers an interpretation.[37] Kauffman argues that large-scale evolution takes place on very rugged, weakly correlated fitness landscapes. Starting from some initial condition where low-fitness multicellular organisms were present (the Ediacarans?), there was a rapid exploration of the landscape. This initial exploration led to an increase of diversity of improved alternative morphologies: the phyla. As the rate of fitter mutations that alter early developmental processes (which define body plans) slowed, variant species founding orders, classes, and the lower taxonomic groups became established. Kauffman's argument is completed by the (quite reasonable) assumption that mutants affecting early development have more profound effects than those affecting late development. In this scenario, by the Permian extinction, early developmental pathways would be expected to have become largely frozen. Fitter variants altering basic body plans would be very rare, but variants affecting late development would remain common, allowing radiations of new families to occur.

A study by the paleontologist Gunter Eble seems to give support to Kauffman's conjecture. Eble's idea can be formulated as follows: if S is the cumulative number of assumed improvements (new body plans) and G is the cumulative number of attempts, then Kauffman's prediction is that S will grow with G in a logarithmic fashion.[38] Graphically, this means that S grows rapidly at the beginning and then slows down (in a particular way defined by a logarithmic relation $S \approx \log(G)$). The cumulative number of attempts can be approximated by the cumulative number of lower-level entities (e.g., genera) viewed as successive experiments in the generation of higher-level entities (such as phyla, classes, and orders). In Figure 9.18 we see the result of Eble's analysis: there is a well-defined logarithmic increase involving rapid innovation at the beginning and a marked slowdown in later stages. Because most animal phyla lack skeletons and therefore leave almost no fossils, we cannot know whether this picture is statistically representative. Still, this result is certainly promising and suggests that previous models can be meaningfully tested. In fact, Eble tested Kauffman's model for classes and orders as well (with statistical confidence) and found that it does not hold at those levels. This suggests that models based on complexity theory can be not only tested but modified in light of new empirical data. Interestingly, this type of phenomenon—rapid diversification followed by locking—is very common to complex adaptive systems evolving on rugged landscapes. It seems to be common in real-world systems as well, although no serious quantitative studies have yet been performed.

The Cambrian explosion and its underlying mechanisms are a great scientific challenge, both for biologists and for those who try to understand the origins of innovation and diversity in nature. Some of the most fascinating discoveries in this area are taking place in molecular biology and embryology, where many of the basic keys to macroevolution are to be found.

We are at the beginning of a new integrative theory of development and evolution. The huge complexity of gene networks means that we will need new theoretical studies and that new mathematical tools will be required to integrate the vast amounts of information concerning genes, gene interactions, developmental pathways, and their relation to the fossil record of life. The Cambrian event can be of extraordinary value in our search for how self-organization, information, and adaptation are related, and how they relate to evolution. It will be a long journey across the vast ocean of time.

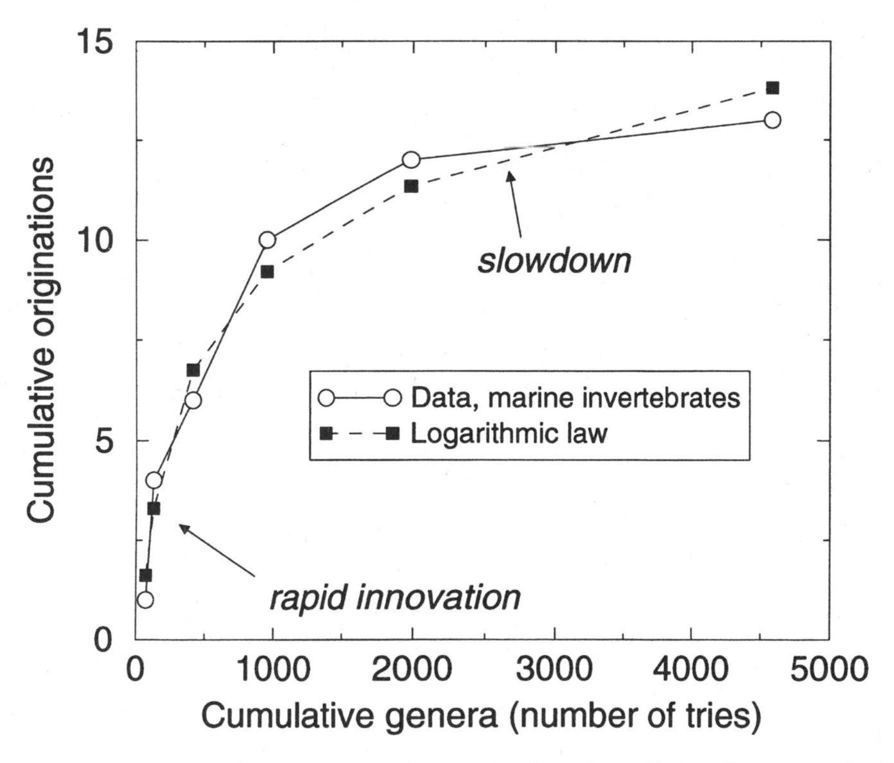

Figure 9.18 Logarithmic increase of innovation in early evollution (Data provided by G. Eble).

TEN

Fractal Cities and Market Crashes

The city was desolate. . . . It lay before us like a shattered bark in the midst of the ocean, her masts gone, her name effaced, her crew perished, and none to tell whence she came, to whom she belonged, how long on her voyage, or what caused her destruction; her lost people to be traced only by some fancied resemblance in the construction of the vessel, and, perhaps, never to be known at all.

—J. L. Stephens

The Chaos of History: Path Dependence

The birth and death of ancient civilizations is a fascinating theme to historians and scientists alike. Why such powerful and successful organizations declined and eventually disappeared is one of the most intriguing problems for the social sciences.[1] One of the most interesting examples of the rise and decay of a civilization is provided by the Mayans, who built a powerful society extending over a vast area of southern Mexico and northern Central America. There flourished the most brilliant civilization of the New World in pre-Columbian times, spanning nearly twelve centuries until 1541. When in 1839 two American explorers, John Stephens and Frederic Catherwood, found the first Maya ruins in the middle of the rainforest in Copán, they were astonished by the terrible contrast between the majesty of those huge and elegant monuments and the wild jungle surrounding them. The city that Stephens discovered was already empty before Columbus's journey.

And the streets and houses were silent, like "a shattered bark in the midst of the ocean." At other sites in Mexico, ruins like Monte Albán, Palenque, and Mitla were waiting to be discovered after centuries of abandonment.

Human societies are complex adaptive systems with many features in common with complex biosystems. All the large-scale patterns arising from human activities show trends through time indicating that to a large extent, our social structures, our cities, and our economies are not static but evolving. Cities and civilizations grow and die following hidden patterns where historical accidents and nonlinearities are at work. Let us start with a very interesting story about technological innovation, where some familiar phenomena are at work.

The standard keyboard configuration used by almost all computers and typewriters is known as the QWERTY model (because of the first six keys in the top row of letters). If they think it about it at all, most people think that QWERTY is the result of a rational, optimal distribution of keys to allow faster typing. We usually think that most technological products have been created after a long, probably difficult period of design and testing. The story of QWERTY, however, is not a tale of optimality but of a frozen accident.

The letters used most frequently in English texts are R, S, E, A, H, D, I, T, N, and O, which together make up more than 70 percent of letter usage in English. Common sense establishes that these letters should be the letters most easily available to our fingers in the keyboard. But QWERTY strongly contradicts this picture. The awkward locations of several of these letters clearly make the typewriter a suboptimal device. Why, for instance, do the rarely used *j* and *k* get prime real estate under the two strongest fingers of the right hand? How did this come about?

The typewriter was invented by C.L. Sholes around 1860 and improved over the next decade.[2] More than thirty models were made in the search for a satisfactory design. The problem with the initial designs was that because each key controlled a lever that printed a letter, these levers became locked if one typed too fast. To solve this problem, Sholes decided to put some common letters in difficult locations through the keyboard. A safer disposition of keys encouraged smooth typing but at the cost of lower speed. As Duncan James notes,[2] "ergonomically unsound, the QWERTY layout is an anachronism on the modern word-processor, but attempts to introduce a better system have, so far, failed."

The question is why this suboptimal arrangement was so successful that all standard keyboards now follow it.

QWERTY was not alone. At the time of Sholes's design, other types of typewriters were begin created by other engineers. Their designs were diverse (Figure 10.1), and some of them were superior in speed, using a different type of technology and organization. But the next decades were those of an exploding industry of typewriters, and several companies decided to use some of the available designs. During the period from 1880 to 1900, many clever and often extraordinary machines were launched onto the market by designers and manufacturers based in the northeast United States, all fiercely competing against one another. At the time these industries started to flourish, QWERTY was already in the market, taking advantage of a number of contingencies, like the fact that some people were being trained in typewriting with QWERTY machines. Despite only a small number (a few thousand) of QWERTY machines in use in the early 1880s, the initial advantage of this system was strongly amplified by the increasing returns generated by the economic forces, and QWERTY finally won the race. Some other designs were technologically better (others bizarre), but today we can see them only in science museums around the world (see Figure 10.1). None of them survived.

The dominancc of the QWERTY keyboard is an example of a very common phenomenon called *increasing returns*. Conventional eco-

Figure 10.1 An old typewriter.

nomic theory assumes that economic actions trigger negative feedback, eventually leading to an equilibrium where market shares and prices reach a steady state. According to this thesis any market should approach a single equilibrium state providing the most efficient use of resources, regardless of the market's initial state. But as the economist Brian Arthur has pointed out,[3,4] this nice picture often does violence to reality. Instead of being stabilized by negative feedback, economic markets seem to be driven by positive feedback that lead to complex fluctuations and a variety of possible steady states. These multiple solutions arise from path-dependent processes, where historical accidents plus increasing returns play a crucial role in defining the final outcome.

Brian Arthur, an economist at the Santa Fe Institute, has argued for years against the dominant equilibrium view of economics based on decreasing returns. He illustrates the importance of such path-dependent processes using a well-known example: the videocasette recorder. Initially, two competing formats fought to dominate the VCR market: VHS and Betamax. Both were sold at similar prices, and each experienced increasing returns as its market share increased. Only one of those two competitors survived. Small gains in the initial state led to further improvements in their market share. The small advantage VHS held over Beta was quickly amplified, and as can happen with two biological species in competition, symmetry was broken in favor of the one that took advantage of random fluctuations. There is no rationality in this final decision, just nonlinearity and broken symmetry. The process leading to the final state is strongly path-dependent.

The economy contains many examples of such increasing returns. Arthur and his colleagues have explored these ideas through a very useful mathematical model known as the Pólya urn problem.[4] In 1931 George Pólya solved a particular version of this problem, which was later generalized by Arthur, Yuri Ermoliev, and Yuri Kaniovski. Let us consider the original formulation by Pólya (Figure 10.2, top). Imagine an infinite urn that at the outset contains two balls of different colors, black and white, say. The rules are very simple: at each step, take at random one of the balls in the urn. If it is black, say, we return it to the urn together with another black ball. If it is white, we return it and add a white ball. The same operation is repeated again and again. Let

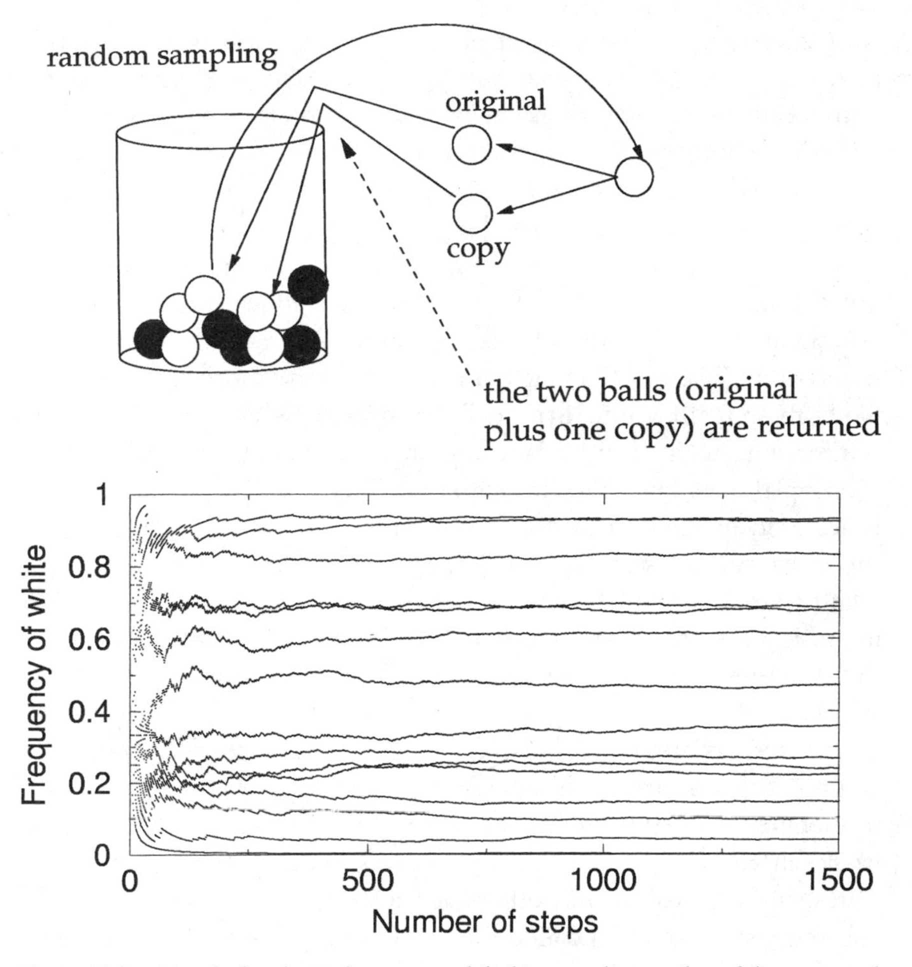

Figure 10.2 (a) rule for the Polya urn model; (b) several examples of the temporal evolution of the frequency of white balls: we can see that the probabilities stabilize to fixed values, but any value between zero and one is possible. An infinite set of attractors is available.

us ask the following question: what will be the distribution of colors after a large number of steps? Three natural solutions seem possible: one of the colors always dominates, its proportion approaching 100 percent; no equilibrium state emerges; or the urn will always reach an equal-frequency distribution.

Surprisingly, none of these things happens. If we follow the system in time, and make many runs of it, we discover that the system reaches well-defined steady states (Figure 10.2, bottom). But their number is *infinite*: any possible ratio of black and white balls is equaly likely to occur. There is not just a multiplicity of solutions but an infinite number of them. Arthur has shown that this basic dynamical pattern underlies a number of economic processes and explains some relevant though striking features of how the economy self-organizes in space and time. In particular, Arthur has shown that industrial location displays path-dependence as new firms are added to a given region where some initial firms were settled. These new firms are added one at a time by "spinning off" from parent firms. Assuming that each firm stays close to its parent location and that all existing firms are more or less equally likely to spin off a new firm, a Pólya process is at work. This self-reinforcing mechanism leads to regional settlements where historical events play a leading role (although not all patterns are possible: sometimes a given region can dominate, sometimes several regions share the industry). This means that large aggregations of industry might result from increasing returns rather than geographic superiority. No intrinsic geographic advantage (such as access to raw materials) caused the rapid expansion of the American electronics industry in Silicon Valley.

Arthur has often mentioned the deep analogies between economic systems and nonlinear physics. As happens with magnetic materials (remember the Ising model) where self-reinforcing processes are at work, an economy can lock into one of several (perhaps many) end states. In fact, Arthur and others at the Santa Fe Institute have advocated a view of the economy as a complex evolving system, with many of the features displayed by biological systems, including complex fluctuations, punctuated equilibrium, adaptation, and coevolution.[5]

Fluctuations in economic indices strongly suggest the presence of self-similar dynamics, together with characteristic features often found in turbulent fluids. These self-similar properties in the economy were early demonstrated by Benoit Mandelbrot. By analyzing long-term fluctuations of financial market records, he showed that it is often virtually impossible to distinguish a daily price record from a monthly records. Figure 10.3, for instance, plots the logarithm of daily closing prices of an economic index. We also plot a 60-day daily price record, a 60-week weekly price record and a 60-month monthly price record. It is very difficult to distinguish them.

It is still not clear what rules underlie large-scale market dynamics, but some trends start to appear. The market is reasonably predictable at

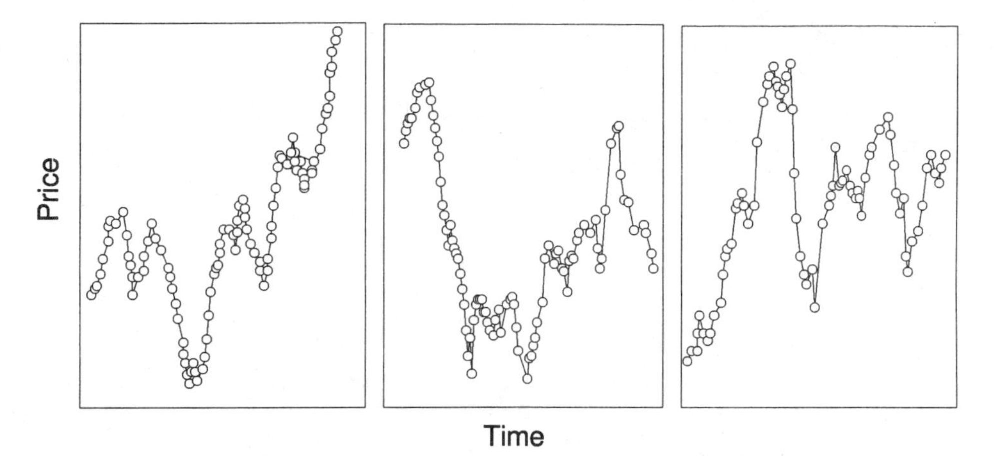

Figure 10.3 Fluctuations in closing prices of the DAX index. Here the time series involves a 60-day daily price record, a 60-week weekly price record and a 60-month monthly price record. All the plots look similar and reveal that fluctuations at all scales are present.

short time scales. The proof of this is the success of two chaos theorists, Doyne Farmer and Norman Packard, who started a private investment firm (the Prediction Company) with very good results. Markets are nonlinear systems, and the nonlinearities make them predictable at some scale but intrinsically unpredictable in the long run.

What about long-term trends? This is a hot area of research (known as *econophysics*) where many economists, mathematicians, and physicists, but also biologists, have found a common language for new theoretical approaches and metaphors.[6] Physicists such as Eugene Stanley, Didier Sornette, and Jean-Paul Bouchaud have suggested that market fluctuations share some nontrivial features with critical phenomena in physics. One very interesting result is that large financial crashes are analogous to the critical points studied in Chapter 2. Focusing on the extreme behavior of stock markets, namely the two largest financial crashes of the twentieth century, Didier Sornette and coworkers have shown that markets reveal some of their structure and internal organization when close to such highly stressed situations.[7,8]

Two large crashes occurred in 1929 and 1987. In spite of the immense amount of detailed information concerning trade, fluctuations in economic indices, and the complex network of interactions among different parts of the economic web, no clear interpretation of the

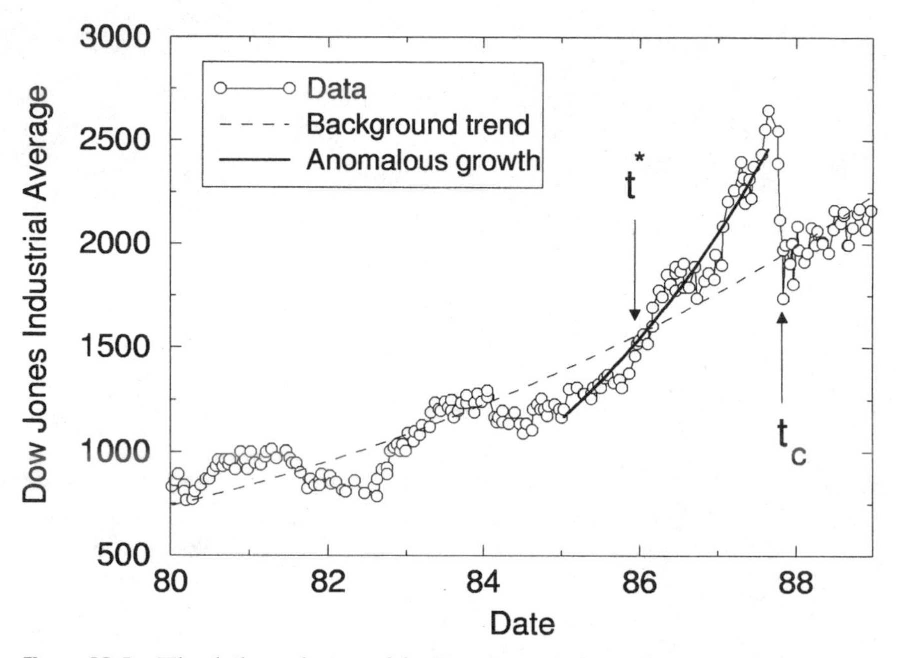

Figure 10.4 The daily evolution of the Dow Jones Industrial Average from January 1980 to December 1988. An anomalous growth is observed at some point (here indicated as t^*) followed by the later, 1987 market crash at another point t_c.

origins of these large financial crashes has emerged. The only clear fact is that several positive feedback mechanisms amplified the descent. The available time series (Figure 10.4) reveal a number of fascinating features suggesting that the collective organization of market traders led to a critical point (t_c in Figure 10.4) where, as we know from Chapter 2, strong, even catastrophic, fluctuations (or avalanches) are likely to occur. An analysis of this time series shows that several years before both crashes, some type of precursor patterns (starting at t^* in Figure 10.4) were detectable showing characteristic fluctuations associated with the approach to instability. And each crash was followed by characteristic "aftershock" patterns.

These results are consistent with the observation that crashes occur during periods of generalized economic euphoria (we can see this at work in Figure 10.4, as indicated by the continuous increasing line starting at t^*). Sornette and his colleagues suggest that in fact, the market, as a complex adaptive system, anticipates the crash in a subtle, self-organized, and cooperative fashion, hence releasing precursor fingerprints observable in stock market prices. They also suggest that the market as a whole can exhibit emergent behavior not shared by any of its constituents. According to Sornette and Johansen,[7] "We have in mind the emergence of intelligent behavior at a macroscopic scale that individuals at the microscopic scale have no idea of." They claim that the so-called efficient market hypothesis, according to which traders extract and incorporate consciously (by their action) all information contained in market prices, does not work. Instead, they present a more cooperative, emergent picture of how markets are organized. The intrinsic instability close to criticality would explain why the many specific explanations of the crashes have failed. Essentially, anything may precipitate a crash once the system is ripe. The crash is self-induced, and exogenous shocks serve only as triggering factors.

This should prevent us from thinking about the market in a simplistic way. Out-of-control markets are intrinsically unstable and unpredictable. If critical dynamics may lead to interesting side effects (such the emergence of biodiversity in nature), in others extreme events can also lead to poverty and depression. A better understanding of how complex economies evolve can only help the world move toward sustainable development. The laws of complexity are at work at all levels, including, as we will see in the next section, how our cities are built.

Fractal Cities and Zipf's Law

The oldest known city map was found by archeologists in ancient Sumerian ruins. A clay tablet from about 1500 B.C. shows a city formed by a set of concentric circles around Babylon. The city, defined as an ordered, structured pattern, is a common motive from Sumer to the Renaissance, but in fact, the Greeks early made a distinction between urban settlements involving regular patterns and those growing in an organic way. The latter were the result of many individual decisions made according to local rules.[9]

Urban populations continue to grow worldwide. Many of the cities projected to be megacities in the next century will experience growth rates of about 3% per year in the next decade. This global increase is the result of two processes: the internal growth of the cities and migration. While the former becomes more important at higher population levels, the latter is the leading force in the first stages of urban growth. During the whole life of a city, movements of population from one site to another and individual decisions to settle in this or that area shape the small- and large-scale features of the urban center. The result can be a complex spatial net of growth centers of many sizes. A map of neighboring cities (Figure 10.5) strongly reminds us of organic entities expanding in space like dendrites in neural tissue.

The history of every city creates patterns arising from particular uses of the land and a distinctive integration of the economic, cultural, and political past into the present. Although an observer embedded in the bulk of a city will find striking differences among cities in North and South America, Europe, and Africa, these differences mask some common large-scale features that have been identified worldwide. With cities as with other living systems, universal laws seem to be at work.

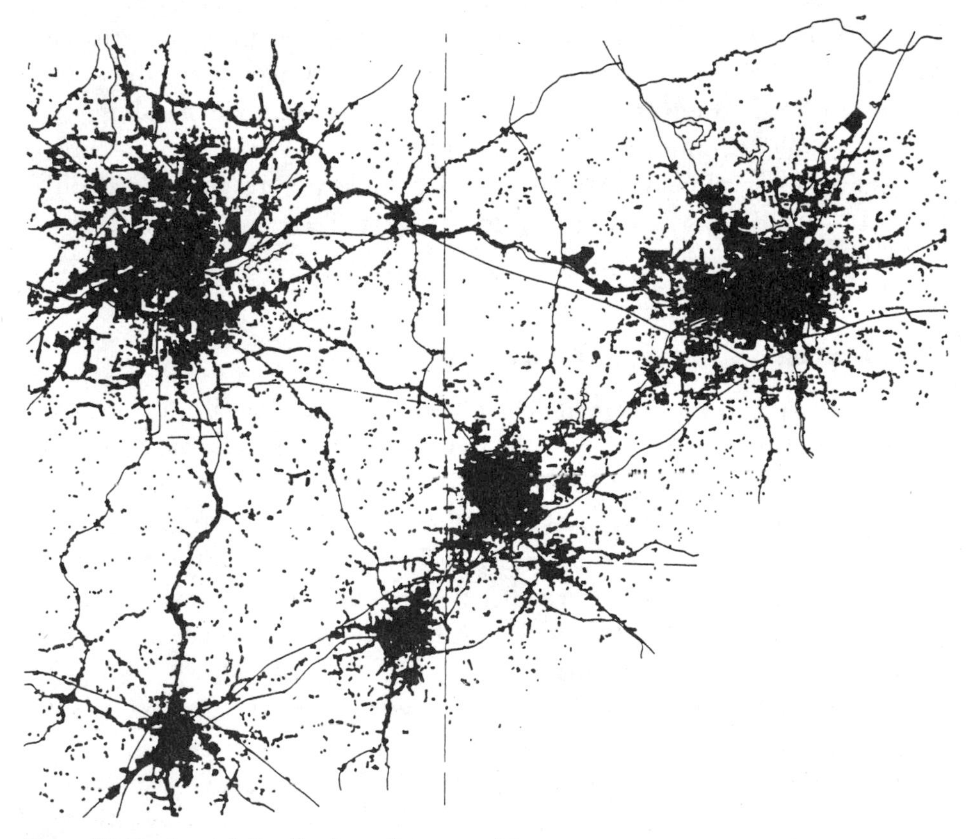

Figure 10.5 Spatial distribution of nearest cities.

One of the best known is Zipf's law, which states that the fraction of cities with n inhabitants shows a definite power-law dependence ($f(n) \propto n^{-r}$ with $r \approx 2$) that does not depend on cultural, social, or historical factors or on the short- or long-term economic or political plans of different countries (Figure 10.6). The fraction of cities with area a also follows a power law: $f(a) \propto a^{-s}$ with $s \approx 1.85$. The relation to Zipf's law is obvious: as a city's population grows, so does the urbanized area. In fact, field studies have led to the *population-area law*, which states that $n \propto a^{\beta}$ with $\beta \approx 1$, that is, urban population grows in proportion to the urbanized area.

We also find several regularities at the level of single cities. As a given city grows (such as Berlin; see Figure 10.7), the density of urbanized areas decays in a well-defined exponential way from the most populated nucleus, often known as the *central business district*.[10]

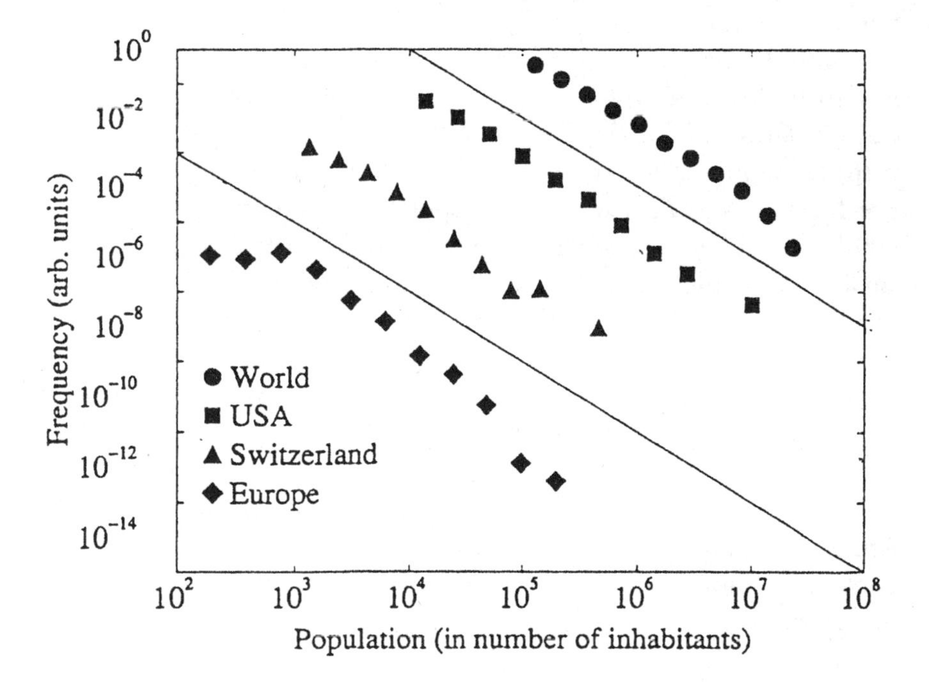

Figure 10.6 Frequency distribution of cities obtained from different data sets.

The morphology of the urban boundary is known to be fractal, again suggesting that the city behaves as an evolving complex system with nonlinear properties.

These coherent macroscopic patterns in human demography have led to theoretical models aimed at understanding how cities grow and evolve in time. One of these models,[11,12] was developed by Damian Zanette and Susanna Manrubia, then at the Fritz-Haber Institut in Berlin. It presents an oversimplified picture of reality, but as often happens with complex systems, we will show that the observed regularities are due to universal behavior. This model shows power laws and fractal patterns, but as we will see, it is not a self-organized critical system.*

The model starts with an initial condition whereby the population is randomly distributed over a two-dimensional lattice. Two rules are then applied to each lattice site (Box 1). The first involves diffusion between neighboring locations. If a given site is more populated than its nearest neighbors, the population will diffuse outwards. This diffusion process is controlled by a parameter D between zero and one that indicates how fast this migration takes place. The second rule introduces amplification processes (through another parameter p, also between zero and one) that permit wild fluctuations of local populations, including eventual drops to very low levels (or even zero). This model generates complex patterns in space, and in Figure 10.8a we show a sequence of population fluctuations for a 100×100 lattice. There are several localized urban centers, most of them small and fluctuating, although they survive over very long periods of time (in the model, as in the real world, even very large "cities" can sometimes decay and be replaced by new urban nuclei). Figure 10.8b shows a snapshot of the self-organized urban structure initiated with a single seed placed at the center of the lattice (here an additional rule has been added involving slow addition of population: newly arriving people tend to place themselves close to already existing populations[12]). Over time, the old center shows a remarkable contrast in morphology with the new, more fragmented developing periphery.

---

*Although self-organized criticality is cited at several places in this book, different types of dynamical mechanisms have been identified in relation to the origin of power laws.

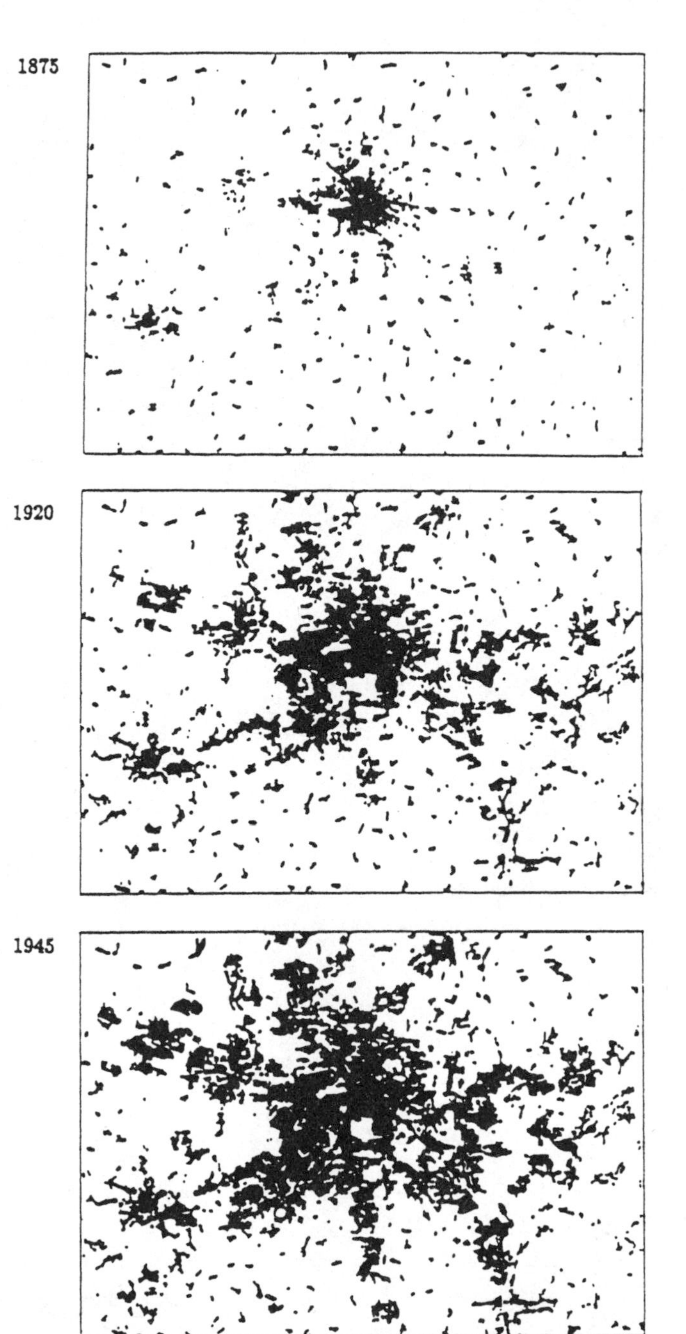

Figure 10.7 The growth of Berlin at three different stages.

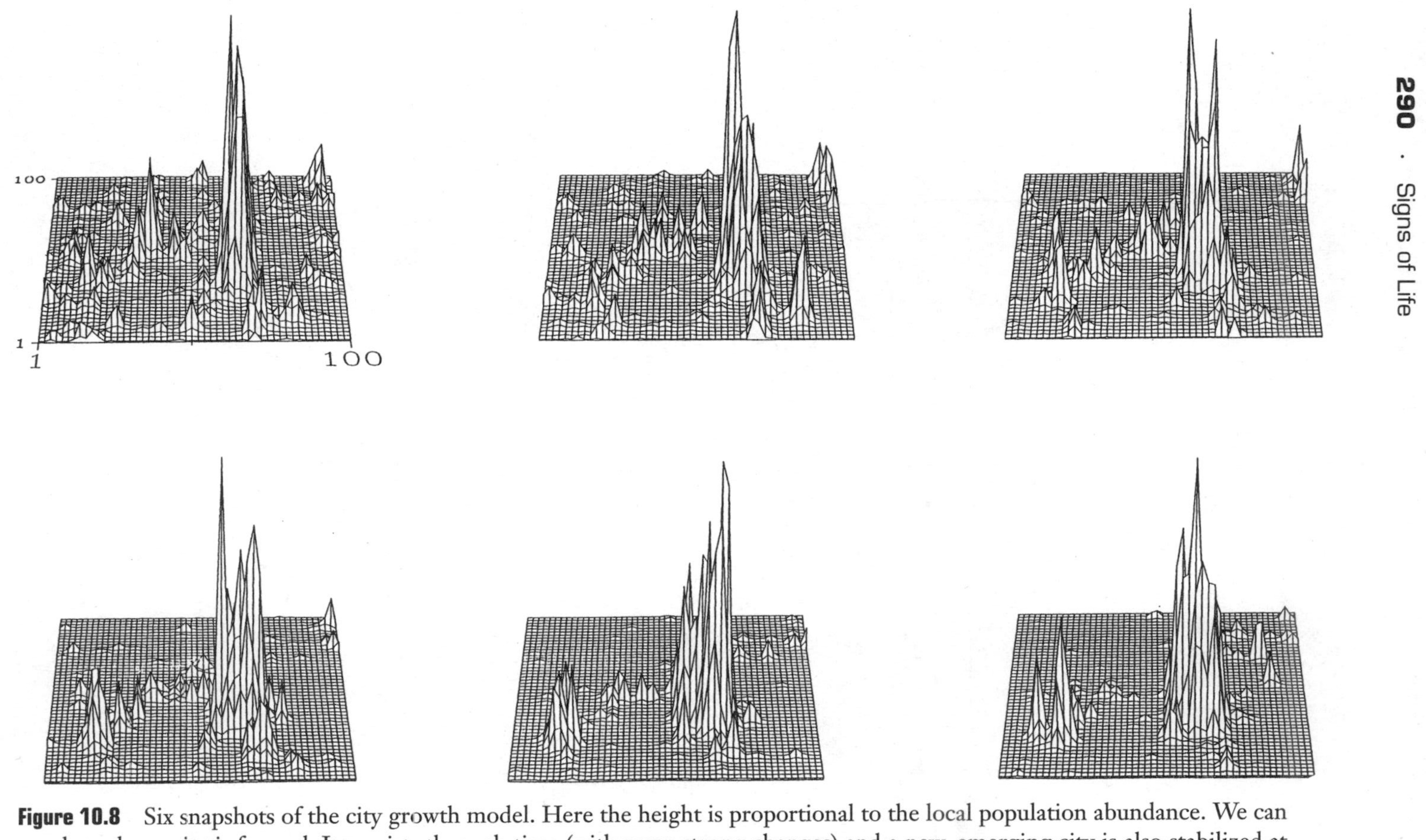

Figure 10.8 Six snapshots of the city growth model. Here the height is proportional to the local population abundance. We can see that a large city is formed. It persists through time (with some strong changes) and a new, emerging city is also stabilized at the end of the simulation.

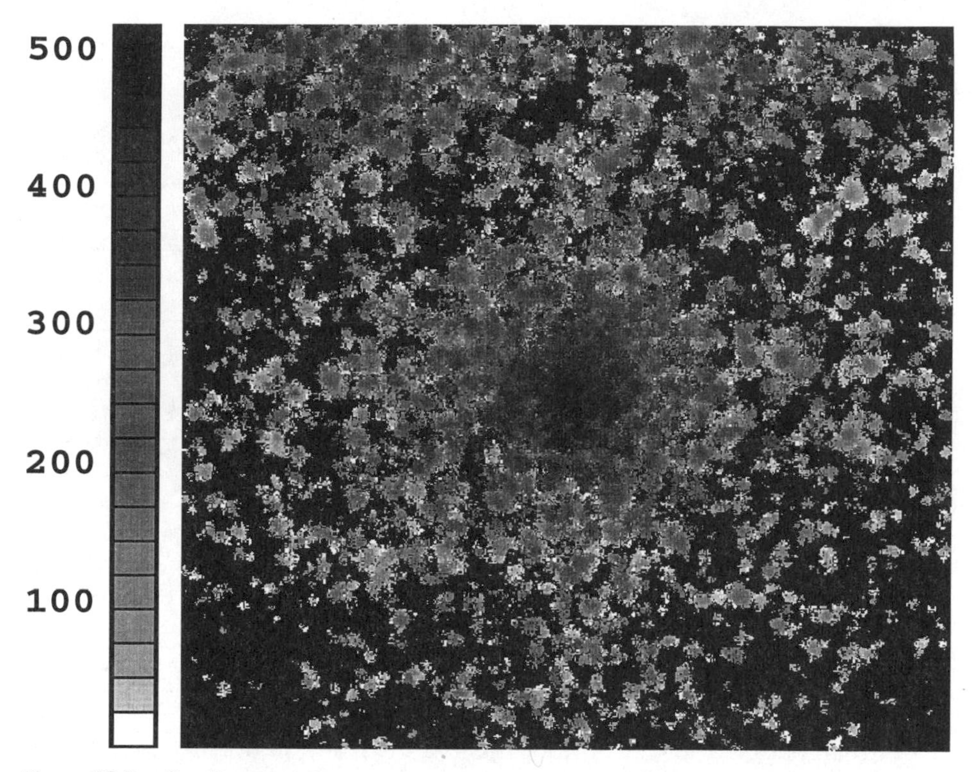

Figure 10.8 *Continued*

Stochastic Urban Growth Model

The city growth model runs on a square lattice of size $L \times L$ with periodic boundary conditions. Each cell at the position i, j at time t has a total population $m(i, j, t)$. The following dynamical rules define the evolution of the model:

1. **Diffusion.** At each time step, each cell loses a fraction D of its contents, which is evenly distributed among its four neighboring cells according to

$$m\left(i, j, t + \frac{1}{2}\right) = Dm(i, j, t) + \frac{D}{4} \sum_{\text{neighbors}} m(i, j, t),$$

   where the summation is over the four nearest neighbors.
2. **Reaction.** With probability p, each cell multiplies its contents by a factor p^{-1}, and with the complementary probability $1 - p$ its resources are set to zero:

$$m(i,j,t+1) = p^{-1} m(i,j;t+1/2) \text{ with probability } p,$$

$$m(i,j,t+1) = 0 \text{ with probability } (1-p).$$

This model leads to the scaling law $f(m) \approx m^{-z}$ with $z = 2$ irrespective of the parameters α and p [Figure 10.8]. This can be theoretically understood by assuming that a stationary distribution $m(i,j)$ exists and that the global process is mainly led by the reaction step. In fact, under these assumptions the new distribution $f'(m')$ of the rescaled variable $m' = m/p$ reads

$$f(m)dm = 1pf(mp)dm = pf(m)dm,$$

and in the stationary regime, $f = f'$. In this case, the previous identity is satisfied by the function $f(m) \propto m^{-2}$. This result shows that the chosen parameters play no role, supporting the universality of the suggested mechanism.

This model is simple, but it allows us to explore our statistical observations concerning city size distributions and the fractal nature of urban patterns. If we consider the size distribution of clusters in the lattice model (roughly speaking, the size of our virtual cities), then Zipf's law is recovered whatever values of D and p are used in the simulation (Figure 10.9). In spite of any underlying differences in the rates of change and migration specifically described by the these two parameters, the overall result of the process is completely universal. In this model, the parameters p and D stand for different social conditions such as economy, urban planning, and rapid movement from or toward cities. One would associate values of p close to 1 and of D close to 0 to areas with an old urban tradition (like Western Europe, where the central cities are very crowded and diffusion to the suburbs is extremely slow). Cities in formation, which are still absorbing population from outside and where migration rates generally are much higher, would be better represented with lower p and higher D, as in Africa, for instance. Here some other terms should probably be added, which we consider in the next section. In the most general case, as the parameters p and D vary with time, either randomly or in a complex deterministic fashion, they do not change the general features of the pattern.

Most models of urban growth involve multiparametric characterizations of interactions among "agents" (either citizens or compa-

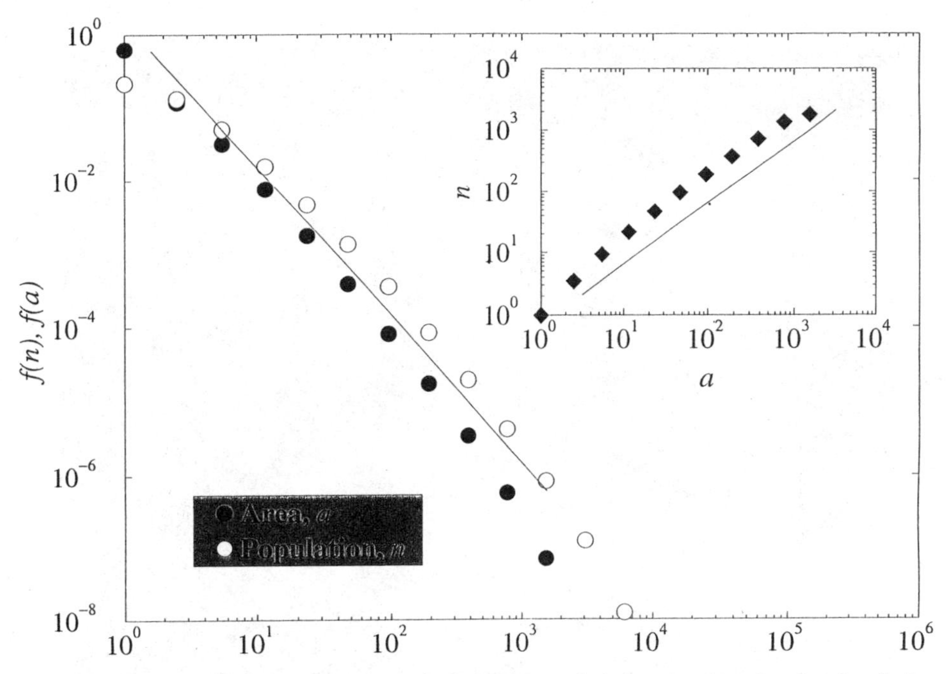

Figure 10.9 Distribution of city population $f(n)$ and city areas $f(a)$ for the simulation model. Both distributions are in agreement with observed data. Inset: relation between area and population, predicting that the number of inhabitants grows proportional to its area, consistently with real cities.

nies). Most detailed models include rules emphasizing the economy-dependent nature of urban settlements. They are usually applied to specific situations or remain highly qualitative in their basic conclusions. Yet our results suggest that very generic mechanisms are at work and that a well-defined theory can be formulated for urban growth phenomena on a large scale: the properties of human settlements emerge from the universal properties of their intrinsic dynamics.

Traffic at the Edge of Chaos

Few words are more often linked with "chaos" than "traffic." Figure 10.10 shows a drawing by Gustave Doré illustrating the chaotic traffic in the old city of London. A tremendous traffic jam has formed, and individuals and vehicles can hardly move. But we all know that traffic jams do not appear if the density of cars is low enough. Is there a critical density where changes suddenly emerge?

Figure 10.10 Old-fashioned picture of a traffic jam by Dore.

Traffic is a man-made transportation system that shares a number of features with other complex systems. It involves many different locally interacting units. It typically occurs on a network, and because of the complexity of their dynamics, networks tend to react to external control decisions in counterintuitive ways. In this section we will analyze a simple model[13] of traffic flow that shows a phase transition close to a critical density. This density separates two well-defined regimes: fluid, ordered traffic flow and turbulent, disordered flow.

Kai Nagel and Steen Rasmussen, two Los Alamos physicists, have studied this problem in depth by means of a one-dimensional system[14] where the simulated cars travel on a ring (i.e., those that reach the end appear at the beginning, which can be quite frustrating for humans but good for simulation purposes). A number of sites are available (we are treating the line as a series of discrete sites), and the cars are defined by their position x on the line as well as by their velocity ν. A maximum velocity ν_{max} is allowed, and the distance between nearest cars, g, will be called the *gap*. The rules are as follows:

—Acceleration of free vehicles: Each car has a velocity $\nu < \upsilon_{max}$, and when $gap\nu + 1$ it accelerates to $\nu + 1$ (i.e., if there is enough space between us and the car in front, our speed is increased).

—Slowing down: Due to the behavior of other cars, the speed can be reduced. If $gap\nu - 1$, then the speed is reduced to gap. In other words, if our speed is too fast, we have to slow down so as not to collide with the next car.

—Randomization: Noise is always present, and it is introduced by means of a simple rule representing stochastic effects due to the intrinsic complexity of drivers. Here each car reduces its speed with some probability p (here we take $p = 0.5$) to $\nu - 1$.

—Movement: Each car advances ν sites at each time step.

In Figure 10.11 we can see several runs of this model for different values of the density ρ. At low densities, all cars can travel at the maximum speed with no major problems. The flow of vehicles is fluid and appears as ordered, straight parallel lines in the space–time diagram, where each car is a small black square. The high-density regime, on the other hand, gives us the expected turbulent motion: the cars stop here and there with no order. But at the critical boundary between order and disorder, complex structures emerge: although the traffic is fluid in many parts of the line, jams appear and fractal structures show up. If we look at the jams closely, we can see that they display a self-similar structure. These fractal waves propagate backwards and last a very long time at the critical boundary.

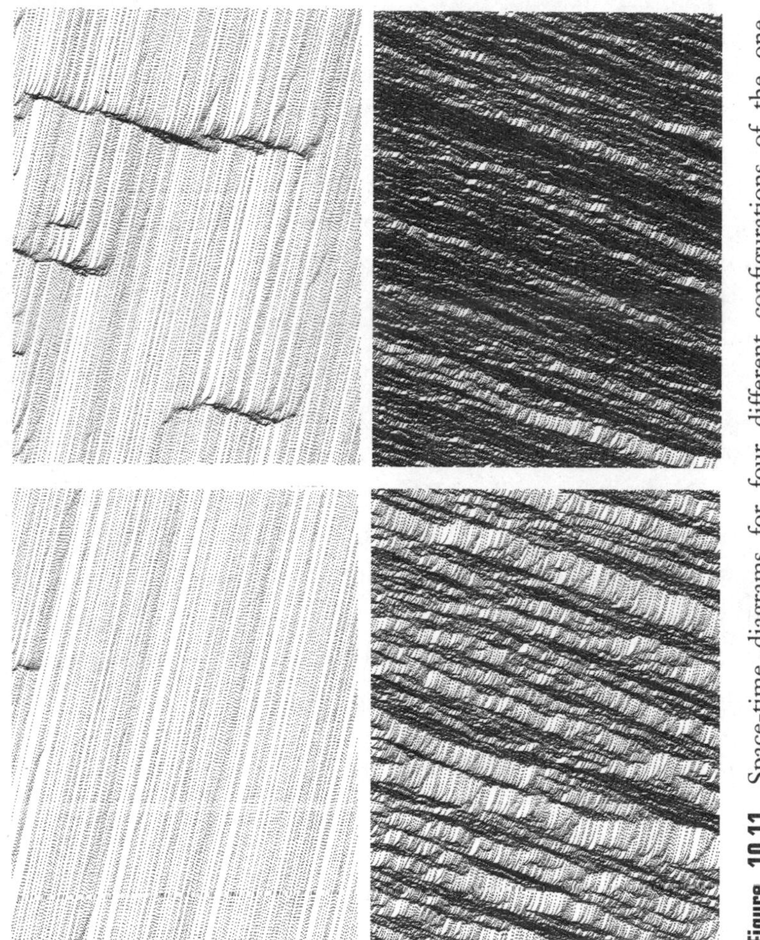

Figure 10.11 Space-time diagrams for four different configurations of the one-dimensional car traffic model. Here space is goes from left to right and time from up to down. The subcritical system (upper left) is observed at low densities, where cars simply describe straight lines (i.e., no jams are formed) in the space-time diagram. An example of a critical system is shown in the upper right plot, where large jams are formed in the middle of an ordered pattern. Two examples of the supercritical (jammed) phase are shown in the lower row, for a slightly larger density than critical (left) and very high density (right).

Several measures can be used to characterize this transition. Three of them are shown in Figure 10.12. Plotting vehicle flow against density shows what we expected: at low densities, the flow increases as more cars are added, but once jams start to emerge the flow can no longer increase. At the disordered phase the flow decreases. This diagram matches data analysis of traffic flows on real highways. Travel times typically increase with density, and the interesting point has to do with fluctuations: if we measure the *variance* of the travel time (i.e., how unpredictable is the time required for a car to travel through the whole line), it also shows a maximum at criticality.

This maximal fluctuation at the critical point should be familiar: we know from Chapter 2 that phase transitions involve complex fluctuations. This time the fluctuations have important practical consequences. As Nagel and Rasmussen discussed in an earlier paper on "traffic at the edge of chaos," as advanced traffic management systems drive larger parts of the transportation system toward the regime of maximum flow (the critical state), these systems will become highly unpredictable. In other words, a side effect of optimality is that the predictability of travel times sharply decreases, and our efforts to obtain better control of the transportation system through traffic management actually produce more unpredictable traffic dynamics.

The problem stems from a linear view of the transportation system that assumes that individuals can be treated as isolated entities characterized by some well-defined average values like mean speed. But traffic patterns are an emergent result of interactions among many agents. At the critical regime, individual microscopic decisions can have a large influence on the macroscopic dynamics. It is not clear whether we will ever solve the intrinsic problem of optimality versus unpredictability. But clearly, this problem is not restricted to highways: the Internet also displays well-defined fractal properties associated with the self-organization of the system. As happens in many other critical systems, the traffic of information displays self-similar behavior, with high fluctuations known as "Internet storms." This is probably the result of the conflict between the constant input of information coming from many users and the global saturation that forces users to leave the net (or reduce their activity). Simple models displaying phase transitions can account for these measurements (Box 2). Future work will be necessary to understand how highly interactive information-based systems lead to such storms on the web and whether this congestion can be avoided.

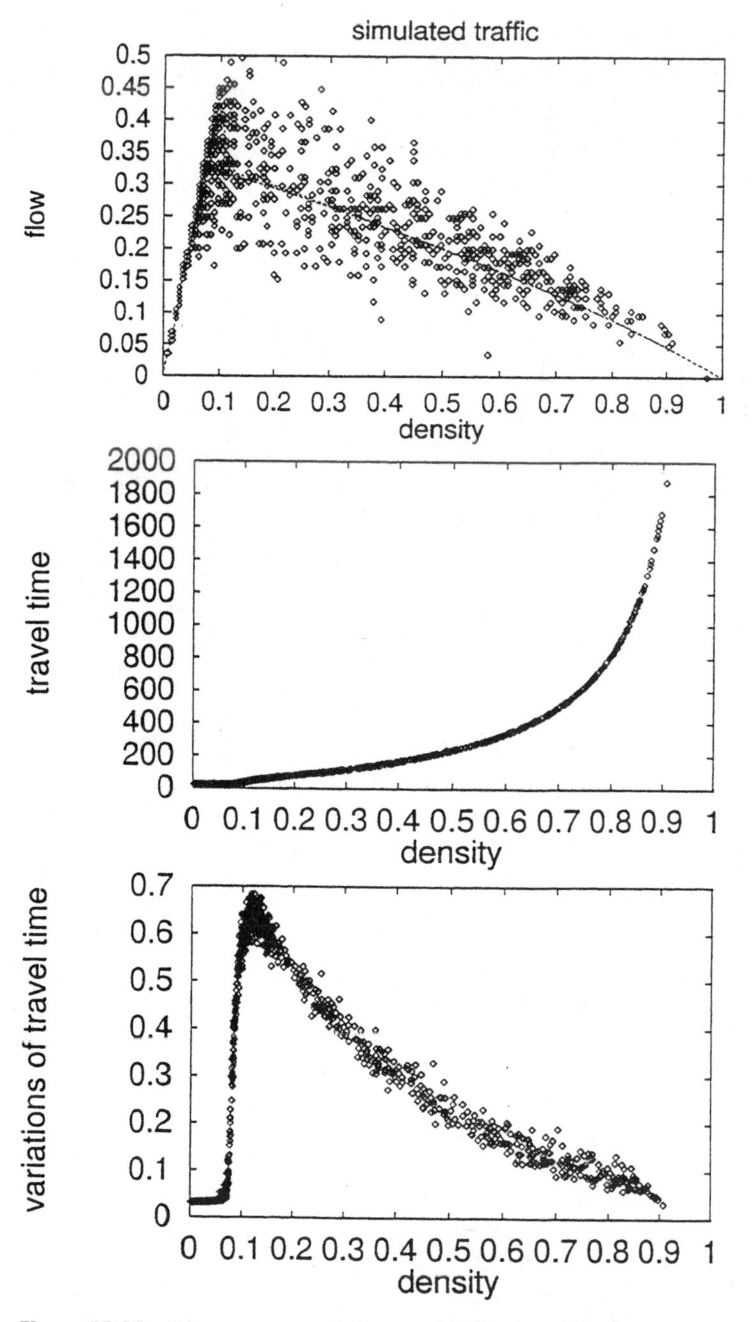

Figure 10.12 Flow rate, travel time and variance of fluctuations as a function of density for the traffic model. We can see that the maximum flow is obtained for high rates of variability. This means that maximal efficiency is associated with maximum unpredictability.

We conjecture that solving these problems will need a change of philosophy in the standard management rules. Central control is probably ineffective in these types of systems. A more distributed and adaptive set of rules, always in direct interaction with the internal web dynamics, is more likely to effect cooperation (and not conflict) among users.

A Model of Internet Dynamics

Most of the reported statistical features of Internet traffic can be recovered from a simple model of information traffic on a two-dimensional $L \times L$ lattice. Inspired by a recent model by Toru Ohira and Ryusuke Sawatari,[15] a parallel array of computers has been modeled by Sergi Valverde and Ricard Solé,[16] This array (Figure 10.13a) involves two types of nodes: *hosts* and *routers*. The first are nodes that can generate and receive messages, and the second can only store and forward messages.

Only a fraction ρ of the nodes are hosts, and the rest are routers, which are randomly distributed through the lattice. Each node maintains a queue of unlimited length where the arriving message packets are stored. The rules are defined as follows:

—Creation: The hosts create packets with given probability λ. Only another host can be the destination of a packet, which is also selected randomly. Finally, this new packet is appended at the tail of the host queue.

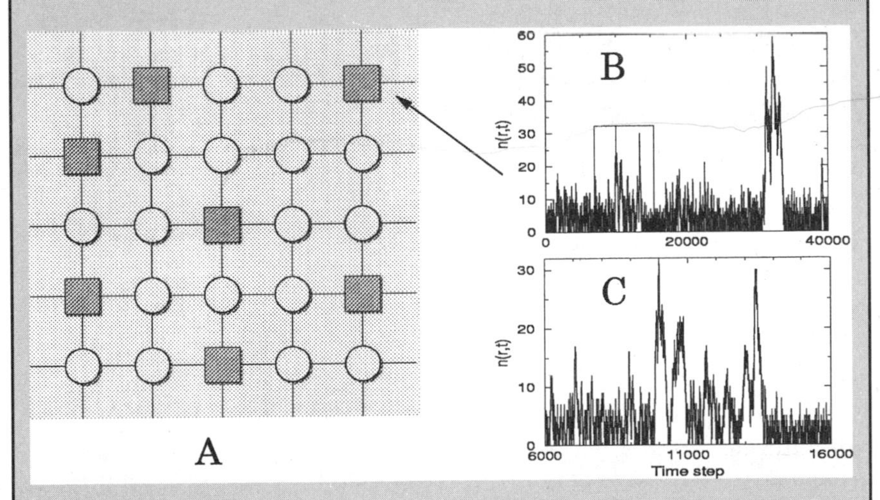

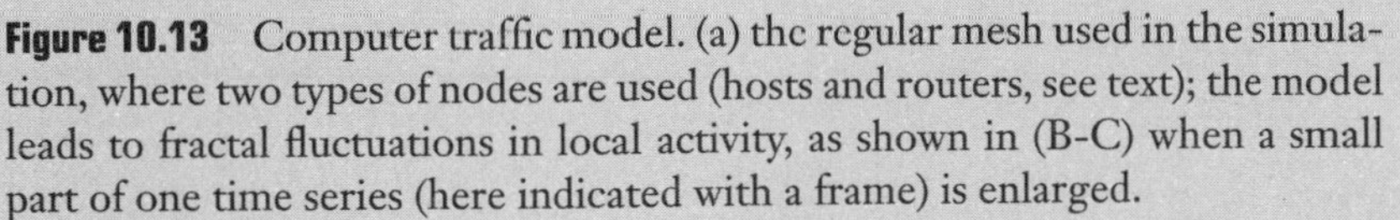
Figure 10.13 Computer traffic model. (a) the regular mesh used in the simulation, where two types of nodes are used (hosts and routers, see text); the model leads to fractal fluctuations in local activity, as shown in (B-C) when a small part of one time series (here indicated with a frame) is enlarged.

—Routing: Each node picks up the packet at the head of its queue and decides which outgoing link is best suited to the packet destination. Here, the objective is to minimize the communication time for any single message, taking into account only shortest paths and avoiding congested links as well. First, the selected link is the one that points to a neighbor node that is nearest to the packet destination. Second, when two choices are possible, the less congested link is selected. The measure of congestion of a link is simply defined as the number of packets forwarded through that link. Once the node has made the routing decision, the packet is inserted at the end of the queue of the node selected and the counter of the outgoing link is incremented by one.

These rules are applied to each site, and each $L \times L$ updatings define our time step. The model exhibits complex fluctuations of activity: figure 10.13b shows the time series dynamics of the number of packets $n(r, t)$ at a given node, for a lattice with $L = 256$, $\rho = 0.08$, and $\lambda = 0.055$. The parameter choice is related to a phase transition point. This point defines the boundaries separating a fluid phase of no congestion from a congestion phase where packets accumulate and the total congestion increases without bounded. In Figure 10.13c an enlarged piece of the previous series is plotted, showing the presence of self-similar (fractal) properties.

Although this model is not self-organized (the transition point must be tuned through a specific choice of λ and ρ), it can be made self-organized by including simple behavioral rules. If hosts can self-adjust their probability of releasing messages by depending on local congestion, then the whole system becomes self-organized into the critical state.

The Collapse of Civilizations

We have seen that the economy is an evolving complex system, sharing a number of relevant features with biological systems. Generic features appear to be common to all markets, suggesting that some constraints to their internal structure are at work. Coevolution is an inevitable outcome of the interactions among agents, and it is interesting that stock market crashes and long-range correlations are observable, as are mass extinctions in the fossil record.

The network-like organization of markets prompts us to ask whether our previous models might provide a theoretical framework for the economy. Models based on rugged landscapes (see Chapter 9) have been used as a powerful metaphor for the evolution of technologies. The economy has certainly a web structure and displays features of complex, self-organized systems, and Stuart Kauffman has found deep links between economies and large-scale evolutionary systems.[17] In particular, using the rugged landscape approximation, he has shown that biological and economic webs share a nontrivial set of regularities in evolving new properties. The Cambrian explosion and the sudden emergence of innovations in technology share (at least at some level) common patterns of search in rugged landscapes. Initial bursts of innovation are followed by extinction of less-fit forms. In biology, a large number of forms that emerged in the Cambrian "big bang" became extinct, and no remains of them exist in our biosphere. In the technological context (to cite just one example), in the nineteenth century an explosion of bicycle designs was followed by a large "extinction" of most of the designs in favor of a very small number of them.

Another fascinating phenomenon is also linked with models of macroevolution and extinction: the collapse of civilizations.[18] Because archaeologists have traditionally devoted most of their attention to the rise of civilizations, we still have little understanding of their collapse. Except for the Mayan and Roman cases, little is said about collapse in the literature. Collapse, loosely related to extinction events in macroevolution, involves the breakdown of institutional organizations after some period of social and economic instability. This is a very important issue, and perhaps new models might shed some light on it. Is it inevitable? Are emerging societies prone to instability and eventually to collapse? How important is the intrinsic web of relationships among different structures in relation to external perturbations? Is the increasing exchange of information among entities within a society a source of stability or instability? These questions are difficult to answer, but the huge amounts of macroeconomic data, as well as an increasing understanding of how real economic webs emerge and become organized at different scales, may soon provide a consistent theoretical framework.

Studies of past civilizations have generated a number of interesting hypotheses.[1] Some authors have suggested that the increasing complexity of social organizations leads to a degree of interdependence that reduces diversity and flexibility. If resources decline, then collapse occurs as a result of an avalanche in the system. Many of the theoretical approaches to collapse employ biological metaphors. Some of them are inspired by the Darwinian approximation, and others involve a punctuated-equilibrium framework. Using the lexicon of general systems theory, cybernetics, and information theory, this evolutionary perspective was enriched in the 1970s when some authors explicitly suggested that complex societies spontaneously evolve toward a *hypercoherent* state where "failure in one part of the system affects all the other parts in a domino theory of disaster." This coherent response is in fact what we found in the web model of macroevolution. Applied to civilizations, it means that social institutions become so integrally connected that failure in one subsystem affects many others and brings the whole hierarchy crashing down.

Let us mention, however, that collapses are seldom wholly catastrophic. Rather, the breakdown occurs institution by institution. Eventually, however, collapse takes place, and old traditions, large urban centers, and organized economic webs disappear. Whether external causes were responsible for the Anasazi,[19] Mayan, and Roman collapses or were merely the trigger points of a collective response remains an open question. But we can conjecture that given the long duration of these societies and the likely occurrence of several episodes of, say, climatic change, something more than external events (although necessary for a complete explanation) is necessary to account for collapse. Network dynamics and information exchange among different parts of the system must be part of the explanation as well. One clearly important element is the exchange of resources between the organized urban center (the stable core, providing the source of order through stable structures) and the periphery, which provides resources to the center with some expectation of benefits in return. If conflicts occur between these two parts and the flow of resources and information becomes altered, the system can break down. Collapse in this scenario entails the dissolution of the central core that had facilitated the transmission of resources and information.

The maintenance of order in complex societies demands flexibility and resilience to external stress. This can be understood as a type of collective adaptation in Kauffman's sense of a system evolving toward the best place in parameter space. We think it is much more useful, however, to think of complex societies as spontaneously self-organizing into hierarchical patterns where conflict arises between the need for intrinsic order and the spontaneous tendency to growth and competition. The ordered "core" is maintained by intrinsic rules that tend to freeze information exchange in order to maintain structure and information.

But the different parts of the system (whether economic agents or competing companies) are also pursuing their self-interest and thus constantly push the system toward the unstable regime. Whether this chaotic regime is reached will depend on the web structure.

We still do not know why civilizations collapse. We need some new explorers able to penetrate the jungles of complexity and find the theories that will help us locate ourselves in our complex world. Complexity shows us that we live in a fascinating but counterintuitive universe, a nonlinear and unpredictable world operating by rules still to be discovered. These laws will show us how our ecosystems are organized as they are and how fragile they can be. They will tell us more than we can imagine about our brains and societies. The study of complexity and emergent phenomena is opening the door of a library never before explored, full of amazing books with unexpected insights. We need to read them, both because they will shed light on our understanding of nature and because they will help us preserve it. Let us remember, when walking through the ruins of those great ancient cities, imagining their past grandeur, that they all had something in common: a long time ago, their citizens believed that those cities and those civilizations would last forever.

Notes

Chapter 1

1. Poincaré, H. (1913). *The Foundation of Science: Science and Method.* English translation, 1946, The Science Press. Lancaster PA, 397.

2. Anderson, P.W. (1994). *A Career in Theoretical Physics.* Singapore: World Scientific Publishing.

3. Einstein, A., Podolsky, P., and Rosen, N. (1935). Can quantum-mechanical description of physical reality be considered complete? *Phys. Rev.* 47, 777–780.

4. Bell, J.S. (1966). On the problem of hidden variables in quantum theory. *Rev. Mod. Phys.* 38, 447–452.

5. Aspect, A., Dalibard, J., and Roger, G. (1982). *Phys. Rev. Lett.* 49, 1804–1807.

6. Aspect, A., Grangier, P., and Roger, G. (1982). *Phys. Rev. Lett.* 49, 91–94.

7. Silberstein, M. (1998). Emergence and the mind–body problem. *J. Consc. Stud.* 5, 464–482.

8. Crutchfield, J. (1993). The Calculi of Emergence: Computation, Dynamics, and Induction. Santa Fe Working Paper no. 94-03-016.

9. Höfer, T,. Sherratt, J.A., and Maini, P.K. (1995). *Dictyostelium discoideum*: cellular self-organisation in an excitable biological medium. *Proc. Roy. Soc. Lond. B*, 259, 249–257.

10. Goodwin, B.C. (1994). *How the Leopard Changed Its Spots.* London: Weidenfeld and Nicolson.

Chapter 2

1. Jacob, F. (1988). *The Statue Within.* Basic Books, New York.

2. Coveney, P. and Highfield, R. (1990). *The Arrow of Time*, W.H. Allen, London.

3. Mandelbrot, B.B. (1991). *The Fractal Geometry of Nature.* Freeman, New York.

4. Schroeder, M. (1991). *Fractals, Chaos and Power Laws.* Freeman, New York.

5. Stanley, H.E. (1971). *Introduction to Phase Transitions and Critical Phenomena.* Oxford University Press, New York.

6. Bruce, A. and Wallace, D. (1989). Critical point phenomena: universal physics at large length scales. In: *The New Physics*, P. Davies (editor); 236–267. Cambridge University Press, Cambridge.

7. Back, C.H. et al. (1995). Experimental confirmation of universality for a phase transition in two dimensions. *Nature* 378, 597–600.

8. Chaikin, P.M. and Lubensky, T.C. (1995). *Principles of Condensed Matter Physics*. Cambridge University Press, Cambridge.

9. Gardner, M. (1978). White and brown music, fractal curves and one-over-f noise. *Scientific American*, January, 110–121.

10. Bak, P., Tang. C., and Wiesenfeld K. (1987). Self-organized criticality: an explanation for 1/f noise. *Phys. Rev. Lett.* 59, 381–384.

11. Bak, P. (1996). *How Nature Works*. Springer, New York.

12. Turcotte, D.L. (1997). *Fractals and Chaos in Geology and Geophysics* (2nd edition). Cambridge University Press, Cambridge.

13. Arthur, B. (1990). Positive feedbacks in the economy. *Scientific American*, February, 92–99.

Chapter 3

1. Brent, R. (1999). Functional genomics: learning to think about gene expression. *Curr. Biol.* 9, R338–341.

2. Ko, E.P., Yomo, T., and Urabe, I. (1994). Dynamic clustering of bacterial population. *Physica* D 75, 81–88.

3. For discussions of epigenesis in contemporary biology, see the following references: Waddington, C.H. (1956). *The Principles of Embryology*. London: Allen and Unwin. Webster, G and Goodwin, B (1996). *Form and Transformation: Generative and Relational Principles in Biology*. Cambridge: Cambridge University Press. Strohman, R.C. (1997). The coming Kuhnian revolution in biology. *Nature Biotech.* 15, 194–200.

4. Ben-Jacob, E., Schochet, O., Tenenbaum, A., Cohen, I., Czirok, A., and Vicsek, T. (1994). Generic modelling of cooperative growth. Generic modelling of cooperative growth patterns in bacterial colonies. Nature *368*, 46–49.

5. Kaneko, K., and Yomo, T. (1997). Isologous diversification: a theory of cell differentiation. *Bull. Math. Biol.* 59, 139–196.

6. Rubin, H. (1990). The significance of biological heterogeneity. *Cancer and Metastasis: Reviews* 9, 1–20.

7. Furusawa, C., and Kaneko, K. (1998). Emergence of rules in cell society: differentiation, hierarchy, and stability. *Bull. Math. Biol.* 60, 659–687.

8. Kaneko, K., and Yomo, T. (1999). Isologous differentiation for robust development of cell society. *J. Theoret. Biol.* 199, 243–256.

9. Kaneko, K. (1990). Clustering, coding, switching, hierarchical ordering, and control in networks of chaotic elements. *Physica* D 41, 137–172.

10. Kaneko, K. (1994). Relevance of clustering to biological networks. Physica D 75, 55–73.

11. Kacser, H., and Burns, J.A. (1973). The control of flux. *Symp. Soc. Exp. Biol.* 27, 65–104.

12. Kacser, H. (1957). Some physico-chemical aspects of biological organisation. Appendix to *The Strategy of the Genes* by C.H. Waddington, 191–249. London: Allen and Unwin.

13. The application of the principle of robust dynamics to embryonic development generally can be found in Goodwin, B.C., Kauffman, S.A., and Murray, J.D. (1993). Is morphogenesis an intrinsically robust process? *J. theoret. Biol.* 163, 135–144.

The relevance of robust design in the molecular organization underlying bacterial chemotaxis has been described by Barkal, N., and Liebler, S. (1997). Robustness in simple biochemical networks. *Nature* 387, 913–917.

14. Kauffman, S.A. (1995). *At Home in the Universe. The Search for the Laws of Self-Organisation and Complexity.* New York: Oxford University Press.

15. Jacob. F., and Monod, M. (1961) Genetic regulatory mechanisms in the synthesis of proteins. *J. Molec. Biol.* 3. 318.

16. Alberts, A., Bray, D., Lewis, J., Raff, M., Roberts, K., and Watson, J.D. (1983). *Molecular Biology of the Cell.* New York: Garland.

17. Harris, S.E., Sawhill, B.K., Wuensche, A., and Kauffman, S.A. (1997). Biased eukaryotic gene regulation rules suggest genome behaviour is near edge of chaos. Santa Fe Institute Working Paper 97–05-039.

18. Cooke, J., and Zeeman, E.C. (1976). A clock and wavefront model for control of the number of repeated structures during animal morphogenesis. *J. Theoret. Biol.* 58, 455–476.

19. Thom, R. (1970). Topological models in biology. In *Towards a Theoretical Biology*, ed. C.H. Waddington, vol. 3, 89–116. Edinburgh: Edinburgh University Press.

20. Palmeirin, I., Henrique, D., Ish-Horowitcz, D., and Pourquié, O. (1997). Avian hairy gene expression identifies a molecular clock linked to vertebrate segmentation and somitogenesis. *Cell* 91, 639–648.

21. McGrew, M.J., Dale, J.K., Fraboulet, S., and Pourquié, O. (1998). The *lunatic Fringe* gene is a target of the molecular clock linked to somite segmentation in avian embryos. *Curr. Biol.* 8, 979–982.

22. Forsberg, H., Crozet, F., and Brown, N.A. (1998). Waves of mouse *Lunatic fringe* expression, in four-hour intervals, precede somite boundary formation. *Curr. Biol.* 8, 1027–1030.

23. Descriptions of phyllotaxis can be found in the following papers:

Cummings, F.W. and Strickland, J.C. (1998). A model of phyllotaxis. *J. Theoret. Biol.* 192, 531–544.

Douady, S. and Couder, Y. (1996). Phyllotaxis as a dynamical self organising process. Parts I, II, and III. *J. Theoret. Biol.* 178, 255–274; 275–294; 295–312.

Goodwin, B.C. (1994). *How the Leopard Changed Its Spots; The Evolution of Complexity.* London: Weidenfeld and Nicolson.

24. Meinhardt, H. (1995). *The Algorithmic Beauty of Sea Shells.* Berlin: Springer.

25. Meinhardt, H. (1982). *Models of Biological Pattern Formation.* London: Academic Press.

Meinhardt, H. (1992). Pattern-formation in biology—a comparison of models and experiments. Reports on Progress in Physics 55, 797–894.

26. Turing, A.M. (1952). The chemical basis of morphogenesis. *Phil. Trans. Roy. Soc.* B 237, 37–72.

27. Gunji, Y. (1989). Molluscan pigment pattern generation by a dynamic structure with intrinsic time, illustrating subjective autonomy. In *Dynamic Structures in Biology*, eds. B. Goodwin, A. Sibatani, and G. Webster, 219–235. Edinburgh: Edinburgh University Press.

Chapter 4

1. Saunders, P.T., Koeslag, J.H., and Wessels, J.A. (1998). Integral rein control in physiology. *J. Theoret. Biol.* 194, 163–173.

2. Watson, A.J., and Lovelock, J. E, (1983). Biological homeostasis of the global environment. *Tellus* 35B, 284–289.

3. Goodwin, B.C. (1965). Oscillatory behaviour in enzymatic control processes. In *Advances in Enzyme Regulation*, vol. 3 (ed. G. Weber). 425–438. Oxford: Pergamon Press.

4. Ruoff, P., and Rensing, L. (1996). The temperature-compensated Goodwin model simulates many circadian clock properties. *J. Theoret. Biol.* 179, 275–285.

5. Aronson, B.D., Johnson, K.A., Loros, J.J., and Dunlap, J.C. (1994a). Negative feedback defining a circadian clock: autonomous regulation of the clock gene *frequency*. *Science* 263, 1578–1584.

6. Zeng, H., Hardin, P.E., and Rosbash, M. (1994). Constitutive overexpression of the *Drosophila* period protein inhibits period mRNA cycling. *EMBO J.* 13, 3590–3598.

7. Aronson, B.D., Keith, A.J., and Dunlap, J.C. (1994b). Circadian clock locus frequency: protein encoded by a single open reading frame defines period length and temperature compensation. *Proc. Nat. Acad. Sci. U.S.A.* 91, 7683–7687.

8. Knobil, E. (1980). The neuroendocrine control of the menstrual cycle. *Recent Prog. Hormone Res.* 30, 1–46.

Knobil, E. (1989). The circhoral hypothalamic clock that governs the 28-day menstrual cycle. In *Cell to Cell Signalling: From Experiments to Theoretical Models* (ed. A. Goldberger), 353–358. London: Academic Press.

9. Filicori, M. (1989). The critical role of signal quality: Lessons from pulsatile GnRH pathophysiology and clinical applications. In *Cell to Cell Signalling: From Experiments to Theoretical Models* (ed. A. Goldberger), 395–405. London: Academic Press.

10. Rapp, P.E., Mees, A I., and Sparrow, C.T. (1981). Frequency encoded biochemical regulation is more accurate than amplitude dependent control. *J. Theoret. Biol.* 90, 531–544.

11. Goldbeter, A., and Yue-Xian Li (1989). Frequency coding in intercellular communication. In *Cell to Cell Signalling: From Experiments to Theoretical Models* (ed. A. Goldberger), 415–432. London: Academic Press.

12. Glass, L., and Mackey, M. (1988). *From Clocks to Chaos: The Rhythms of Life*. Princeton: Princeton Univ. Press.

Bélair, J., Glass, L., an der Heiden, U., and Milton., J. (eds.) (1995). *Dynamical Disease; Mathematical Analysis of Human Disease.* New York: American Inst. of Phys.

13. Mackey, M.C. and Glass, L. (1977). Oscillation and chaos in physiological control systems. *Science* 197, 287–289.

14. Winfree, A.T. (1987). *When Time Breaks Down. The Three-Dimensional Dynamics of Electrochemical Waves and Cardiac Arrhythmias.* Princeton: Princeton Univ. Press.

15. Poon, C.-S., and Merrill, C.K. (1997). Decrease of cardiac chaos in congestive heart failure. *Nature* 389, 492–495.

16. Ivanov, P.Ch., Rosenblum, M.G., Peng, C.-K., Mietus, J., Havlin, S., Stanley, H.E., and Goldberger, A.L. (1996). Scaling behaviour of heartbeat intervals obtained by wavelet-based time-series analysis. *Nature* 383, 323–327.

Chapter 5

1. Sacks, O. (1985). *The Man Who Mistook His Wife for a Hat.* Picador, London.

2. Rose, S. (1992). *The Making of Memory.* Anchor Books, New York.

3. Von Neumann, J. (1954). *The Computer and the Brain. Yale U. Press.*

4. Bear, M.F., Connors, B.W., and Paradiso, M.A. (1996). *Neuroscience.* Williams & Wilkins, Baltimore.

5. Hopfield, J. (1982). Neural networks and physical systems with emergent collective computational abilities. *Proc. Natl. Acad. Sci. USA* 79, 2554–2558.

6. Peretto, P. (1992). *An Introduction to the Modeling of Neural Networks.* Cambridge U. Press, Cambridge.

7. Mikhailov. A.S. (1990). *Foundations of Synergetics* I. Springer, Berlin.

8. Başar, E. (editor) (1990). *Chaos in Brain Function*, Springer, Berlin.

9. Schuster, H.G. and Wagner, P. (1990). A model for neuronal oscillations in the visual cortex. *Biol. Cybern.* 64, 77–82.

10. Babloyantz, A. and Destexhe, A . (1986). Low dimensional chaos in an instance of epilepsy. *Proc. Natl. Acad. Sci. USA* 83, 3513–3517.

11. Freeman, W.J. (1991). The physiology of perception. *Scientific American* 264, 78–85.

12. Skarda, C.A., and Freeman, W.J. (1987). How brains makes chaos in order to make sense the world. *Behav. Brain Sci.* 10, 161–195.

13. Haken, H. (1996). *Principles of Brain Functioning*. Springer, Berlin.

14. Kelso, J.A.S. (1992). A phase transition in human brain and behavior. *Phys. Lett.* A 169, 134–144.

15. Crick, F. (1995). *The Astonishing Hypothesis.* Touchstone Books, London.

Chapter 6

1. Hölldobler B., and Wilson, E.O. (1994). *Journey to the Ants.* Harvard: Belknap Press.

2. Hölldobler B., and Wilson, E.O. (1990). *The Ants.* Springer-Verlag, Berlin.

3. Gordon, D.M. (1999). *Ants at Work.* Free Press, New York.

4. Gordon, D.M., Goodwin, B.C., and Trainor, L.E.H. (1992). A parallel distributed model of ant colony behaviour. *J. Theor. Biol.* 156, 293–307.

5. Bonabeau E., Theraulaz G., Deneubourg J.-L., Aron S., and Camazine S. (1997). Self-organization in social insects. *Trends in Ecol. Evol.* 12, 188–193.

6. Deneubourg, J.L. (1977). Application de l'ordre par fluctuation à la description de certaines étapes de la construction du nid chez les termites. *Insectes Sociaux* 24, 117–130.

7. Franks, N.R., Bryants, S, Griffiths, R., and Hemerik, L. (1990). Synchronization of the behavior within the nests of *Leptothorax acervorum*. *Bull. Math. Biol.* 52, 597–612.

8. Cole, B.J. (1991). Short-term activity cycles in ants: generation of periodicity by worker interaction. *Am. Nat.* 137, 244–259.

9. Cole, B.J. (1991). Is animal behavior chaotic? Evidence form the activity of ants. *Proc. R. Soc. London* B 244, 253–259.

10. Solé, R.V., Miramontes, O., and Goodwin, B.C. (1993). Oscillations and Chaos in Ant Societies, *J. Theor. Biol.* 161, 343–357.

11. Solé, R.V., Miramontes, O., and Goodwin, B.C. (1993). Emergent behavior in insect societies: global oscillations, chaos and computation. In: *Interdisciplinary Approaches to Nonlinear Complex Systems*, H. Haken and A.S. Mikhailov (eds.), Springer Series in Synergetics, Springer, Berlin.

12. Goss, S. and Deneubourg, J.L. (1988). Autocatalysis as a source of synchronized rhythmical activity in social insects. *Insectes Sociaux* 35, 310–315.

13. Cole, B.J. (1996). Mobile cellular automata models of ant behavior: movement activity of *Leptothorax allardycei*, *Am. Nat.* 148, 1–15.

14. Solé, R.V. and Miramontes, O. (1995). Information at the edge of chaos in Fluid Neural Networks, *Physica* D80, 171–180.

15. Langton, C.G. (1990). Computation at the edge of chaos: phase transitions and emergent computation. *Physica* D42, 12–37.

16. Schneirla, T.C. (1940). Further studies on the army-ant behavior pattern: Mass-organization in the swarm-raiders. *J. Comp. Psych.* **25**, 51–90.

17. Deneubourg J.-L., Goss S., Franks N.R., and Pasteels J.M. (1989). The blind leading the blind: Modeling chemically mediated army ant raid patterns. *J. Insect Behav.* 2, 719–72.

18. Solé, R.V., Bonabeau, E., Delgado, J., Fernández, P., and Marin, J. (2000). Pattern Formation and Optimization in Army Ant Raids, *Artificial Life*, in press.

19. Wilson, E.O. (1971). *The Insect Societies*, Belknap Press, Harvard.

20. Theraulaz, G., and Bonabeau, E. (1995). Coordination in distributed building, *Science* 269, 686–688.

21. Theraulaz, G. and Bonabeau, E. (1995). Modelling the collective building of complex architectures in social insects with lattice swarms. *J. Theor. Biol.* 177, 381–400.

Chapter 7

1. Whitmore, T.C. (1990). An introduction to tropical rainforests. Clarendon Press, Oxford.

2. Gause, G.F. (1971). The Struggle for Existence. Dover, New York.

3. Brown, J.H., and Heske, E.J. (1990). Control of a dester-grassland transition by a keystone rodent guild. Science 250, 1705–1707.

4. Drake, J.A. (1990). The mechanics of community assembly and succession. *J. Theor. Biol.* 147, 213–233.

5. Paine, R.T. (1966). Food web complexity and species diversity. *Am. Nat.* 100, 65–75.

6. Brown, J.H. (1994). Complex ecological systems, in *Complexity: Metaphors, Models and Reality* (SFI Series) (Cowan, G., Pines, D., and Meltzer, D., eds.), 419–449, Addison Wesley.

7. Kaneko, K. (editor) (1992). Focus issue on coupled map lattices. *Chaos* 2, 279–408.

8. Solé, R.V., Bascompte, J., and Valls, J. (1992). Stability and complexity in spatially-extended two-species competition. *J. Theor. Biol.* 159, 469–476.

9. Solé, R.V., Bascompte, J., and Valls, J. (1992). Nonequilibrium dynamics in lattice ecosystems: chaotic stability and dissipative structures. *Chaos* 2, 387–395.

10. Godfray, H.C.J. (1990). *Parasitoids: Behavioral and Evolutionary Ecology.* Princeton U. Press, Princeton.

11. Solé, R.V. and Valls, J. (1991). Order and chaos in a 2D Lotka–Volterra coupled map lattice. *Phys. Lett.* A 153, 330–336.

12. Solé, R.V., Valls, J., and Bascompte, J. (1992). Spiral waves, chaos and multiple attractors in lattice models of interacting populations. *Phys. Lett.* A 166, 123–128.

13. Hassell, M.P., Comins, H., and May, R.M. (1991). Spatial structure and chaos in insect population dynamics. *Nature* 353, 255–258.

14. Hanski, I. (1999). *Metapopulation Ecology.* Oxford U. Press, Oxford.

15. Bascompte, J., and Solé, R.V. (1996). Habitat fragmentation and extinction thresholds in spatially explicit models. *J. Anim. Ecol.* 65, 465–473.

16. May, R.M. (1972). Will a large complex system be stable? *Nature* 238, 413–414.

17. Pimm, S.A. (1984). The complexity and stability of ecosystems. *Nature* 307, 321–326.

18. Levin, S.A. (1999). *Fragile Dominion.* Princeton U. Press, Princeton.

19. Keitt, T.M., and Marquet, P.A. (1996). The introduced Hawaiian avifauna reconsidered: evidence for self-organized criticality? *J. Theor. Biol.* 182, 161–167.

20. Ellner, S., and Turchin, P. (1995). Chaos in a noisy world: new methods and evidence from time-series analysis. *Am. Nat.* 145, 343–375.

21. Solé, R.V., Alonso, D., and McKane, A. (2000). Scaling in a network model of a multispecies ecosystem. *Physica* A 286, 337–344.

22. Solé, R.V. and Manrubia, S.C. (1995). Are rainforests self-organized in a critical state?. *J. Theor. Biol.* 173, 31–40.

23. Katori, M., Kizaki, A., Terui, Y., and Kubo, T. (1998). Forest dynamics with canopy gap expansion and stochastic Ising model. *Fractals* 6, 81–86.

24. Yodzis, P. (1989). *Introduction to Theoretical Ecology*, Harper and Row, New York.

25. de Ruiter, P.C., Neutel, A., and Moore, J.C. (1995). Energetics, patterns of interaction strengths and stability in real ecosystems. *Science* 269, 1257–1260.

Chapter 8

1. Sacks, O. (1990). *Awakenings.* Harper and Collins, New York.

2. Morse, S. S. (1993). *Emerging Viruses.* Oxford U. Press, New York.

3. Holland, J. et al. (1982). Rapid evolution of RNA genomes. *Science* 215, 1577–1585.

4. Lodish, H. et al. (1995). *Molecular Cell Biology.* Scientific American Books, New York.

5. Eigen, M. (1996). *Steps Towards Life.* Oxford, New York.

6. Eigen, M. McCaskill, and Schuster, P. (1987). The Molecular Quasispecies. *Advan. Chem. Phys.* 75, 149–263.

7. Solé, R.V., Ferrer, R., Gonzalez-Garcia, I., Quer, J., and Domingo, E. (1999). Red Queen dynamics, competition and critical points in a model of RNA virus quasispecies. *J. Theor. Biol.* 198, 47–59.

8. Domingo, E., Flawell, R.A., and Weissman (1976). *Gene* 1, 3–25.

9. Hofbauer, J., and Sigmund, K. (1988). *The Theory of Evolution and Dynamical Systems*, Cambridge U. Press.

10. Maynard Smith, J., and Shatzmary, E. (1995). *The Major Transitions in Evolution*, Oxford U. Press.

11. Küppers, B.O. (1988). *Molecular Theory of Evolution.* Springer, Berlin.

12. Boerjlist, M.C., and Hogeweg, P. (1991). Spiral wave structure in prebiotic evolution: hypercycles stable against parasites. *Physica* D 48, 17–26.

13. Pearson, J.E. (1993). Complex patterns in a simple system. *Science* 261, 189–192.

14. Lee, K.J., McCormick, W.D., Pearson, J.E., and Swinney, H.L. (1994). Experimental observation of self-replicating spots in a reaction–diffusion system. *Nature* 369, 215–218.

15. Cronhjort, M. (1995). Models and computer simulations of origins of life and evolution. Ph.D. thesis, Royal Institute of Technology, Stockolm.

16. Chacón, P., and Nuño, J.C. (1995). Spatial dynamics of a model of prebiotic evolution. *Physica* D 81, 398–410.

17. Farmer, J.D., Kauffman, S.A., and Packard, N.H. (1986). Autocatalytic replication of polymers. *Physica* D 22, 50–67.

18. Luque, B. (1999). Phase transitions in random networks. Ph.D. thesis, Universitat Politecnica de Catalunya, Barcelona.

19. Fontana, W. (1992). Algorithmic chemistry. In: *Artificial Life* II (C.G. Langton, J.D. Farmer, S. Rasmussen, and C. Taylor, eds.). Addison-Wesley, Reading, Mass.

20. Adami, C. (1998). *Introduction to Artificial Life.* Springer, New York.

Chapter 9

1. Conway Morris, S. (1997). *The Crucible of Creation: The Burgess Shale and the Rise of Animals.* Oxford University Press, Inc., New York.

2. Gould, S.J. (1989). *Wonderful Life: The Burgess Shale and the Nature of History.* Norton, New York.

3. Valentine, J.W., Jablonski, D., and Erwin, D.H. (1999). Fossils, molecules and embryos: new perspectives on the Cambrian explosion. *Development* 126, 851–859.

4. Darwin, C. (1975). *On the Origin of Species.* Harvard U. Press.

5. Kimura, M. (1994). Population genetics, molecular evolution and the neutral theory (selected papers). Chicago U. Press, Chicago.

6. Ward, P. (1992). *On Methuselah's Trail: Living Fossils and the Great Extinctions.* Freeman, New York.

7. Hallam, A., and Wignall, P.B. (1997). *Mass Extinctions and Their Aftermath.* Oxford U. Press, Oxford.

8. Raup, D.M. (2000). *The Nemesis Affair.* Norton, New York.

9. Raup, D.M., and Sepkoski, J.J. (1984). Periodicity of extinctions in the geologic past. *Proc. Natl. Acad. Sci, USA* 81, 801–805.

10. Raup, D.M. (1993). *Extinctions: Bad Genes or Bad Luck?* Oxford U. Press, Oxford.

11. Newman, M.E.J. (1997). A model of mass extinction. *J. Theor. Biol.* 189, 235–252.

12. Maynard Smith, J. (1989). Causes of extinction. *Phil. Trans. R. Soc. Lond.* B325. 241–252.

13. Van Valen, L. (1973). A new evolutionary law. *Evol. Theory* 1, 1–30.

14. Solé, R.V., Manrubia, S.C., Benton, M., Kauffman, S.A., and Bak, P. (1997). Self-similarity of extinction statistics in the fossil record. *Nature* 388, 764–767.

15. Newman, M.E.J., and Eble, G.J. (1999). Power spectra of extinction in the fossil record. *Proc. R. Soc. London* B 266, 1267–1270.

16. Solé, R.V., and Bascompte, J. (1996). Are critical phenomena relevant to large-scale evolution? *Proc. R. Soc.* B 263, 161–168.

17. Newman, M.E.J. (1996). Self-organized criticality, evolution and the fossil extinction record. *Proc. R. Soc.* B 263, 1605–1610.

18. Burlando, B. (1993). The fractal geometry of evolution. *J. Theor. Biol.* 163, 161–172.

19. Nottale, L., Chaline, J., and Grou, P. (2000). Les arbres de l'évolution. Hachette, Paris.

20. Kauffman, S., and Johnsen, S. (1991). Coevolution at the edge of chaos: coupled fitness landscapes, poised states and coevolutionary avalanches. *J. Theor. Biol.* 149, 467–482.

21. Sneppen, K., Bak, P., Flyvbjerg, H., and Jensen, M.H. (1995). Evolution as a self-organized critical phenomenon. *Proc. Natl. Acad. Sci. USA* 92, 5209–5213.

22. Kauffman, S.A., and Levin, S. (1987). Towards a general theory of adaptive walks on rugged landscapes. *J. Theor. Biol.* 128, 11–??.

23. Kauffman, S.A. (1994). *At Home in the Universe.* Oxford U. Press, Oxford.

24. Bak, P., Flyvbjerg, H., and Lautrup, B. (1992). Coevolution in a rugged fitness landscape. *Phys. Rev.* A46, 6724–6732.

25. Ratcliffe, D. (1979). The end of the large blue butterfly. *New Scientist* 8, 457–458.

26. Jablonski, D. (1991). Extinctions: a paleontological perspective. *Science* 253, 754–757.

27. Owen-Smith, N. (1987). Pleistocene extinctions: the pivotal role of megahervibores. *Paleobiology* 13, 351–362.

28. Solé, R.V., Bascompte, J., and Manrubia, S.C. (1996). Extinctions: bad genes or weak chaos? *Proc. R. Soc. Lond* B263, 1407–1413.

29. Solé, R.V., Manrubia, S.C., Pérez-Mercader, J., Benton, M. and Bak, P. (1998). Long-range correlations in the fossil record and the fractal nature of macroevolution. *Advances in Complex Systems* 1, 255–266.

30. Solé, R.V., Manrubia, S.C., Benton, M., Kauffman, S.A., and Bak, P. (1999). Criticality and scaling in evolutionary ecology. *Trends in Ecology and Evolution* 14, 156–160.

31. Lenski, R.E., and Travisano, M. (1994). Dynamics of adaptation and diversification: a 10,000 generation experiment with bacterial populations. *Proc. Natl. Acad. Sci. USA* 91, 6808–6814.

32. Elena, S., Cooper, V.S., and Lenski, R.E. (1996). Punctuated evolution caused by selection of rare beneficial mutations. *Science* 272, 1802–1804.

33. Rainey, P.B., and Travisano, M. (1998). Adaptive radiation in a heterogeneous environment. *Nature* 394, 69–72.

34. Adami, C. (1998). *Introduction to Artificial Life*. Springer, New York.

35. Ray, T. (1992). An approach to the synthesis of life. In: *Artificial Life II*, C. Langton et al., editors. 371–408. Addison-Wesley, Redwood City, CA.

36. Lenski, R.E., Ofria, C., Collier, T.C., and Adami, C. (1999). Genome complexity, robustness and genetic interactions in digital organisms. *Nature* 400, 661–664.

37. Kauffman, S.A. (1998). Cambrian explosion and Permian quiescence: implications of rugged fitness landscapes. *Evolutionary Ecology* 3, 274–281.

38. Eble, G.J. (1998). The role of development in evolutionary radiations. In: *Biodiversity Dynamics*, M. McKinney and J. Drake, editors. 132–161. Columbia U. Press, New York.

Chapter 10

1. Yoffee, N., and Cowgill, C.L. (1988). *The Collapse of Ancient States and Civilizations*. University of Arizona Press, Tucson.

2. James, D. (1993). *Old typewriters*. Shire Publications, London.

3. Arthur, B. (1990). Positive feedbacks in the economy. *Scientific American*, February, 92–99.

4. Arthur, B. (1994). *Increasing Returns and Path Dependence in the Economy*. Michigan U. Press.

5. Anderson, P., Arrow, K., and Arthur, B., eds. (1987). *The Economy as an Evolving Complex System*. Addison-Wesley, Reading, MA.

6. Mantegna, R.N., and Stanley, H.E. (2000). *An Introduction to Econophysics*. Cambridge U. Press.

7. Sornette, D., and Johansen, A. (1997). Large financial crashes. *Physica* A 245, 411–422.

8. Vandevalle, N., Boveroux, Ph., Minguet, A., and Ausloos, M. (1998). The crash of October 1987 seen as a phase transition: amplitude and universality. *Physica* A 255, 201–210.

9. Batty, M., and Longley, P. (1994). *Fractal Cities*, Academic, San Diego.

10. Makse, H.A., Havlin, S., and Stanley, H.E. (1995). Modelling urban growth patterns. *Nature* 377, 608–612.

11. Zanette, D., and Manrubia, S.C. (1997). Role of intermittency in urban development: A model of large-scale city formation. *Phys. Rev. Lett.* 79, 253–256.

12. Manrubia, S.C., Zanette, D.H., and Solé, R.V. (1999). Transient dynamics and scaling phenomena in urban growth, *Fractals* 7, 1–8.

13. Nagel, K., and Schreckenberg, J. (1992). A cellular automaton model for freeway traffic. *J. Phys.* I France 2, 2221–2224.

14. Nagel, K., and Rasmussen, S. (1994). Traffic at the edge of chaos. In: *Artificial Life IV*, R.A. Brooks and P. Maes (eds.), 222–230. MIT Press (Cambridge, MA, 1994).

15. Ohira, T., and Sawatari, R. (1998). Phase transition in a computer network traffic model. *Phys. Rev.* E 58, 193–195.

16. Solé, R.V., and Valverde, S. (2000). Information transfer and phase transitions in a model of Internet Traffic. Santa Fe Institute Working Paper 00-03-020.

17. Kauffman, S. (1995). *At Home in the Universe*. Viking, Great Britain.

18. Solé, R.V. (1999). Statistical Mechanics of Macroevolution and Extinction, In: *Statistical Mechanics of Biocomplexity* (M. Rubi et al., eds.) Lecture Notes in Physics, 217–250, Springer, Berlin.

19. Gumerman, H., and Gell-Mann, M. (eds.) (1994). *Understanding Complexity in the Prehistoric Southwest.* Addison-Wesley, Reading.

Index